Lecture Notes in Applied and Computational Mechanics 65

Series Editors

Prof. Dr.-Ing. Friedrich Pfeiffer
Lehrstuhl B für Mechanik
Technische Universität München
Boltzmannstraße 15
85748 Garching
Germany
E-mail: pfeiffer@amm.mw.tu-muenchen.de

Prof. Dr. Peter Wriggers
FB Bauingenieur- und Vermessungswesen
Inst. Baumechanik und Numer. Mechanik
Universität Hannover
Appelstr. 9 A
30167 Hannover
Germany
E-mail: wriggers@ikm.uni-hannover.de

For further volumes:
http://www.springer.com/series/4623

Numerical Analysis of Vibrations of Structures under Moving Inertial Load

Czesław I. Bajer and Bartłomiej Dyniewicz

Springer

Authors
Czesław I. Bajer
Polish Academy of Sciences
and Warsaw University of Technology
Warsaw
Poland

Bartłomiej Dyniewicz
Polish Academy of Sciences
Warsaw
Poland

ISSN 1613-7736
e-ISSN 1860-0816
ISBN 978-3-642-29547-8
e-ISBN 978-3-642-29548-5
DOI 10.1007/ 978-3-642-29548-5
Springer Heidelberg New York Dordrecht London

Library of Congress Control Number: 2012935520

Printed on acid-free paper

Springer is part of Springer Science+Business Media (www.springer.com)

Preface

Computer methods and simulations allow engineers to study complex problems in detail. They can take into account various factors that influence the investigated phenomenon. Moreover, they can examine problems throughout a wide range of parameters. Such cases occur when the existing or designed structures are to carry heavier loads and should optimally resist external forces that involve static displacements or vibrations. Generally, vibrational or wave problems in structural dynamics require a detailed study of numerous cases.

Moving inertial loads are applied to structures in civil engineering, robotics, and mechanical engineering. Some fundamental books exist, as well as thousands of research papers. Well known is the book by L. Frýba, *Vibrations of Solids and Structures Under Moving Loads*, which describes almost all problems concerning non-inertial loads. Unfortunately, this wide literature is rarely reflected in computer codes. Well known commercial packages enable the analysis of complex mechanical problems, with material and geometrical non-linearities, but they fail in the case of moving loads.

This book presents broad description of numerical tools successfully applied to structural dynamic analysis. Unfortunately none of the classical methods can be directly applied to non-classical problems. Moving mass problems are an example of such a group of problems. It can be generally considered as problems with distributed parameters. Physically we deal with non-conservative systems. Mathematically they are described by linear partial differential equations with variable coefficients. We will focus our discussion on the moving inertial particle rather than on the structure carrying the massless load. The discrete approach formulated with the use of the classical finite element method (FEM) results in elemental matrices, which can be directly added to global structure matrices. The classical approach is considered in the simplest case in our book as the finite element method applied to space with another method applied to integration of the time derivatives. A more general approach is carried out with the space-time finite element method. It can be considered as an extension, to the time domain, of the well known finite element method: the spatial finite element gains an additional time dimension. In such a case, a trajectory of the moving concentrated parameter in space and time can be simply defined. What is more, elemental characteristic matrices can easily be derived and the formulation is relatively clear. The crucial point, however, is the uniform treatment of the space and time dependent terms in the differential equations. Discussion and the experience gained then allow of a better understanding of a formulation in the case of the Newmark method, central difference method, and other time integration methods commonly used in structural dynamics.

We consider structures described by pure hyperbolic differential equations such as strings and structures described by hyperbolic–parabolic differential equations such as beams and plates. More complex structures such as frames, grids, shells, and three-dimensional objects, can be treated with the use of the solutions given in this book.

The problems treated in the monograph can be related to problems of mathematical physics. The resulting matrices that describe the influence of the moving inertial particle can be directly implemented in computer codes.

This monograph would not have been possible without the support of the project Lider/26/40/L-2/10/NCBiR/2011 and project of Foudation for Polish Science – *START*.

Warsaw,
February 2012

Czesław I. Bajer
Bartłomiej Dyniewicz

Contents

1 Introduction 1
1.1 Literature Review 5
1.2 Solution Methods 7
1.3 Approximate Methods 9
1.4 Review of Analytical-Numerical Methods in Moving Load Problems 12
1.4.1 d'Alembert Method 13
1.4.2 Fourier Method 14
1.4.3 Lagrange Formulation 17
1.5 Examples 18

2 Analytical Solutions 21
2.1 A Massless String under a Moving Inertial Load 22
2.1.1 Case of $\alpha \neq 1$ 23
2.1.2 Case of $\alpha = 1$ 25
2.2 Discontinuity of the Solution 26
2.3 Conclusions 29

3 Semi-analytical Methods 31
3.1 String 32
3.1.1 Fourier Analysis 32
3.1.2 The Lagrange Equation 37
3.2 Bernoulli–Euler Beam 46
3.2.1 Fourier Solution 47
3.2.2 The Lagrange Equation of the Second Kind 50
3.2.3 Conclusions 55
3.3 Timoshenko Beam 55
3.3.1 Fourier Solution 56
3.3.2 The Lagrange Equation 56
3.3.3 Examples 59
3.3.4 Conclusions and Discussion 61

3.4 Bernoulli–Euler Beam vs. Timoshenko Beam 66
3.5 Plate 67
3.6 The Renaudot Approach vs. The Yakushev Approach 70
3.6.1 The Renaudot Approach 71
3.6.2 The Yakushev Approach 72

4 Review of Numerical Methods of Solution 77
4.1 Oscillator 79
4.1.1 String Vibrations under a Moving Oscillator 79
4.1.2 Beam Vibrations under a Moving Oscillator 83
4.2 Inertial Load 84
4.2.1 A Bernoulli–Euler Beam Subjected to an Inertial Load 85
4.2.2 A Timoshenko Beam Subjected to an Inertial Load 89

5 Classical Numerical Methods of Time Integration 95
5.1 Integration of the First Order Differential Equations 97
5.2 Single-Step Method SSpj 102
5.3 Central Difference Method 105
5.3.1 Stability of the Method 107
5.3.2 Accuracy of the Method 108
5.4 The Adams Methods 109
5.4.1 Explicit Adams Formulas (Open) 110
5.4.2 Implicit Adams Formulas (Closed) 112
5.5 The Newmark Method 114
5.6 The Bossak Method 117
5.7 The Park Method 118
5.8 The Park–Housner Method 118
5.8.1 Stability of the Park–Housner Method 119
5.9 The Trujillo Method 121

6 Space–Time Finite Element Method 123
6.1 Formulation of the Method—Displacement Approach 129
6.1.1 Space–Time Finite Elements in the Displacement Description 135
6.2 Properties of the Integration Schemes 138
6.2.1 Accuracy of Methods 140
6.3 Velocity Formulation of the Method 140
6.3.1 One Degree of Freedom System 140
6.3.2 Discretization of the Differential Equation of String Vibrations 144
6.3.3 General Case of Elasticity 149
6.3.4 Other Functions of the Virtual Velocity 151
6.4 Space–Time Element Method and Other Time Integration Methods 154

6.4.1 Convergence ... 154
6.4.2 Phase Error ... 157
6.4.3 Non-inertial Problems ... 158
6.5 Space–Time Finite Element Method vs. Newmark Method ... 160
6.6 Simplex Elements ... 161
6.6.1 Property of Space Division ... 162
6.6.2 Numerical Efficiency ... 167
6.7 Simplex Elements in the Displacement Description ... 169
6.7.1 Triangular Element of a Bar Vibrating Axially ... 169
6.7.2 Space–Time Finite Element of the Beam of Moderate Height ... 170
6.7.3 Tetrahedral Space–Time Element of a Plate ... 172
6.8 Triangular Elements Expressed in Velocities ... 176

7 Space–Time Finite Elements and a Moving Load ... 181
7.1 Space–Time Finite Element of a String ... 182
7.1.1 Discretization of the String Element Carrying a Moving Mass ... 182
7.1.2 Numerical Results ... 184
7.1.3 Conclusions ... 188
7.2 Space–Time Elements for a Bernoulli–Euler Beam Carrying a Moving Mass ... 188
7.2.1 Numerical Results ... 190
7.3 Space–Time Element of Timoshenko Beam Carrying a Moving Mass ... 198
7.3.1 Conclusions ... 203
7.4 Space–Time Finite Plate Element Carrying a Moving Mass ... 204
7.4.1 Thin Plate ... 204
7.4.2 Thick Plate ... 213
7.4.3 Plate Placed on an Elastic Foundation ... 215
7.5 Problems with Zero Mass Density ... 218

8 The Newmark Method and a Moving Inertial Load ... 223
8.1 The Newmark Method in Moving Mass Problems ... 223
8.2 The Newmark Method in the Vibrations of String ... 226
8.3 The Newmark Method in Vibrations of the Bernoulli–Euler Beam ... 229
8.4 The Newmark Method in Vibrations of a Timoshenko Beam ... 230
8.5 Numerical Results ... 230
8.6 Accelerating Mass—Numerical Approach ... 233
8.6.1 Mathematical Model ... 233
8.6.2 The Finite Element Carrying the Moving Mass Particle ... 235
8.6.3 Accelerating Mass—Examples ... 238
8.7 Conclusions ... 239

9 Meshfree Methods in Moving Load Problems 241
9.1 Meshless Methods (Element-Free Galerkin Method) 241
9.2 Results 243

10 Examples of Applications 247
10.1 Dynamics of the Classical Vehicle–Track System 249
10.2 Dynamics of the System Vehicle—Y-Type Track 253
10.3 Dynamics of Subway Track 262
10.4 Vibrations of Airport Runways 266

Appendix 271

A Computer Programs 271
A.1 String—Space–Time Element Method 271
A.2 Timoshenko Beam—Newmark Method 274
A.3 Mindlin Plate—Space–Time Element Method 277
A.4 Kirchhoff Plate — Space-Time Element Method 283

References 285

Index 293

Chapter 1
Introduction

Computer methods are commonly used now in engineering design, manufacturing, and applications. They replace experimental methods of verification, especially if experiments are expensive, time consuming, or difficult to perform. Static analysis, plastic deformations, optimization, and free vibrations are fields sufficiently well explored, and now possess efficient numerical procedures implemented in commercial software. But the case of moving loads is not represented in such computer codes. Design engineers use simplifications and approximations known from analytical solutions. These are often adequate if the load does not change the dynamical properties of the structure, i.e. is massless. In the case of an inertial load we do not have adequate tools.

In this book we will present numerical methods which enable us to solve problems of the vibrations of structures subjected to inertial moving loads. Only simple and particular cases of problems with moving inertial load can be solved analytically. Such problems usually require numerical computations at the final stage or, if we use discrete methods, during the whole analysis. Analytical and semi-analytical solutions are indispensable when we verify our numerical results. Therefore we will present semi-analytical solutions as a base for a better understanding of both the differential equations that govern the motion of these structures, and features and properties of solutions. Engineers, researchers and students will find here matrices and algorithms ready for use, material which will enable them understanding mechanical problems, and an elaboration of the software procedures for basic or more complex structural elements.

Inertial loads moving on strings, beams, and plates at sub- or super-critical speed are of special interest. Theoretical solutions are applied to many practical problems: train–track interaction, vehicle–bridge interaction, pantograph collectors in railways, magnetic rails, guideways in robotic solutions, etc. Such problems have been widely treated in the literature. Attempts at solving such problems began in the middle of the 19th century. However, up to now we have not had a complete and closed analytical solution. The term describing the concentrated mass motion is the reason for the difficulties. Systems of differential equations of variable coefficients, which, except in a few cases, do not have analytical solution, are serious roadblocks

C.I. Bajer and B. Dyniewicz: Numerical Analysis of Vibrations of Structures, LNACM 65, pp. 1–20.
springerlink.com

to obtaining closed-form solutions. These types of equations are finally solved by numerical means.

Structures subjected to moving loads are often encountered in engineering practice. Such are the bridges and viaducts loaded with vehicles [147], flyovers for traditional or magnetically lifted trains, road or airfield plates, sliding robot manipulators, machine tools, weapon firing barrels, ropes of transporting systems, and current collectors for power supply systems for rail vehicles (Figure 1.1). They are exposed to much larger displacements than when under static loads or slowly sliding loads. This becomes obvious if we look at the undeformed structure at rest, which is suddenly subjected to a force. Such a structure starts to vibrate around its

Fig. 1.1 Examples of problems with moving mass.

equilibrium position in the unladen state, through the state of static equilibrium under load, to obtain an amplitude equal to twice the static deflection under load. Therefore, the rapid entry of a loading force has a similar effect. The passage of the load, if takes place cyclically with a certain frequency, will increase the deflection. If its frequency is associated with the passage speed in such a way that at the exit of one load the next will enter, then we obtain the dependence of the maximum deflection of the structure under load on the speed of the travel of the load $w_{max}(\mathrm{v})$.

The maximum deflection occurs at the point located generally around 0.5–0.7 of the span, depending on the speed of travel. Proceeding further, we can examine the speed at which the maximum deflection of the journey will be greatest. This speed is called the critical speed. In the case of a string, the critical velocity corresponds to the wave propagation velocity c. The critical speed is the important feature from a practical point of view. It determines the most unfavorable value of the deflection, to which the structure must be made resistant. For this reason, the study of a structure under a moving load is an important engineering problem. Unfortunately, the existing commercial packages do not perform computational simulations of such tasks. As we will see, the problem is difficult and this can be ascribed to the lack of appropriate computerized procedures.

Here, attention should be given to the classification of the loads. The simplest case is shown in Figure 1.2*a*. It is irrelevant that the force applied directly to the structure is replaced with an oscillator, which will have non-zero mass. Although the mass effect will be visible in the results, it will not be the result of a task with the inertial load [112]. Moreover, although the impact of the mass of the oscillator will increase with increasing spring stiffness, then the solution does not tend uniformly to the solution for the case of an inertial load. Additional degree of freedom oscillations introduce additional artificial effects in the form of resonance: the increase or decrease in amplitude at certain speeds. A mass load is shown in Figure 1.2*b*. The mass motion affects the outcome at non-zero displacements $w(x,t)$ when $0 \leq t < l/\mathrm{v}$. Otherwise, the participation of factors causing displacement is required, such as the pulse force (Figure 1.2).

Let us take a railway wheel with mass 500 kg. Together with the axle and the axle box its inertia exceeds a ton. The rail has a linear mass density of 60 kg/m. The influence of this concentrated mass significantly changes the dynamic properties of a structure (Figure 1.3). We claim that a significant part of the rail wheel should be

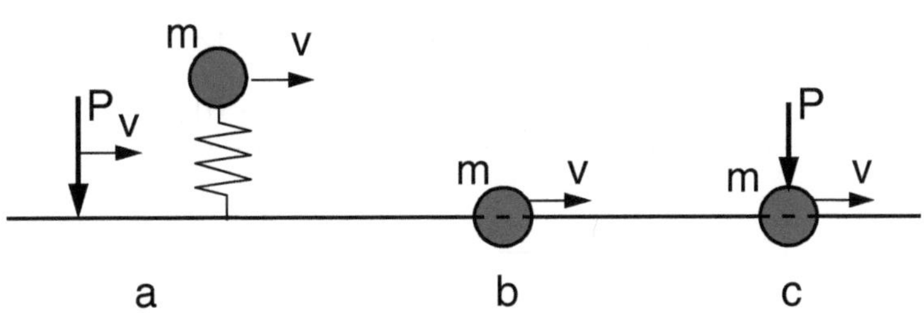

Fig. 1.2 Loads: a) massless, b) inertial, c) inertial and gravity.

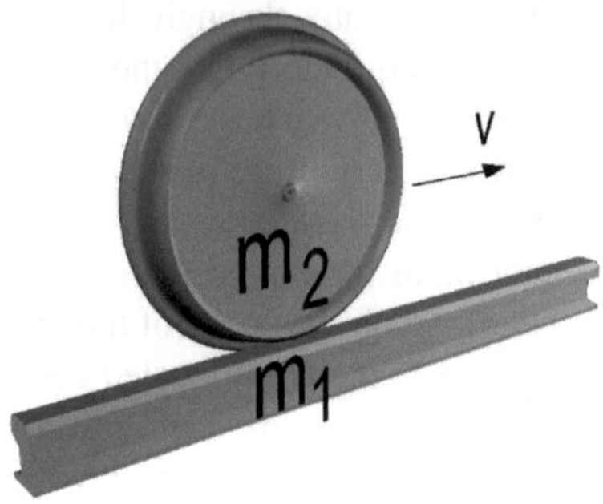

Fig. 1.3 The wheel inertia influencing the rail motion

unsprung. In engineering practice we wish to take into account a real structure, with all the atypical elements for an analytical model, i.e. ballast as an elastic foundation, sleepers as periodic supports, elastic pads, influence of several wheelsets, coupling and interacting with a boogie, etc. Accurate results are fundamental for decisions at the design stage. An accurate estimation of the dynamic influence is essential for proper modelling. Accurate results are important not only for increasing the durability and reliability of systems: predicting the level of the dynamic response of structures under a moving load facilitates the protection of the environment, especially populated urban centres or historic places.

Existing finite element (FEM) modelling software allows us to perform computations in several stages grouped into a batch procedure. In such a case, one can split the dynamic problem into a sequence of static problems with structures subjected to gravitational and inertial forces. Such a solution corresponds to the simple lumping of a moving mass into nodal points, as depicted in Figure 1.4. Such an approach is correct only at extremely low speeds of the mass, practically quasistatic. Problems containing beams as a supporting structure in the case of a moderate speed of a mass

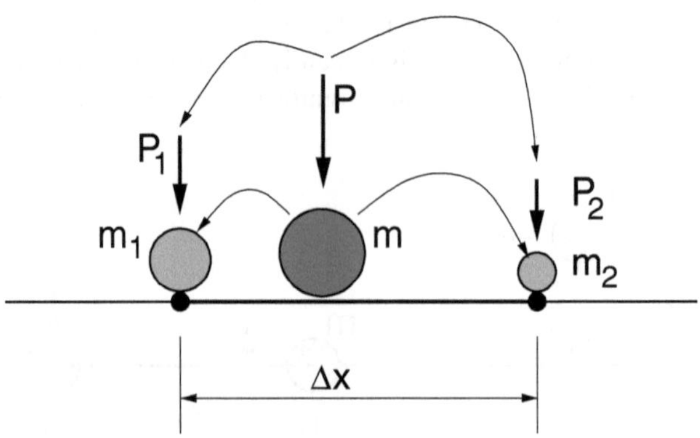

Fig. 1.4 Ad-hoc mass lumping to nodes.

usually give limited solutions. Unfortunately they are neither convergent nor stable. Even a small variation in the parameters produces significant alterations of the results. The pure hyperbolic differential equation describing the string motion results in divergent solutions.

Travelling loads are generally unlikely to be solved by commercial codes. Most of the existing systems for dynamic simulations usually do not allow us to solve even simple problems comprising travelling massless point forces, travelling distributed non-inertial loads, or even travelling and elastically joined moving substructures. Inertial moving loads are completely unimplemented by computer systems. The intuitive approach to discrete analysis with ad-hoc lumping of forces and masses to neighbouring nodes always fails. Sometimes, especially in the case of beams, numerical solutions are limited, but significantly inaccurate. We emphasize here that the travelling mass problem is not trivial, even if at first sight it seems to be.

1.1 Literature Review

In the literature, numerous historical reviews concerning the moving load problem exist (for example Panovko [106], Yakushev [146], Dmitrijev [43]). In most cases the moving massless constant force was considered as a moving load. This type of problem results in closed solutions. Unfortunately, the problem of inertial loads is still open. Saller in [123] considered a moving mass for the first time. He proved, in spite of essential simplifications, the significant influence of the moving mass on the beam dynamics. In the 1930s, two contributions appeared, important for researchers working in the field of moving loads. Inglis [70] applied simplifications and the solution was expressed by only the first term of a trigonometric series. The time function fulfilled a second order differential equation with variable coefficients. This equation was derived by considering the acceleration under the moving mass, expressed by the so-called Renaudot formula. In fact it is the derivative computed with the chain rule. The final solution of the differential equation with variable coefficients was proposed as an infinite series. It results in an approached solution.

Schallenkamp [124] proposed another approach to the problem of a moving mass. However, his attempt only allows us to describe the motion under the moving mass. The method of separation of variables by the expansion of the unknown function into a sine Fourier series was applied. Boundary conditions in the beam were taken into account in a natural way. The ordinary differential equation, which describes the motion under the moving mass, was expressed in generalized coordinates by using the second Lagrange equations. The generalized force was derived from the virtual work principle. Schallenkamp's consideration is relatively complex and converges slowly since the final solution is expressed in terms of a triple infinite series.

The works of Inglis and Schallenkamp can be considered as the basis for the analysis of the problem of a moving mass in the succeeding works of Bolotin [36, 35],

Morgajewskij [101] and others. An excellent and important monograph in this field was written by Szcześniak [133]. One can find there hundreds of references concerning moving loads on beams and strings. In [138] the authors consider a simply supported beam modelled by the Bernoulli–Euler theory. The equation of motion is written in an integral-differential form with Green's function terms. In order to solve this equation, a dual numerical scheme was used. A backward difference technique was applied to treat the time parameter and numerical integration was used for the spatial parameter. This method of solution, though applied to higher velocities, still requires complex mathematical operations. Each solution enables us to determine only the displacements under the moving load and does not give solutions for a wide range of parameters x and t. Only one closed analytical solution can be found in the literature. Smith [127] proposed a purely analytical solution for the inertial moving load, however, only in the case of a massless string. The basic motion equation, without the term which describes the inertia of the string, was transformed to the hyper-geometrical equation. It has an analytical solution in terms of infinite series. Frýba [56] applied the same approach and found a closed analytical solution for the particular case. However, the formula given in [56] has mistakes.

Recent papers have contributed analyses of complex problems of structures subjected to moving inertial loads [144] or oscillators [29, 97, 112]. Variable speeds were analysed in [3, 58, 99]. The equivalent mass influence is analysed in [57]. An infinitely long string subjected to a uniformly accelerated point mass was also treated [121] and analytical solution of the problem concerning the motion of an infinite string on a Winkler foundation subjected to an inertial load moving at a constant speed was given [74].

In one of our papers [48], we considered small vibrations of the massless and massive string subjected to a moving inertial load. We proposed an analytical–numerical solution of the problem. The final equation has the form of a matrix differential equation of the second order. Numerical integration results in a solution over a wide range of the velocity: under-critical and over-critical. It exhibits a discontinuity of the mass trajectory at the end support point. This new feature had not yet been reported in the literature. A closed-form solution in the case of a massless string was analysed and its discontinuity was proved mathematically. Fully numerical results obtained for the inertial string had a similar property. Since small vibrations are analysed, the discontinuity effect discussed in the paper was of purely mathematical interest.

The results are compared with the approximate numerical solutions obtained by the finite element method (FEM). The string is subjected to a moving oscillator. In the case of a rigid spring, we approach the analytical solution. However, in the case of higher speeds (greater than 20% of the critical wave speed), the accuracy of the FEM solution is poor.

A review of the literature devoted to numerical methods applied to moving mass problems will be given in Chapter 4.

1.2 Solution Methods

In the early period of the rapid development of computer methods, falling in the eighties of the last century, researchers analysed and described the basic properties of discrete methods of calculation. These included:

- the impact of the finite element mesh density on the results, the estimation of the error of approximation,
- reducing the size of the task by using techniques of static and dynamic condensation, the division into subsystems, etc.
- creation and study of the properties of new, more accurate finite element models, mainly bending elements, the analysis of the locking of degrees of freedom, overstiffening, the inclusion of complex constitutive relations,
- the development of methods for the integration of the differential equations of motion, characterized by unconditional stability, low computational cost, and appropriately matched characteristics of spurious damping.

The capabilities of the known techniques were combined (finite difference and finite element method) and new methods were formulated (the boundary element method, moving elements, meshless methods). The limited computational ability of computers still forced work on improving the performance of the computing algorithms. With the increasing power of processors and reductions in the costs of memory, the effort of software developers has shifted to improve the utilisation of existing computational programs: improved data input methods and attractive forms of visualisation of the results. Computer programs were widely used in engineering practice.

Today, computer modelling generally involves the phenomenon of change over time. Both knowledge and computer tools allow you to take into account many factors influencing the processes in structures with complex shape. At the same time, less and less importance is attached to the evaluation of the correctness of the results, and attention is focused on a faithful reproduction of the geometry. Geometric modelling, an appropriate choice of the type of finite elements, and then imaging the stress fields, are the activities which usually limit the operation engineer, i.e., the user of the computing package. Less time is devoted to understandng the numerical and mechanical properties of the models created. Hence, in many cases, the results obtained are difficult to interpret. Effects arising from properties of the numerical model emerge. Sometimes they may be mistakenly regarded as the characteristic features of the phenomenon studied. Differences in the results obtained with two different commercial packages are no longer a cause for concern. Knowledge and experience is slowly being replaced by knowledge about the flaws in package design and how to overcome the technical difficulties encountered. Packages designed for crashworthiness analysis can be a good example. The explicit time integration method of differential equations used in the computations turns out to be unstable in the case of lightweight discrete elements, or those with small dimensions, or which are relatively rigid. In this case, there is a way to prevent instability by artificially increasing the weight of selected points. A stable solution is obtained, and then the resulting graphs are scaled to partially offset the impact of such operations and to match the experimental results. Such activities become more random, and the results less reliable. There are a large number of degrees of freedom in the numerical model, reaching the millions, and one can attain an attractive visualisation of the results, but engineers no longer pay due attention to the substantive aspects of the task, leaving such issues to the software developers.

Insight into the reports related to computer modelling shows that we usually seek to illustrate the calculations and provide a narrative summary of the quantitative findings. Very rarely are users able to substantively address results of analysis to the phenomena of structural mechanics and mentally categorise the problem with respect to simple, well-known phenomena such as wave propagation, reflection, stability of motion, relevant feedback, etc. Only the insights and skills in isolating such aspects can lead to improved research, by means of at least methodological changes in the task. Otherwise you are condemned to the path of successive trials and errors. Improvement of the job is done largely by accident. We can not expect good engineering effects in such a case.

In subsequent chapters, we will select a group of tasks related to the dynamics and vibrations caused by moving loads. In particular, we will show the applicability of the space-time finite element method to the dynamics of structures. There are characteristics and properties of this method which are attractive from the standpoint of a researcher or engineer. We will also briefly discuss selected classical computational methods. We will discuss their differences and similarities with the space-time finite element method. The information provided will help readers gain some useful experience in the computer calculations of the dynamics of structures and the interpretation of the results.

The current level of development of computational and simulation techniques definitely altered the balance between the engineer's own creative contribution to the computational process, led by a deliberate plan, developed algorithm, a computer program written by his own mathematical knowledge, versus the use of algorithmic solutions catalogued in computer packages, about which usually little is know. In the first case we must have appropriate knowledge, especially mathematical, and the appropriate ability to use and program computer methods, as well as in the interpretation of the results. In the second, we have to rely on available commercial computing packages, which usually do not fully correspond to our needs. In both cases, we need knowledge of calculation methods, the mathematical models and numerical methods most suited to our situation, and their properties and even their curiosities.

The most frequent problem we encounter in writing our own programss or using ready-made computational computer tools is the selection of the material data. For example, it is easy to take the elastic modulus of steel or concrete, but the real challenge is the value of the damping coefficients of these materials in the chosen construction. Without experimental studies it is impossible to determine the stiffness of the attachment joints, the stiffness of the elastic rubber spacers in the accepted range of work, or the stiffness of the rail anchors. What is worse, the data obtained in a particular case, is only to a limited extent suitable for use in other cases.

The potentially wide range of topics for this book was limited to groups of problems mainly related to dynamic phenomena in transportation and robotics. In this area, much has changed in recent years.

The observed increased speed of vehicles in rail traffic in passenger transportation and the increased capacity in freight traffic involves increases in the dynamic load of the wheel–rail system. At the same time, the aim is to reduce noise and improve the safety of travel. We are looking for new solutions for both track and vehicle design. A tendency to use the optimum parameters of the system can be observed. Resonant states and dynamic critical states cause overloading. One phenomenon of this type is the self-excited vibration. This is directly related to the problem of stability of the interaction and the motion of rolling wheelsets. Such phenomena in rail traffic should be strictly eliminated.

1.3 Approximate Methods

The computational methods used in practice do not allow of an accurate representation of the phenomena occurring in the processes described. Despite the effort and engagement of the knowledge from many disciplines (Figure 1.5) we can not describe all the phenomena which occurr. For this reason, we will try to separate them from each other and focus our attention on one single phenomenon. But in this case we are forced to go back even further and accept the solution of simpler problems,

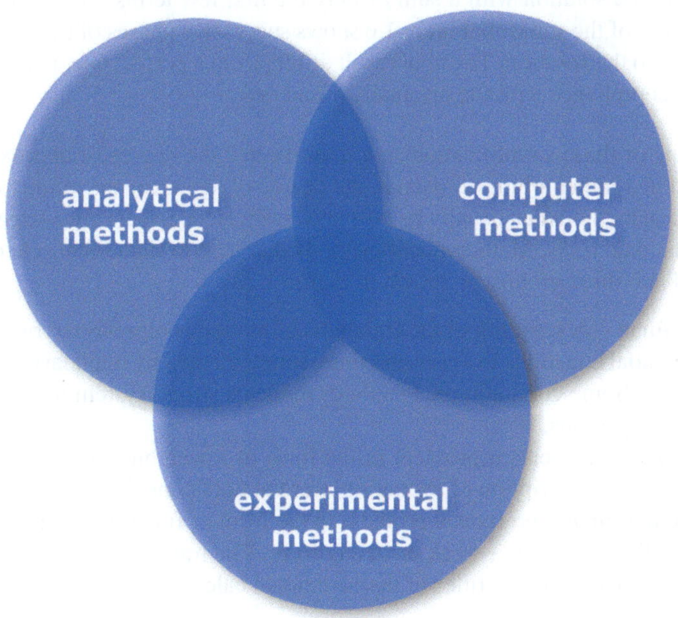

Fig. 1.5 Fields of knowledge involved in modelling.

which significantly deviates from our initial expectations. We make simplifications every step of the calculation. one can pay attention to some of them.

Statics or dynamics? This selection requires a decision as to whether any time-varying phenomena, especially the dynamic effects arising from the motion of the construction and acting inertia forces, are large enough that we must decide on a dynamic analysis. In simple analyses, we assume that the processes varying in time are slow and treat the problem as quasi-static. Such is the case with subsidence, the material plastic flow in strain rate-dependent phenomena or the fatigue of periodically loaded material.

Linear or nonlinear description? In the phenomena of a linear system, the response is linearly dependent on the cause. The question arises to what extent the relationship between cause and effect can be regarded as linear. Can we replace the nonlinear characteristics with segments of linear dependencies? If so, how do we estimate the resulting error?

What method of calculation? As a rule, we simplify the mathematical description of the phenomenon and adapt it to the available computational tools. We agree that the differential equations are not exactly satisfied throughout the area, but only at the selected points. We assume that beyond these points the solution is sufficiently smooth. We replace the solution, a continuous function: we represent the solution and its derivatives by derivatives continuous which are piece-wise continuous. We discretize the test area. Other ways of proceeding need to replace the infinite series representing the solution with a sum of only the first few terms. We discretize the assumed form of the solution. Figure 1.6 shows successive phases of transition from the problem to the solution. The final result differs from a perfect one, i.e. from a result in full compliance with the originally posed task.

While many of these simplifications are intuitive and the degree of approximation is assessed in a rather arbitrary manner, the degree of approximate validity of these mathematical methods can usually be estimated well. Hence, there are a large variety of computational tools. Mathematical methods that lead to numerical schemes can be divided into three groups, as shown in Figure 1.7.

Strong form – description of the equations of motion is represented by a system of differential equations in space and time, supplemented by boundary and initial conditions. Some computational methods reduce a strong system to a system of algebraic equations.
Weak form – presents a problem in the form of a weighted integral equation. This leads to a description of averaging over the field considered.
Variational form – presents a problem in the form of functional whose stationarity conditions lead to a weak or strong form. The transition from one form to another is done using the rules of the variational calculus.

Strong forms of description of phenomena have been used for a long time. Newton's second law is an example. Solutions of the tasks described by strong forms consist

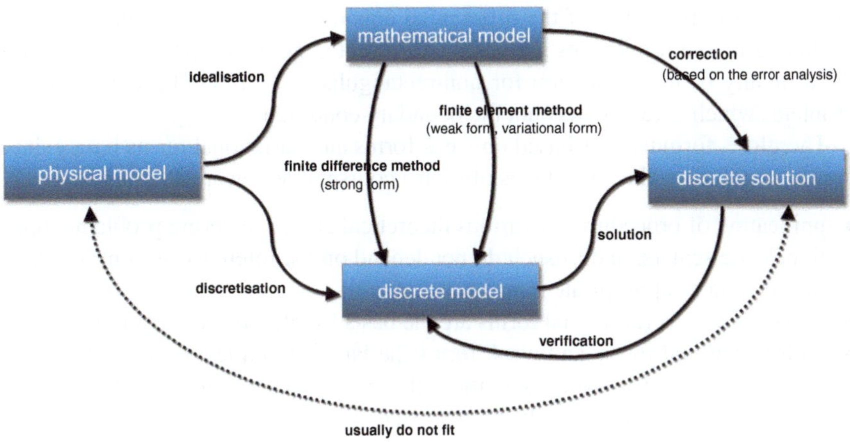

Fig. 1.6 Dependencies in the solution process.

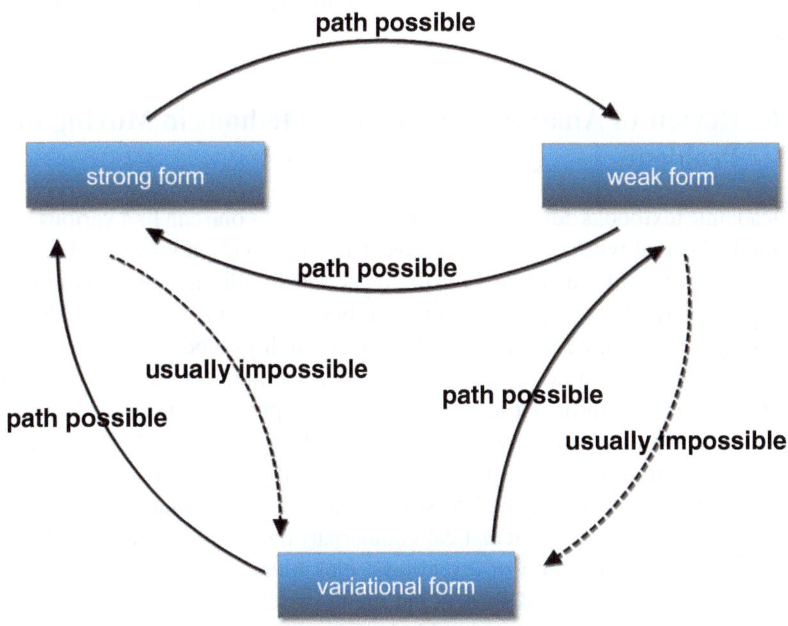

Fig. 1.7 The transition between different mathematical formulations.

of a direct discretization of the differential equation, such as the finite difference method. Suitable differences replace the differential quotients. The basic defect is the difficulty of its application for non-rectangular or non-circular areas, and for problems which take into account the boundary conditions.

Therefore, formulations based on weak forms and variational forms have gained greater popularity. The advantages of such a course of action are as follows:

- unification of procedures in various theoretical and engineering problems; functionals are scalars, and as such do not depend on the reference system, which in turn facilitates appropriate transformations,
- weak forms and variational forms are the basis for effective computer methods,
- with variational forms and weak forms the basic principles of mechanics, such as energy conservation, conservation of mass, conservation of momentum and angular momentum can easily be expressed,
- the error estimation is facilitated, and so is the determination of the stability and the convergence of the numerical method employed.

Variational methods of solutions dominate methods based on strong forms. The latter are slowly beginning to address the conservation problems involved with historic sites.

1.4 Review of Analytical-Numerical Methods in Moving Load Problems

In academic textbooks devoted to differential equations one can find various solution methods. Special types of equations have dedicated solution schemes. In structural mechanics we limit our interest for the most part to elliptic equations in terms of the spatial variables and parabolic or hyperbolic equations in terms of time. Only a few equations can be solved analytically. A small number of them have closed solutions. Others must be solved numerically at the final stage, since solutions are usually given in the form of a series. Nowadays computational effort decreases when we use mathematical software, for example Mathematica, Matlab, Maple, etc. Numerical computations can be performed with the use of algebra packages such as Lapack. The Octave package allows us to automate computations. However, the physical model and its mathematical counterpart must still be elaborated by engineers. Proper simplification of a real task, classification of the phenomena, elaboration of the mathematical model, and finally an effective and accurate solution, must be accomplished by a human. Mathematical and mechanical knowledge must be properly advanced to verify solutions. In relatively simple cases we can compare the results of our solution procedures with known classical analytical solutions. In more complex problems we must rely on our experience.

Below we will present some solution methods as a background, to perform some simple tasks. The method of d'Alembert and Fourier methods are commonly applied. We will indicate the basic steps for the exemplary case of the string equation.

We will also show an alternate use of the Lagrange equation to derive the differential equations of motion of a structure.

1.4.1 d'Alembert Method

In 1713–15 B. Taylor elaborated the equation of small transverse vibrations of an infinitely thin, uniform string of length l, fixed at both ends, inclined from the equilibrium state. Only in 1747 did d'Alembert express this mechano-geometric formulation in the form of a second order partial differential equation:

$$\frac{\partial^2 y}{\partial x^2} = \frac{1}{c^2}\frac{\partial^2 y}{\partial t^2}\,, \tag{1.1}$$

where x and y are the coordinates of the material point of a string, t is time, and c is a constant depending on the tension and mass density of the string.

The general solution of (1.1) can be obtained by introducing new variables $\xi = x - ct$ and $\eta = x + ct$. Applying the chain rule we obtain

$$\frac{\partial}{\partial x} = \frac{\partial}{\partial \xi}\frac{\partial \xi}{\partial x} + \frac{\partial}{\partial \eta}\frac{\partial \eta}{\partial x} = \frac{\partial}{\partial \xi} + \frac{\partial}{\partial \eta}\,, \tag{1.2}$$

$$\frac{\partial}{\partial t} = \frac{\partial}{\partial \xi}\frac{\partial \xi}{\partial t} + \frac{\partial}{\partial \eta}\frac{\partial \eta}{\partial t} = -c\frac{\partial}{\partial \xi} + c\frac{\partial}{\partial \eta}\,. \tag{1.3}$$

We use (1.2) and (1.3) to compute the left and right sides of (1.1). We obtain

$$\frac{\partial^2 y}{\partial x^2} = \left(\frac{\partial}{\partial \xi} + \frac{\partial}{\partial \eta}\right)\left(\frac{\partial y}{\partial \xi} + \frac{\partial y}{\partial \eta}\right) = \frac{\partial^2 y}{\partial \xi^2} + 2\frac{\partial^2 y}{\partial \xi \partial \eta} + \frac{\partial^2 y}{\partial \eta^2}\,, \tag{1.4}$$

$$\frac{\partial^2 y}{\partial t^2} = \left(-c\frac{\partial}{\partial \xi} + c\frac{\partial}{\partial \eta}\right)\left(-c\frac{\partial y}{\partial \xi} + c\frac{\partial y}{\partial \eta}\right) = c^2\frac{\partial^2 y}{\partial \xi^2} - 2c^2\frac{\partial^2 y}{\partial \xi \partial \eta} + c^2\frac{\partial^2 y}{\partial \eta^2}\,, \tag{1.5}$$

respectively. Finally (1.1) reduces to

$$\frac{\partial^2 y}{\partial \xi \partial \eta} = 0\,. \tag{1.6}$$

This partial differential equation has the general solution

$$y(x,t) = f(\xi) + g(\eta) = f(x-ct) + g(x+ct)\,, \tag{1.7}$$

where f and g are arbitrary functions, with f representing a wave travelling to the right and g a wave travelling to the left.

The solution of the initial value problem of a string located at the position $y(x,0) = y_0(x)$ as a function of distance along the string x, and with vertical speed $\partial y/\partial t|_{t=0} = v_0(x)$, can be found as follows.

From the initial conditions and the general solution (1.7) we have,

$$y_0(x) = f(x) + g(x) . \tag{1.8}$$

Differentiating the displacement y with respect to t we have

$$v_0(x) = f'(x)\frac{\partial(x-ct)}{\partial t} + g'(x)\frac{\partial(x+ct)}{\partial t} = -cf'(x) + cg'(x) . \tag{1.9}$$

Integration gives us

$$\int_a^x v_0(s)ds = -cf(x) + cg(x) . \tag{1.10}$$

Solving (1.8) and (1.10) simultaneously for f and g we have

$$f(x) = \frac{1}{2}y_0(x) - \frac{1}{2c}\int_a^x v_0(s)ds , \tag{1.11}$$

$$g(x) = \frac{1}{2}y_0(x) + \frac{1}{2c}\int_a^x v_0(s)ds . \tag{1.12}$$

Plugging them into (1.8), we obtain the solution of the wave equation with specified initial conditions

$$y(x,t) = \frac{1}{2}y_0(x-ct) + \frac{1}{2}y_0(x+ct) + \frac{1}{2c}\int_{x-ct}^{x+ct} v_0(s)ds . \tag{1.13}$$

Later, d'Alembert proposed looking for solutions in the form of a product of two functions of one variable, i.e. with the method of separation of variables: $y = f(x)g(t)$. At the beginning of the 19th century this idea was extensively developed by J.-B. Fourier.

1.4.2 Fourier Method

The method of Fourier is one of the most widely used methods for solving partial differential equations. This method is described in detail in many books about vibrations and waves. The following is a brief description of the Fourier method for one-dimensional structures.

Let us consider the one-dimensional wave equation

$$\frac{\partial^2 w(x,t)}{\partial t^2} - c^2\frac{\partial^2 w(x,t)}{\partial x^2} = 0 . \tag{1.14}$$

In the case of an axially vibrating rod, the wave speed c is given by the equation

$$c = \sqrt{\frac{E}{\rho}} , \tag{1.15}$$

where E is the Young modulus and ρ is the mass density. We take a rod clamped on its left side, so the boundary conditions are

$$w(0,t) = 0 \quad \left.\frac{\partial w(x,t)}{\partial x}\right|_{x=0} = 0 \,. \tag{1.16}$$

Also we assume the initial conditions

$$w(x,0) = w_0(x), \quad \left.\frac{\partial w(x,t)}{\partial t}\right|_{t=0} = v_0(x) \,. \tag{1.17}$$

The solution of this initial-boundary problem can be obtained in the classical way of integration of a differential equation by grouping variables and proposing the solution in the form of a series of harmonic functions

$$w(x,t) = X(x)T(t) \,. \tag{1.18}$$

Substituting (1.18) into the equation of motion (1.14) we obtain

$$\ddot{T}(t)X(x) - c^2 T(t)X''(x) = 0 \,, \tag{1.19}$$

or in separated form

$$\frac{\ddot{T}(t)}{T(t)} = c^2 \frac{X''(x)}{X(x)} \,. \tag{1.20}$$

Since the left-hand side of equation (1.20) is only a function of time, while the right-hand side is only a function of the spatial variable, to ensure the equality at any time and at any point of the rod, both sides must be constant. Let us denote this constant by $-\omega^2$. This gives us

$$\frac{\ddot{T}(t)}{T(t)} = c^2 \frac{X''(x)}{X(x)} = -\omega^2 \,. \tag{1.21}$$

The equation (1.21) can be written in the following form

$$\ddot{T}(t) + \omega^2 T(t) = 0 \,, \tag{1.22}$$

$$X''(x) + \frac{\omega^2}{c^2} X(x) = 0 \,. \tag{1.23}$$

Applying the boundary conditions (1.16) to (1.18) and using the fact that $T(t) \not\equiv 0$, we obtain

$$X(0) = 0, \quad X'(x)\big|_{x=0} = 0 \,. \tag{1.24}$$

The solution of equation (1.21) satisfying the conditions (1.24) is called the solution of the boundary problem. Solutions of the boundary problem are called eigenfunctions, and the values of ω called eigenvalues. The solution of (1.21) is in the following form

$$X(x) = A\sin\left(\frac{\omega}{c}x\right) + B\cos\left(\frac{\omega}{c}x\right), \tag{1.25}$$

for $\omega \neq 0$, where A and B are constants. According to the boundary conditions (1.24), the constants A and B take the form

$$B = 0, \quad A\frac{\omega}{c}\cos\left(\frac{\omega}{c}l\right) - B\frac{\omega}{c}\sin\left(\frac{\omega}{c}l\right) = 0. \tag{1.26}$$

The existence of nonzero solutions is proved by equating to zero the determinant of the characteristic system of equations (1.26)

$$\begin{vmatrix} 0 & 1 \\ \frac{\omega}{c}\cos\frac{\omega l}{c} & -\frac{\omega}{c}\sin\frac{\omega l}{c} \end{vmatrix} = 0. \tag{1.27}$$

After transformation, we obtain the characteristic equation in the form

$$\frac{\omega}{c}\cos\frac{\omega l}{c} = 0. \tag{1.28}$$

Since $\omega = 0$ leads to the trivial solution $X(x) \equiv 0$, the eigenvalues are solutions of the characteristic equation (1.28)

$$\cos\frac{\omega l}{c} = 0. \tag{1.29}$$

Thus

$$\frac{\omega_n}{c}l = (2n-1)\frac{\pi}{2}, \tag{1.30}$$

where n is an integer describing the number of the term. Equation (1.30) defines an infinite sequence of eigenvalues ω_n. Each eigenvalue corresponds to the eigenfunction

$$X_n(x) = A_n\sin\left(\frac{\omega_n}{c}x\right) + B_n\cos\left(\frac{\omega_n}{c}x\right). \tag{1.31}$$

Assuming in equation (1.26) $\omega = \omega_n$ and $A = A_n$, $B = B_n$, we obtain

$$B_n = 0, \quad n = 1, 2, \ldots, \tag{1.32}$$

consequently

$$X_n(x) = A_n\sin\left(\frac{\omega_n}{c}x\right) = A_n\sin\left\{(2n-1)\frac{\pi}{2}\frac{x}{l}\right\}. \tag{1.33}$$

Let us specify a particular integral of equation (1.14). For the nth particular integral the function $T_n(t)$ satisfies the equation (1.22), where $\omega = \omega_n$. We have a solution for T_n in the following form

$$T_n(t) = K_n\cos(\omega_n t) + L_n\sin(\omega_n t). \tag{1.34}$$

In this way an infinite series of particular integrals has been specified

$$w_n(x,t) = X_n(x)T_n(t) = [K_n \cos(\omega_n t) + L_n \sin(\omega_n t)]\, A_n \sin\left(\frac{\omega_n}{c}x\right) , \tag{1.35}$$

where ω_n is given by equation (1.30).

1.4.3 Lagrange Formulation

If the position of the mass is known at two instants of time t_1 and t_2, then its motion during this interval of time can be represented by a curve. A slightly different curve or path is obtained if, at any instant, a small variation in position $\delta\xi$ is allowed with no associated change in time; that is, we assume δt=0. The stipulation is made, however, that at times t_1 and t_2 the two paths coincide, that is,

$$\delta\xi = 0 \quad \text{at} \quad t = t_1 \quad \text{and} \quad t = t_2 \, . \tag{1.36}$$

The problem is to choose the true path from ξ_1 to ξ_2 out of all possible ones.

According to Hamilton's principle for a conservative system, we can write

$$\int_{t_1}^{t_2} \delta\,(E_k - E_p)\,\mathrm{d}t = 0 \; . \tag{1.37}$$

E_k and E_p are the kinetic energy and the potential energy, respectively. The above formula describes the energy balance in the system. Equation (1.37) with respect to E_k=$f(\xi,\dot{\xi})$ is transformed to

$$\int_{t_1}^{t_2} \left(\frac{\partial E_k}{\partial \dot{\xi}}\delta\dot{\xi} + \frac{\partial E_k}{\partial \xi}\delta\xi - \frac{\partial E_p}{\partial \xi}\delta\xi\right)\mathrm{d}t = 0 \; . \tag{1.38}$$

Integration by parts allows us to rewrite one of the terms (1.38) as follows

$$\int_{t_1}^{t_2} \frac{\partial E_k}{\partial \dot{\xi}}\delta\dot{\xi}\,\mathrm{d}t = \frac{\partial E_k}{\partial \dot{\xi}}\delta\xi\bigg|_{t_1}^{t_2} - \int_{t_1}^{t_2} \frac{\mathrm{d}}{\mathrm{d}t}\left(\frac{\partial E_k}{\partial \dot{\xi}}\right)\delta\xi\,\mathrm{d}t \; . \tag{1.39}$$

Substituting (1.39) into (1.38) and employing the assumption (1.36), we can write

$$\int_{t_1}^{t_2} \left[-\frac{\mathrm{d}}{\mathrm{d}t}\left(\frac{\partial E_k}{\partial \dot{\xi}}\right) + \frac{\partial E_k}{\partial \xi} - \frac{\partial E_p}{\partial \xi}\right]\delta\xi\,\mathrm{d}t = 0 \; . \tag{1.40}$$

In the case of a system with many degrees of freedom, the deformation of which is described by n independent displacements, the general form of Lagrange's equation of the second kind is

$$\frac{\mathrm{d}}{\mathrm{d}t}\left(\frac{\partial E_k}{\partial \dot{\xi}_j}\right) - \frac{\partial E_k}{\partial \xi_j} + \frac{\partial E_p}{\partial \xi_j} = 0 \, , \quad j = 1,2,...,n \; . \tag{1.41}$$

1.5 Examples

Let us look at simple examples of a string and a beam vibrating under a moving load. We will treat the case of a massless load as well as the case with non-zero mass. We intuitively expect that differences in the results should occur and these results vary with the ratio of the moving mass to the mass of the structure. The second parameter that influences the process is the velocity of the load. Two types of diagrams are used for the analysis of such a problem: the displacements in time of the contact point moving together with the load, and displacements of the midpoint of the span.

Figure 1.8 depicts the displacements in time of the string motion. The left hand column shows the displacements under a travelling inertial particle and the right hand column depicts the displacements of the midpoint. The case of a massless force is plotted with a continuous line while the inertial load is plotted with a dashed line.

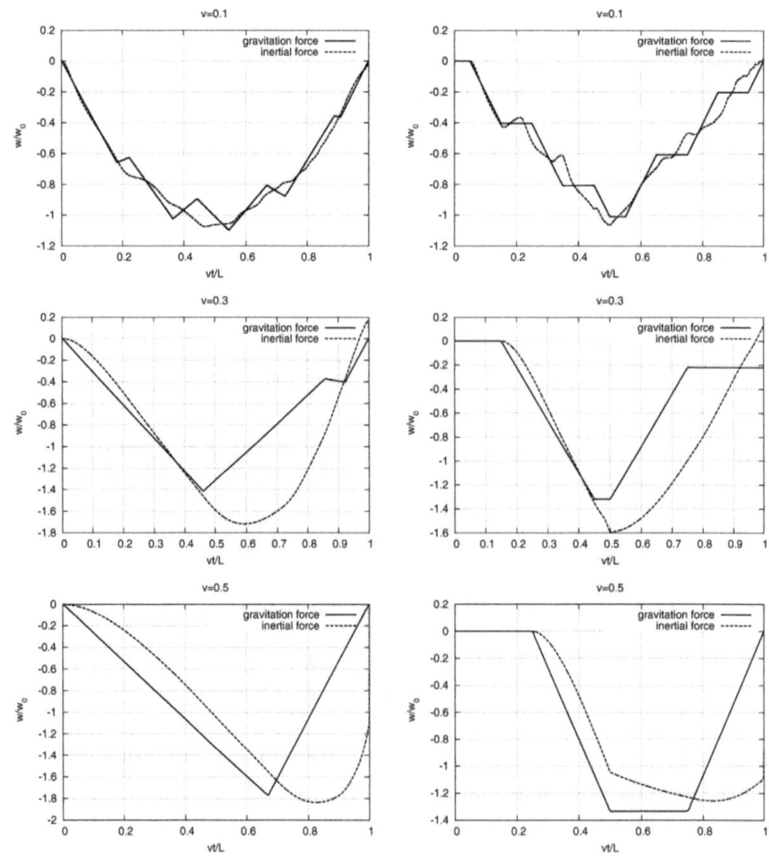

a b

Fig. 1.8 Point load travelling along a string: a) – under a travelling load, b) – at the midpoint.

In the first case, the only force acting on the string is equal to the weight of the corresponding mass. In the case of a massive load, the point mass also increases the inertia of the string. The comparison of both curves exhibits a smooth form in the case of the travelling mass. Moreover, the maximum of the displacement is shifted, exhibiting an elongation of the period of vibrations.

Identical features are observed in the case of a Bernoulli–Euler beam. The simplest model of a beam results in similar diagrams to those of even the more complex Timoshenko model, presented later. Differences in displacements can reach 50% and, although the mass trajectory is smooth, the acceleration values differ (Figure 1.9). Accelerations at the final stage of the process are significantly higher in the case of the moving inertial particle.

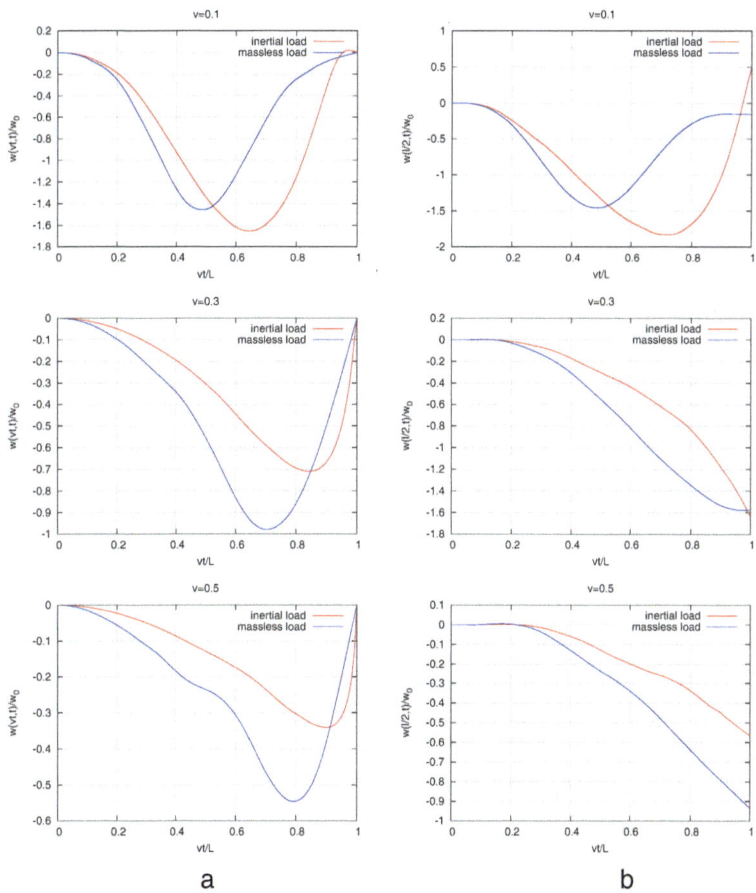

Fig. 1.9 Point load travelling along the Bernoulli–Euler beam: a) – under the travelling load, b) – at the midpoint.

The Timoshenko beam exhibits similar properties to those in the case of a string. We use dimensionless data in the example. The vertical acceleration of the travelling mass is high and noticeable jumps must be taken into account in engineering calculations. Although in real structures described by nonlinear equations we observe a smooth response, we should expect high vertical acceleration to be a physical feature of the problem. We should emphasize a significant difference in the case of both types of a load (Figure 1.10). Generally we can say that in the case of an inertial load, displacements are lower while accelerations are higher.

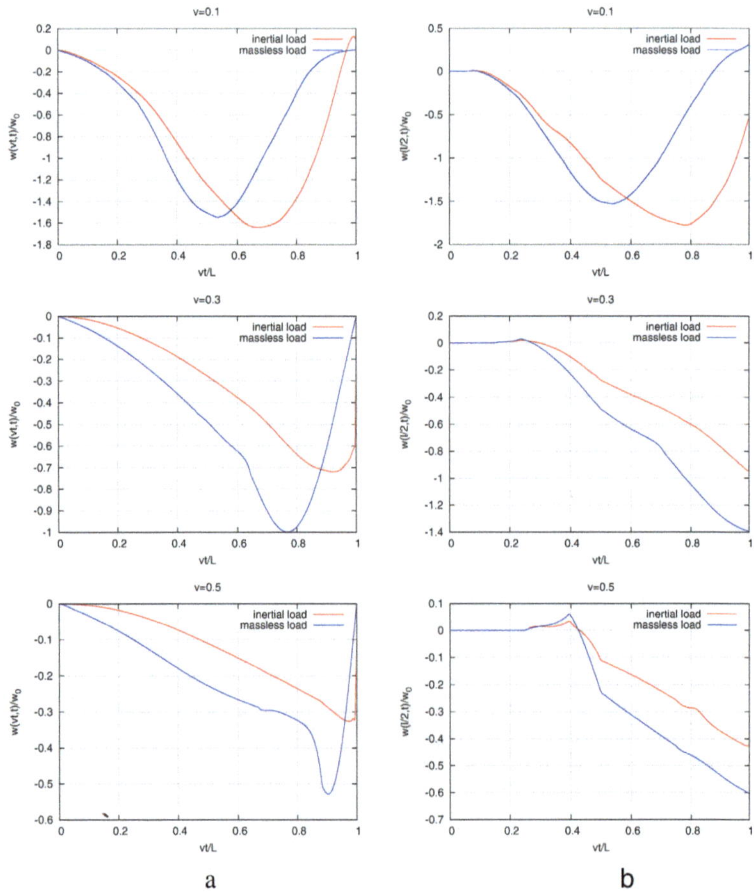

Fig. 1.10 Point load travelling along the Timoshenko beam: a) – under the travelling load, b) – at the midpoint.

Chapter 2
Analytical Solutions

A concentrated load acting on a continuous medium is usually described by a Dirac delta function. The point force or mass whose area of influence is limited, must be described in the entire spatial domain of the structure, for example $0 \le x \le l$. Multiplication of the force by the Dirac delta function $\delta(x)$ leads to such an effect. Then we have the load terms $\delta(x-x_0)P$ or $\delta(x-x_0)m\,\mathrm{d}^2w/\mathrm{d}t^2$ described in the domain of the problem. Unfortunately, the mathematical treatment of the term of the first type is relatively simple. It does not contain the solution variable. The treatment of a term of the second type, which describes the inertial force induced by the material particle, is much more complex. It includes the acceleration of the selected point x_0 as the second derivative of the solution of the differential equation w. What is more, the argument $x_0 = \mathrm{v}t$ moves with velocity v and the inertial term is a function of both x and t:

$$\delta(x-\mathrm{v}t)m\frac{\mathrm{d}^2w}{\mathrm{d}t^2} .$$

Due to the presence of the Dirac delta function in this result, the solutions obtained to these partial differential equations are not solutions in the classical sense, but are called 'weak' or 'distributional' solutions. So we must extend the concept of *solution*, arranging that any limit of an almost uniformly convergent sequence of classical solutions will be regarded as a *generalized solution* in the sense of a distribution. Distributions are therefore defined as the limit of sequences of continuous functions. This is called the sequential theory of distributions [4], in contrast to the functional theory [125]. For each distribution in the sense of L. Schwartz (functional) there is exactly one distribution in the sense of Mikusiński–Sikorski (sequential), and vice versa, so there is a mutual uniqueness [152]. Distributions are thus a generalization of functions. The purpose of the concept of a distribution is to give the correct meaning *qua* mathematical concept to objects such as the Dirac delta $\delta(x)$, which is much used in mathematical physics. An important feature of a distribution is that it ensures the posibility of differentiation, which is not always allowed for an arbitrary set of functions. The starting point for the sequential theory of distributions is the set of functions which are continuous on some fixed interval $A < x < B$ $(-\infty \le A < B \le \infty)$. If a sequence $f_n(x)$ of such continuous functions converges

C.I. Bajer and B. Dyniewicz: Numerical Analysis of Vibrations of Structures, LNACM 65, pp. 21–30.
springerlink.com

almost uniformly to a function $f(x)$, it is also convergent in the sense of distributions to $f(x)$ [4]. Every convergent sequence of distributions can be differentiated term by term (analogous to a uniformly convergent series). Of course, every uniformly convergent sequence is convergent almost uniformly. This allows, in the distribution sense, the differentiation of any function, changing the order of differentiations, and passing to the limit, without any restrictions. Such a statement in classical analysis is in general not true and is only possible under additional assumptions.

2.1 A Massless String under a Moving Inertial Load

Mathematically, a string is the simplest structure to be analysed. Engineering solutions frequently require the application of elements that resist tension but are flexible to bending, for example fibres, ropes, chains, or cables. Moreover, the same equation governs other physical problems such as the longitudinal vibrations of rods. First we will consider a massless string. The analysis will follow the solution of Frýba [56].

Assuming $\rho = 0$, the equation of motion of a string under a moving mass can be put into the following form:

$$-N\frac{\partial^2 w(x,t)}{\partial x^2} = \delta(x-\mathrm{v}t)P - \delta(x-\mathrm{v}t)\,m\frac{\mathrm{d}^2 w(\mathrm{v}t,t)}{\mathrm{d}t^2}\,. \tag{2.1}$$

We impose boundary conditions

$$w(0,t) = 0\,, \qquad w(l,t) = 0 \tag{2.2}$$

and initial conditions

$$w(x,0) = 0\,, \qquad \left.\frac{\partial w(x,t)}{\partial t}\right|_{t=0} = 0\,. \tag{2.3}$$

Here, $w(x,t)$ is the transverse displacement of the string, N is the tensile force, P and m are the point force and the point mass, respectively, and δ is the Dirac delta function.

To solve this equation we use the convolution property

$$w(x,t) = G(x,s) * p(s,t) = \int_0^l G(x,s)\,p(s,t)\,\mathrm{d}s\,, \tag{2.4}$$

where $G(x,s)$ is the Green's function obtained by solving the so-called basic equation by replacing the right-hand side of (2.1) with the Dirac delta function, $\delta(x-s)$. Finally, the displacements of the massless string according to (2.4) are the convolution of $G(x,s)$ and heterogeneity (2.1)

$$p(x,t) = \delta(x-\mathrm{v}t)\left(P - m\frac{\mathrm{d}^2 w(\mathrm{v}t,t)}{\mathrm{d}t^2}\right)\,. \tag{2.5}$$

A simple integration, based on the assumed boundary conditions (2.2) with the following notation

$$x = \mathrm{v}t \quad \text{and} \quad w_1(t) = w(\mathrm{v}t, t) \tag{2.6}$$

according to (2.4), (2.5), (2.6), and (2.2–2.3) results in the differential equation of motion of a moving massless string under a moving inertial load

$$w_1(t) = \left(P - m\frac{\mathrm{d}^2 w_1(t)}{\mathrm{d}t^2}\right)\left[\frac{1}{N}\left(1 - \frac{\mathrm{v}t}{l}\right)\mathrm{v}t\right]. \tag{2.7}$$

We assume dimensionless displacements of the string y and dimensionless time τ

$$y(\tau) = \frac{w_1(t)}{w_0} \quad \text{and} \quad \tau = \frac{\mathrm{v}t}{l}, \tag{2.8}$$

where

$$w_0 = \frac{Pl}{4N} \tag{2.9}$$

is the static deflection in the middle of the string. Substituting (2.8) into (2.7), we obtain the ordinary differential equation, heterogeneous with variable coefficients

$$\tau(1-\tau)\ddot{y}(\tau) + 2\alpha y(\tau) = 8\alpha\tau(1-\tau), \tag{2.10}$$

where

$$\alpha = \frac{Nl}{2m\mathrm{v}^2}. \tag{2.11}$$

The solution of (2.10) depends on the parameter α.

2.1.1 Case of $\alpha \neq 1$

We accept the equation being a solution of equation (2.10)

$$y(\tau) = \tau(1-\tau)\,v(\tau). \tag{2.12}$$

Substituting (2.12) and its second derivative into (2.10) we obtain

$$\tau(1-\tau)\ddot{v}(\tau) + (2-4\tau)\dot{v}(\tau) - 2(1-\alpha)v(\tau) = 8\alpha. \tag{2.13}$$

The homogeneous part (2.13) is the hypergeometric equation [126] of the general form shown below

$$\tau(1-\tau)\ddot{v}(\tau) + [c-(a+b+1)\tau]\dot{v}(\tau) - ab\,v(\tau) = 0. \tag{2.14}$$

In the first step, we solve the homogeneous part of (2.13) where the coefficients a, b, and c have the following form

$$a_{1,2} = \frac{3 \pm \sqrt{1+8\alpha}}{2}, \quad b_{1,2} = \frac{3 \mp \sqrt{1+8\alpha}}{2}, \quad c = 2. \tag{2.15}$$

The solution of the hypergeometric equation, when the number c is a positive integer $c = 1+m$ and $a \neq m$, $b \neq m$, takes the form

$$v_1(\tau) = F(a,b,c,\tau) ,$$

$$\begin{aligned} v_2(\tau) = F(a,b,c,\tau)\ln\tau + \sum_{k=1}^{\infty} \left\{ \frac{(a_k)(b_k)}{(c_k)} [h(k) - h(0)] \frac{\tau^k}{k!} + \right. \\ \left. + \frac{1}{(1-a)(1-b)\tau} \right\} , \end{aligned} \tag{2.16}$$

where $F(a,b,c,\tau)$ is a hypergeometric series:

$$F(a,b,c,\tau) = 1 + \sum_{k=1}^{\infty} \frac{(a_k)(b_k)}{(c_k)} \frac{\tau^k}{k!} , \tag{2.17}$$

and (a_k), (b_k) i (c_k) are the so called Pochhammer symbols:

$$\begin{aligned} (a_k) &= a(a+1)\ldots(a+k-1) , \\ (b_k) &= b(b+1)\ldots(b+k-1) , \\ (c_k) &= c(c+1)\ldots(c+k-1) . \end{aligned} \tag{2.18}$$

The particular solution (2.13) on the basis of its heterogeneity takes the form

$$v_s(\tau) = \frac{4\alpha}{\alpha - 1} . \tag{2.19}$$

According to (2.12) for $\alpha \neq 1$ the equation (2.10) is the following

$$y(\tau) = [A_1 v_1(\tau) + A_2 v_2(\tau) + v_s(\tau)] \tau (1-\tau) . \tag{2.20}$$

Based on the initial conditions (2.3) we calculate the constants A_1 and A_2:

$$A_1 = \frac{-4\alpha}{\alpha - 1} , \qquad A_2 = 0 . \tag{2.21}$$

The constant $A_2 = 0$ greatly simplifies the formula (2.20). Finally, the displacements of the string under a moving load for $\alpha \neq 1$ are of the form

$$\begin{aligned} y(\tau) &= \frac{4\alpha}{\alpha - 1} \tau (1-\tau) [1 - v_1(\tau)] = \\ &= \frac{4\alpha}{\alpha - 1} \tau (1-\tau) [1 - F(a,b,c,\tau)] , \end{aligned} \tag{2.22}$$

where $F(a,b,c,\tau)$ is given by equation (2.17). In Figure 2.1 we can see the impact of the accuracy on the solution near the end supports.

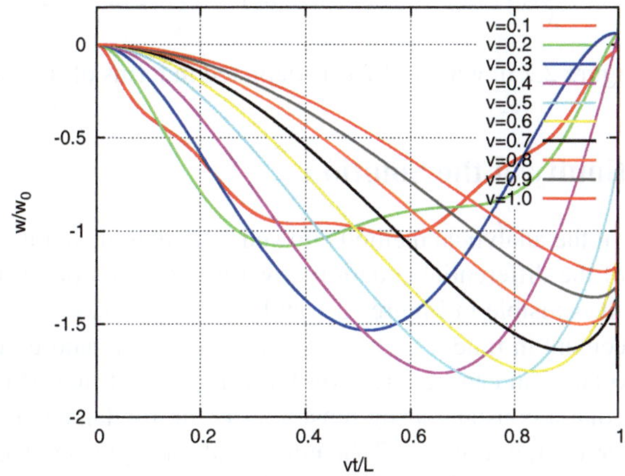

Fig. 2.1 Trajectory of a mass travelling over a massless string at different speeds.

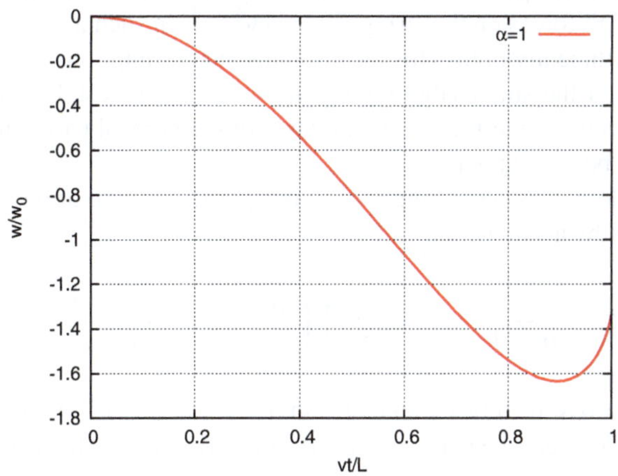

Fig. 2.2 The trajectory of the mass in the case of a massless string (case $\alpha = 1$) described by (2.24).

2.1.2 *Case of* $\alpha = 1$

In this case, equation (2.10) becomes

$$\tau(1-\tau)\ddot{y}(\tau) + 2y(\tau) = 8\,\tau(1-\tau)\,. \tag{2.23}$$

The above equation has a closed-form analytical solution. The displacements of the string when $\alpha = 1$, under the conditions (2.2–2.3) are given below:

$$y(\tau) = \frac{4}{3}\tau(1-\tau) - \frac{4}{3}\tau(1+2\tau\ln(1-\tau) - 2\ln(1-\tau)) \ . \tag{2.24}$$

Let us look at Figure 2.2 in which (2.24) is depicted in terms of time $\mathrm{v}t/l$.

2.2 Discontinuity of the Solution

The advantages of the analytical method allowed us to exhibit an interesting feature of the solution of the differential equation governing the motion of the string near an end support. It is visible in Figures 2.1 and 2.2. The diagrams exhibit jumps of the mass displacement in time. Let us consider the physical nature of these jumps. The simplest explanation can be based on the force equilibrium (Figure 2.3). We must remember that a constant string tension N is the fundamental assumption in our problem. Moreover, in Figure 2.3 the horizontal force pushing the mass to hold the speed v must be seen in the scheme. At the final stage (as depicted in Figure 2.3), the remaining distance d will be traversed in time d/v. In this period, the mass m must be lifted from the position w_B to zero. If the deflection w_B is high enough, compared to the other parameters, the necessary acceleration applied to the mass must result in high forces on the string $F \sim wm\mathrm{v}^2/d^2$. In such a case, F can exceed N if m or v is sufficiently high. This fact violates our assumptions and the condition of applicability of the small vibration equation $(\partial w/\partial x)^2 << 1$. A mathematical proof can be given only in the particular case. In the general case, only numerical simulations can be carried out.

Let us consider a massless string, which is a particular case of our problem. The solution is given by a sum (2.22)

$$y(\tau) = \frac{4\alpha}{\alpha-1}\tau(\tau-1)\sum_{k=1}^{\infty}\prod_{i=1}^{k}\frac{(a+i-1)(b+i-1)}{c+i-1}\frac{\tau^k}{k!} \ , \tag{2.25}$$

where $\tau = \mathrm{v}t/l > 0$ is the time parameter, and $\alpha = Nl/(2m\mathrm{v}^2) > 0$ determines the dimensionless parameter. The parameters a, b and c are given by the equations (2.15).

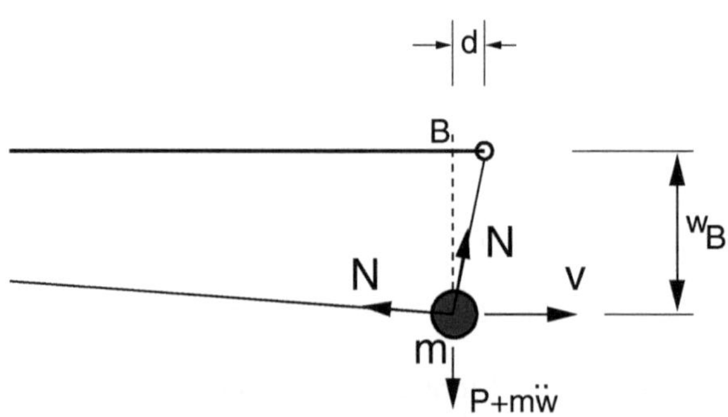

Fig. 2.3 Final stage of the moving mass.

In the case of $\alpha = 1$, the initial problem has a closed-form solution given by (2.24). The mass position near $x = l$, i.e. with $\tau = 1^-$, tends to -4/3.

Below we will consider the case of $\alpha \neq 1$. In Figure 2.4 we can notice the strong influence of the precision on the solution near the end support. Let us consider the solution given by (2.25). The first term $\tau(\tau-1)$ results in zero at $\tau = 1$. The expression

$$\sum_{k=1}^{\infty}\prod_{i=1}^{k}\frac{(a+i-1)(b+i-1)}{c+i-1}\,\frac{\tau^k}{k!} \tag{2.26}$$

tends to ∞ if $\tau \to 1$. We have an indefinite solution at $\tau = 1^-$.

The same result can be obtained on the basis of Abel's theorem. The power series, as a part of (2.25), can be written in the form

$$\sum_{k=1}^{\infty} A_k\,\tau^k\,, \quad A_k = \prod_{i=1}^{k}\frac{(a+i-1)(b+i-1)}{(c+i-1)\,i}\,. \tag{2.27}$$

In this case $\lim\limits_{\tau\to 1^-} A_k\,\tau^k = \infty$ and $y(1^-) = 0\cdot\infty$.

In the case $a+b<c$, the series (2.27) is convergent and there are no singularities. However, this is not our case since it does not fulfill (2.15). In the case $a+b>c$, the series diverges (the sum tends to ∞). We have an indefinite value $0\cdot\infty$ while testing the function.

We can also carry out another scheme of analysis. Below we will include the term $\tau(\tau-1)$ in the sum of (2.25). Thus it can be reduced to the following form:

$$\begin{aligned}(1-\tau)\sum_{k=1}^{\infty}\frac{(a_k)(b_k)}{(c_k)}\frac{\tau^k}{k!} &= \\ = \frac{ab\tau}{c} + \sum_{k=2}^{\infty}\frac{(a_{k-1})(b_{k-1})}{(c_{k-1})}&\left(\frac{(a+k-1)(b+k-1)}{k(c+k-1)} - 1\right)\frac{\tau^k}{(k-1)!}\,,\end{aligned} \tag{2.28}$$

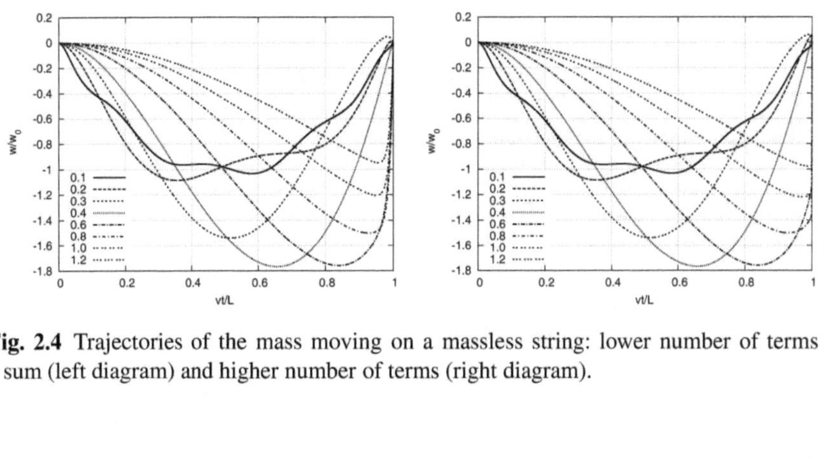

Fig. 2.4 Trajectories of the mass moving on a massless string: lower number of terms in a sum (left diagram) and higher number of terms (right diagram).

where

$$(a_k) = a(a+1)\dots(a+k-1)\,,$$
$$(b_k) = b(b+1)\dots(b+k-1)\,,$$
$$(c_k) = c(c+1)\dots(c+k-1)\,.$$

By using the Rabbe criterion one can show that for $a+b<c+2$ the limit

$$\lim_{\tau\to 1}\left[(1-\tau)\sum_{k=1}^{\infty}\prod_{i=1}^{k}\frac{(a+i-1)(b+i-1)}{c+i-1}\frac{\tau^k}{k!}\right] \tag{2.29}$$

is finite. Now we can estimate the value of the sum (2.28). The sum of the first two or three terms, depending on parameters, including $ab\tau/c$, is positive. The next terms are all positive. This proves that the sum (2.28) is finite and is greater than 0. The scheme of the function (2.25) is depicted in Figure 2.5. The case $a+b=c+1$ is special (our set of parameters), for which the convergence is faster.

Let us look at the boundary condition at $\tau=1$. We can say that it is fulfilled, however. We can imagine a symmetrical problem, with the mass moving from $\tau=2$ towards $\tau=1$ (with opposite direction of the force P) (Figure 2.6). Then we have two analogous problems at $\tau=1$. Both limits result in zero value at $\tau=1$:

$$\frac{1}{2}\left(\lim_{\tau\to 1^-} y(\tau)+\lim_{\tau\to 1^+} y(\tau)\right)=0\,.$$

We can also consider the derivative $\mathrm{d}y/\mathrm{d}\tau$. The resulting formula can be derived. For negative P the result is

$$\lim_{\tau\to 1^-}\frac{\mathrm{d}y}{\mathrm{d}\tau}=\infty\,. \tag{2.30}$$

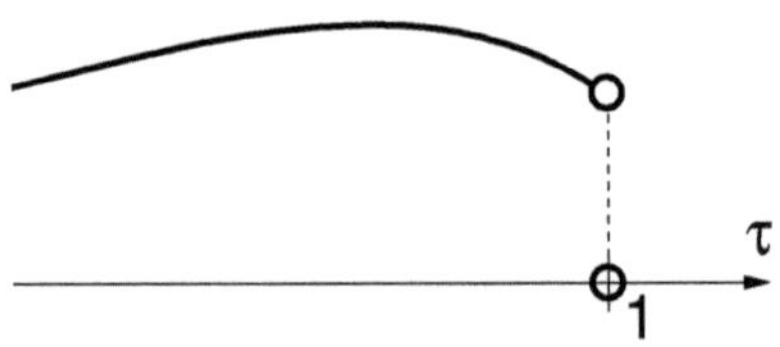

Fig. 2.5 Discontinuity of the function (2.25) at $\tau=1$.

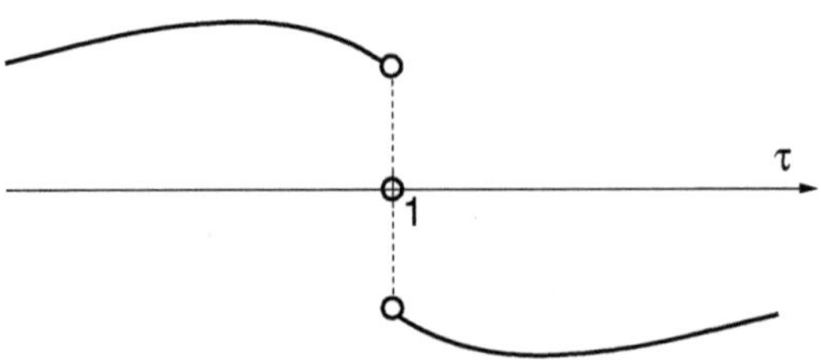

Fig. 2.6 Left and right limits at $\tau=1$.

The above limit shows that the angle of the string line at the final point is infinite. In practice the string vertically approaches the support. Mathematically the function is discontinuous there.

2.3 Conclusions

In this chapter we presented a global analytical formulation of the vibration problem for a string subjected to a moving mass. The numerical solution of the resulting matrix differential equation of the second order is relatively simple and is valid for the whole range of speeds v: sub-critical, critical, and over-critical. The analysis of the results exhibits a jump of the mass in the neighborhood of the end support. The force acting on the mass is, however, limited to the tensile force N. A discontinuity of the mass trajectory at $x = l$ exists in the case of $0 < \mathrm{v} \leq c$. In the case of a massless string this discontinuity is mathematically proved. In case $\mathrm{v} > c$, there is no discontinuity, since for $x \geq \mathrm{v}t$ the deflection $w(\mathrm{v}t,t) = 0$.

Unfortunately, we can not give an answer to the question of whether the string is continuous in the case that the discontinuity of the particle's trajectory occurs. The massless or massive string shape is not determined in the analytical form. We can only expect such a discontinuity on the basis of numerical results. The particle motion is continuous only in the trivial case of $m = 0$. The expression in parenthesis in (2.28) is equal to zero, and since $\alpha = 0$ in (2.25), finally $y(1^-) = 0$ and $y(1) = 0$.

We consider small vibrations. The discontinuity in this case is a feature of mathematical interest rather than any practical one. However, in various analytical or numerical investigations of problems with travelling inertial load, one can find poor convergence of the solutions in the places where the boundary conditions are imposed. Our analysis can explain the anomalies in such cases.

Figure 2.1 shows the mass trajectory moving along the massless string at a speed of from 0.1–1.0 of the wave speed c in the case of the inertial speed. Practically, each curve exhibits ia strong discontinuity of the trajectory.

We can observe the same properties of the solution in the case of an inertial string. A comparative plot is presented in Figure 2.7. We should consider three parameters: the moving mass value, the massless force, and the mass of the string in the case of an inertial string. The plots depend on the ratios of these parameters. The curves for a massless string in Figure 2.7 were compared with the plots computed when the ratio of the moving mass to the string mass was equal to one. We can emphasize that in the case of a lower $m/\rho/A/l$ ratio, the coincidence of each pair of curves is higher. However, an analytical proof of the discontinuity in the case of the inertial string is impossible, because of the complexity of the differential equation and the necessity of a numerical integration stage.

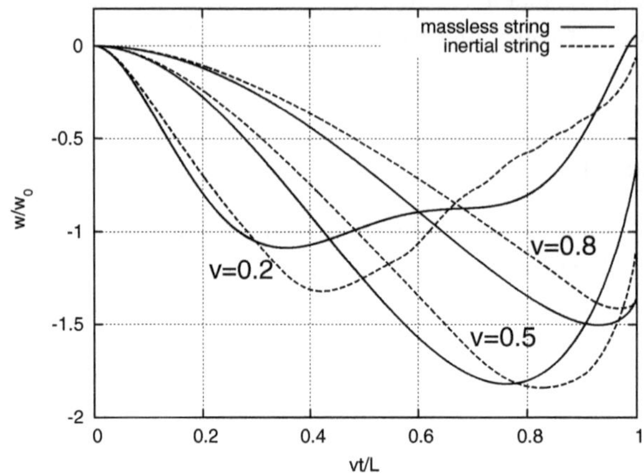

Fig. 2.7 Comparison of particle trajectory for massless and inertial strings.

Chapter 3
Semi-analytical Methods

Problems of the dynamics of moving loads can be divided into three main groups depending on the nature of the load. The first is called the Willis–Stokes [131, 140] problem, describing the motion of an inertial point load travelling along a massless Euler beam. We know its complete analytical solution. The second case is related to the load of a constant amplitude moving along an inertial beam. This task was first solved by Krylov [75]. Further works discussed the influence of the elastic foundation [1, 129] and subcritical and critical velocities of the moving force [53]. Also in the case of a moving force with periodic amplitude, the complete analytical solutions are known [30, 94, 96, 137]. An excellent summary of these works is given by Frýba in his monograph [56]. He discusses in detail the majority of types of such problems. An inertial load moving along a beam is the third problem. Closed analytical solutions of this case do not exist. In the literature we can find approximate analytical solutions [70, 71, 93, 124, 130]. A review of the most important analytical contributions was given by Ayre *et al.* [8]. The authors performed an experiment, of which the results were compared with other approximate analytical solutions published in the literature. Other papers have been devoted to semi-analytical solutions [25, 69, 86, 98, 111, 122, 138, 151]. The equation of motion is transformed analytically to a relatively simple form and then numerical methods are used for its integration. These solutions are much more accurate, but unfortunately they can not be investigated analytically. Our previous research led to the discovery of interesting properties of solutions for the string and Timoshenko beam [48]. The discontinuity of the trajectory of a moving material point near the end supports was exhibited. In the case of a massless string, this feature was mathematically proved. Semi-analytical solutions are not versatile. Altering the imposed boundary conditions requires finding a new method of calculation. Thus semi-analytical solutions are unsuitable for engineering practice. However, they are well-suited for the verification of numerical solutions.

C.I. Bajer and B. Dyniewicz: Numerical Analysis of Vibrations of Structures, LNACM 65, pp. 31–76.
springerlink.com

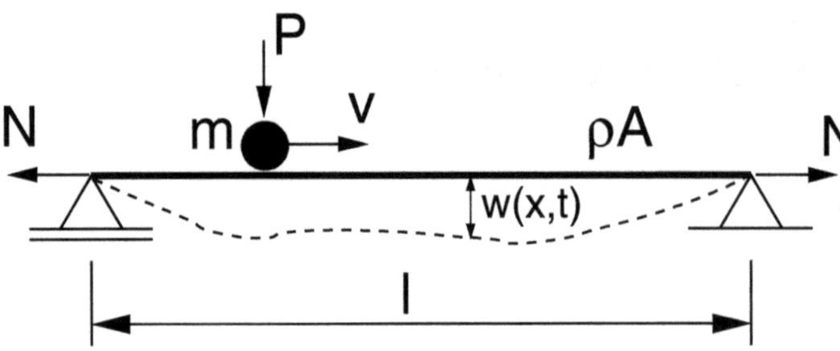

Fig. 3.1 Moving inertial load.

3.1 String

Let us consider a string of length l, cross-sectional area A, mass density ρ, and tensile force N, subjected to a mass m accompanied by a force P (Figure 3.1), moving with constant speed v. The equation of motion of the string under a moving inertial load with constant speed v has the form

$$-N\,\frac{\partial^2 w(x,t)}{\partial x^2} + \rho A\,\frac{\partial^2 w(x,t)}{\partial t^2} = \delta(x-\mathrm{v}t)P - \delta(x-\mathrm{v}t)m\frac{\mathrm{d}^2 w(\mathrm{v}t,t)}{\mathrm{d}t^2}\,. \tag{3.1}$$

We impose boundary conditions

$$w(0,t) = 0, \qquad w(l,t) = 0 \tag{3.2}$$

and initial conditions

$$w(x,0) = 0, \qquad \left.\frac{\partial w(x,t)}{\partial t}\right|_{t=0} = 0\,. \tag{3.3}$$

The first term in equation (3.1) describes the forces that resist the external load in a neighbourhood of the application of the load. The second term introduces the inertia of the material point of the string. The right-hand side of the equation contributes the external force and the inertial force induced by the material inertial point m. The boundary conditions describe the constant zero position of both ends of the string. The initial conditions in turn establish the straight zero initial position and stationary state, i.e. zero velocity, at the initial time.

3.1.1 Fourier Analysis

In order to reduce the partial differential equation to an ordinary differential equation, we apply the Fourier sine integral transformation in a finite range (i.e. a finite length of the string)

$$w(x,t) = \frac{2}{l}\sum_{j=1}^{\infty} V(j,t)\sin\frac{j\pi x}{l}\,, \tag{3.4}$$

$$V(j,t) = \int_0^l w(x,t)\sin\frac{j\pi x}{l}\,\mathrm{d}x\,. \tag{3.5}$$

We can present each of the functions as an infinite sum of sine functions (3.4) with coefficients (3.5). Then the expansion of the moving mass acceleration in a series has a form

$$\frac{\mathrm{d}^2 w(\mathrm{v}t,t)}{\mathrm{d}t^2} = \frac{2}{l}\sum_{k=1}^{\infty}\left[\ddot{V}(k,t)\sin\frac{k\pi \mathrm{v}t}{l} + \frac{2k\pi\mathrm{v}}{l}\dot{V}(k,t)\cos\frac{k\pi\mathrm{v}t}{l} - \right.$$
$$\left. - \frac{k^2\pi^2\mathrm{v}^2}{l^2}V(k,t)\sin\frac{k\pi\mathrm{v}t}{l}\right]\,. \tag{3.6}$$

The integral transformation (3.5) of equation (3.1) with consideration of (3.6) can be performed

$$N\,\frac{j^2\pi^2}{l^2}V(j,t) + \rho A\,\ddot{V}(j,t) =$$
$$= P\sin\frac{j\pi\mathrm{v}t}{l} - m\frac{\mathrm{d}^2 w(\mathrm{v}t,t)}{\mathrm{d}t^2}\int_0^l \delta(x-\mathrm{v}t)\sin\frac{j\pi x}{l}\,\mathrm{d}x\,. \tag{3.7}$$

The integral with the Dirac delta function in the above equation is as follows

$$\int_0^l \delta(x-\mathrm{v}t)\sin\frac{j\pi x}{l}\,\mathrm{d}x = \sin\frac{j\pi\mathrm{v}t}{l}\,. \tag{3.8}$$

Let us consider now (3.6) and (3.8):

$$N\,\frac{j^2\pi^2}{l^2}V(j,t) + \rho A\ddot{V}(j,t) =$$
$$= P\sin\frac{j\pi\mathrm{v}t}{l} - \frac{2m}{l}\sum_{k=1}^{\infty}\ddot{V}(k,t)\sin\frac{k\pi\mathrm{v}t}{l}\sin\frac{j\pi\mathrm{v}t}{l} -$$
$$- \frac{2m}{l}\sum_{k=1}^{\infty}\frac{2k\pi\mathrm{v}}{l}\dot{V}(k,t)\cos\frac{k\pi\mathrm{v}t}{l}\sin\frac{j\pi\mathrm{v}t}{l} +$$
$$+ \frac{2m}{l}\sum_{k=1}^{\infty}\frac{k^2\pi^2\mathrm{v}^2}{l^2}V(k,t)\sin\frac{k\pi\mathrm{v}t}{l}\sin\frac{j\pi\mathrm{v}t}{l}\,. \tag{3.9}$$

Finally, the equation of motion, after a Fourier transformation, can be written

$$\begin{aligned}\rho A\,\ddot{V}(j,t)+\alpha\sum_{k=1}^{\infty}\ddot{V}(k,t)\,\sin\omega_k t\,\sin\omega_j t\,+&\\ +2\alpha\sum_{k=1}^{\infty}\omega_k\dot{V}(k,t)\,\cos\omega_k t\,\sin\omega_j t+\Omega^2V(j,t)-&\\ -\alpha\sum_{k=1}^{\infty}\omega_k^2\,V(k,t)\,\sin\omega_k t\,\sin\omega_j t\;=\;P\,\sin\omega_j t\;,&\end{aligned}\tag{3.10}$$

where

$$\omega_k\;=\;\frac{k\pi v}{l}\,,\qquad \omega_j\;=\;\frac{j\pi v}{l}\,,\qquad \Omega^2\;=\;N\,\frac{j^2\pi^2}{l^2}\,,\qquad \alpha\;=\;\frac{2m}{l}\,.\tag{3.11}$$

The analytical solution of this problem is unknown. We must solve this final system of differential equations numerically. Thus we obtain a semi-analytical solution. Equation (3.10) can be written in matrix form, where **M**, **C** and **K** are square matrices:

$$\mathbf{M}\begin{bmatrix}\ddot{V}(1,t)\\ \ddot{V}(2,t)\\ \vdots\\ \ddot{V}(n,t)\end{bmatrix}+\mathbf{C}\begin{bmatrix}\dot{V}(1,t)\\ \dot{V}(2,t)\\ \vdots\\ \dot{V}(n,t)\end{bmatrix}+\mathbf{K}\begin{bmatrix}V(1,t)\\ V(2,t)\\ \vdots\\ V(n,t)\end{bmatrix}=\mathbf{P}\tag{3.12}$$

or

$$\mathbf{M}\ddot{\mathbf{V}}+\mathbf{C}\dot{\mathbf{V}}+\mathbf{K}\mathbf{V}=\mathbf{P}\,,\tag{3.13}$$

where

$$\begin{aligned}\mathbf{M}=&\begin{bmatrix}\rho A & 0 & \cdots & 0\\ 0 & \rho A & \cdots & 0\\ \vdots & \vdots & \ddots & \vdots\\ 0 & 0 & \cdots & \rho A\end{bmatrix}+\\ &+\alpha\begin{bmatrix}\sin\frac{1\pi vt}{l}\sin\frac{1\pi vt}{l} & \sin\frac{1\pi vt}{l}\sin\frac{2\pi vt}{l} & \cdots & \sin\frac{1\pi vt}{l}\sin\frac{n\pi vt}{l}\\ \sin\frac{2\pi vt}{l}\sin\frac{1\pi vt}{l} & \sin\frac{2\pi vt}{l}\sin\frac{2\pi vt}{l} & \cdots & \sin\frac{2\pi vt}{l}\sin\frac{n\pi vt}{l}\\ \vdots & \vdots & \ddots & \vdots\\ \sin\frac{n\pi vt}{l}\sin\frac{1\pi vt}{l} & \sin\frac{n\pi vt}{l}\sin\frac{2\pi vt}{l} & \cdots & \sin\frac{n\pi vt}{l}\sin\frac{n\pi vt}{l}\end{bmatrix},\end{aligned}\tag{3.14}$$

$$\mathbf{C}=2\alpha\begin{bmatrix}\frac{1\pi v}{l}\sin\frac{1\pi vt}{l}\cos\frac{1\pi vt}{l} & \frac{2\pi v}{l}\sin\frac{1\pi vt}{l}\cos\frac{2\pi vt}{l} & \cdots & \frac{n\pi v}{l}\sin\frac{1\pi vt}{l}\cos\frac{n\pi vt}{l}\\ \frac{1\pi v}{l}\sin\frac{2\pi vt}{l}\cos\frac{1\pi vt}{l} & \frac{2\pi v}{l}\sin\frac{2\pi vt}{l}\cos\frac{2\pi vt}{l} & \cdots & \frac{n\pi v}{l}\sin\frac{2\pi vt}{l}\cos\frac{n\pi vt}{l}\\ \vdots & \vdots & \ddots & \vdots\\ \frac{1\pi v}{l}\sin\frac{n\pi vt}{l}\cos\frac{1\pi vt}{l} & \frac{2\pi v}{l}\sin\frac{n\pi vt}{l}\cos\frac{2\pi vt}{l} & \cdots & \frac{n\pi v}{l}\sin\frac{n\pi vt}{l}\cos\frac{n\pi vt}{l}\end{bmatrix},\tag{3.15}$$

$$
\mathbf{K} = \begin{bmatrix} \frac{1^2\pi^2}{l^2}N & 0 & \cdots & 0 \\ 0 & \frac{2^2\pi^2}{l^2}N & \cdots & 0 \\ \vdots & \vdots & \ddots & \vdots \\ 0 & 0 & \cdots & \frac{n^2\pi^2}{l^2}N \end{bmatrix} - \tag{3.16}
$$

$$
-\alpha \begin{bmatrix} \frac{1^2\pi^2 v^2}{l^2}\sin\frac{1\pi vt}{l}\sin\frac{1\pi vt}{l} & \frac{2^2\pi^2 v^2}{l^2}\sin\frac{1\pi vt}{l}\sin\frac{2\pi vt}{l} & \cdots & \frac{n^2\pi^2 v^2}{l^2}\sin\frac{1\pi vt}{l}\sin\frac{n\pi vt}{l} \\ \frac{1^2\pi^2 v^2}{l^2}\sin\frac{2\pi vt}{l}\sin\frac{1\pi vt}{l} & \frac{2^2\pi^2 v^2}{l^2}\sin\frac{2\pi vt}{l}\sin\frac{2\pi vt}{l} & \cdots & \frac{n^2\pi^2 v^2}{l^2}\sin\frac{2\pi vt}{l}\sin\frac{n\pi vt}{l} \\ \vdots & \vdots & \ddots & \vdots \\ \frac{1^2\pi^2 v^2}{l^2}\sin\frac{n\pi vt}{l}\sin\frac{1\pi vt}{l} & \frac{2^2\pi^2 v^2}{l^2}\sin\frac{n\pi vt}{l}\sin\frac{2\pi vt}{l} & \cdots & \frac{n^2\pi^2 v^2}{l^2}\sin\frac{n\pi vt}{l}\sin\frac{n\pi vt}{l} \end{bmatrix},
$$

$$
\mathbf{P} = P \begin{bmatrix} \sin\frac{1\pi vt}{l} \\ \sin\frac{2\pi vt}{l} \\ \vdots \\ \sin\frac{n\pi vt}{l} \end{bmatrix}. \tag{3.17}
$$

When the coefficients $V(j,t),\ j = 1,...,n$ are computed, the displacements of the string can be determined as a solution of (3.4). This gives the solution for a full range of parameters. We can calculate the displacements at each point of the string and for all values of v. We see that assuming $\rho = 0$ in (3.14) we have the formulation for a massless string.

Numerical computations were carried out for unitary dimensionless data: length, tensile force and mass density. The Newmark scheme was used for integration. In such a way both the problem and the solution were scaled to normalized values. The wave speed in this case is unity ($c = 1$). First we present the moderate convergence rate of the series which constitutes the solution (Figure 3.2). We denote the wave speed in the unloaded string as c ($c^2 = N/\rho/A$). Further diagrams exhibit the vertical deflection of the string w related to the deflection in quasi-static mass motion in the middle of the span w_0. We can notice that the first term is already close to the exact solution. Three or five terms are sufficient for an accurate result in the engineering sense. We must emphasize here that a higher speed of the mass, for example equal to $0.9\,c$ or c requires even a hundred terms and a short time step for time integration of the differential equation, since the solution exhibits small jumps near the final support. An accurate plot for different velocities v is given in Figure 3.3.

Let us look at the diagrams of the displacements of the string at the point under the moving mass. The diagram for various mass values related to the string mass, for the speed $v = 0.2c$, is depicted in Figure 3.4. We see that even a small mass equal to 0.2 of the string mass smooths the trajectory. A more detailed presentation of the string motion is given in Figures 3.5 and 3.6. We can notice the sharp edge of the wave and the reflection from both supports. The case of $v = c$ is interesting since it gives a sharp, practically vertical, surface of the plot at the mass trajectory. The space-time subdomain $x > vt$ of the plot is not affected by the process if $v \geq c$.

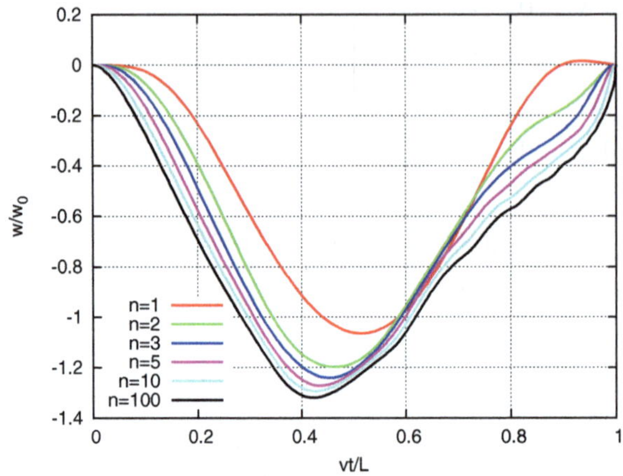

Fig. 3.2 Convergence of the solution for $\mathsf{v} = 0.2c$ depending on the number of terms in the trigonometric series.

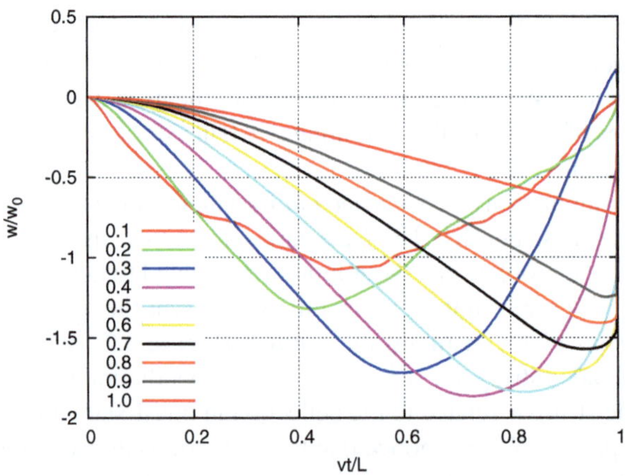

Fig. 3.3 Inertial string — displacements computed semi-analytically at different speeds.

Here we must also focus our attention on the final stage of the path of the inertial particle, which is better exhibited in Figure 3.7. Especially in the case of $\mathsf{v} = 0.5c$ we notice that the travelling mass does not reach the end with the expected position $w = 0$, imposed by the boundary conditions. This problem was discussed in detail in Section 2.2.

The convergence to the discontinuous solution near the end point is depicted in Figure 3.8. The mass trajectory is plotted for increasing numbers of terms at the speed $\mathsf{v} = 0.5\,c$. We notice that the function tends slowly to the jump at $x = l$. All characteristic lines are smooth, since a limited number of terms was assumed in the computations. The convergence rate is slow, and especially near $x = l$, the number of

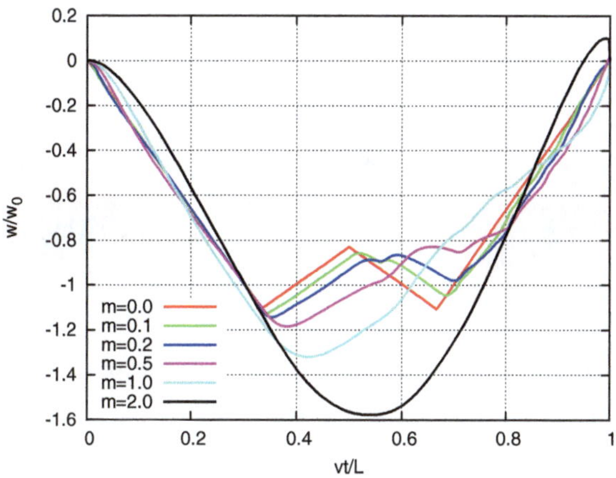

Fig. 3.4 Displacements under the mass for different mass values at the speed $\mathrm{v} = 0.2c$.

terms taken must be at least 50. At high velocities (in our case $\mathrm{v} > 0.8\,c$) a sufficiently small time step for the integration of (3.13) must be applied (even 10^{-5}) to avoid small oscillations of the solution in the last stage.

Figure 3.9 exhibits very good convergence of the deflected axis of the string, even with a small number of terms in a series, for almost entire time of the process. Unfortunately, at the final stage of the process the oscillations of the numerical solution appear. At $t = 0.8$ and $0.9\ l/\mathrm{v}$ the resulting curves are accurate even if plotted with 100 terms. All the curves coincide with those obtained for 400 terms. At $t = 0.99\,l/\mathrm{v}$ the accuracy increases with the the number of terms. Figure 3.9*d* corresponds to the mass position $x = l$. The mass reaches the support. In this particular case we probably do not have convergence. The result is convergent at $x = l - 0$ and $t = l/\mathrm{v} - 0$. Although we cannot prove mathematically the discontinuity of the inertial string matter, we can say that for practical use the differential equation of the string motion under the assumption of small displacements involves the discontinuity of the structure in the neighbourhood of the support. Such a phenomenon is observed in real structures (a track or bridge plates) in the form of high value impacts.

The supersonic motion of the mass results in zero displacement. In the diagram obtained numerically this value oscillates with low amplitude. The amplitude decreases with the increase of the number of terms in a sum (Figure 3.10). From this point of view we can say that the solution is accurate for under and over critical speed of the mass motion.

3.1.2 The Lagrange Equation

Let us consider a string of the length l, cross-sectional area A, mass density ρ, tensile force N, subjected to a mass m accompanied by a force P (as in the Section 3.1 and in Figure 3.1), moving with a constant speed v. We impose the same

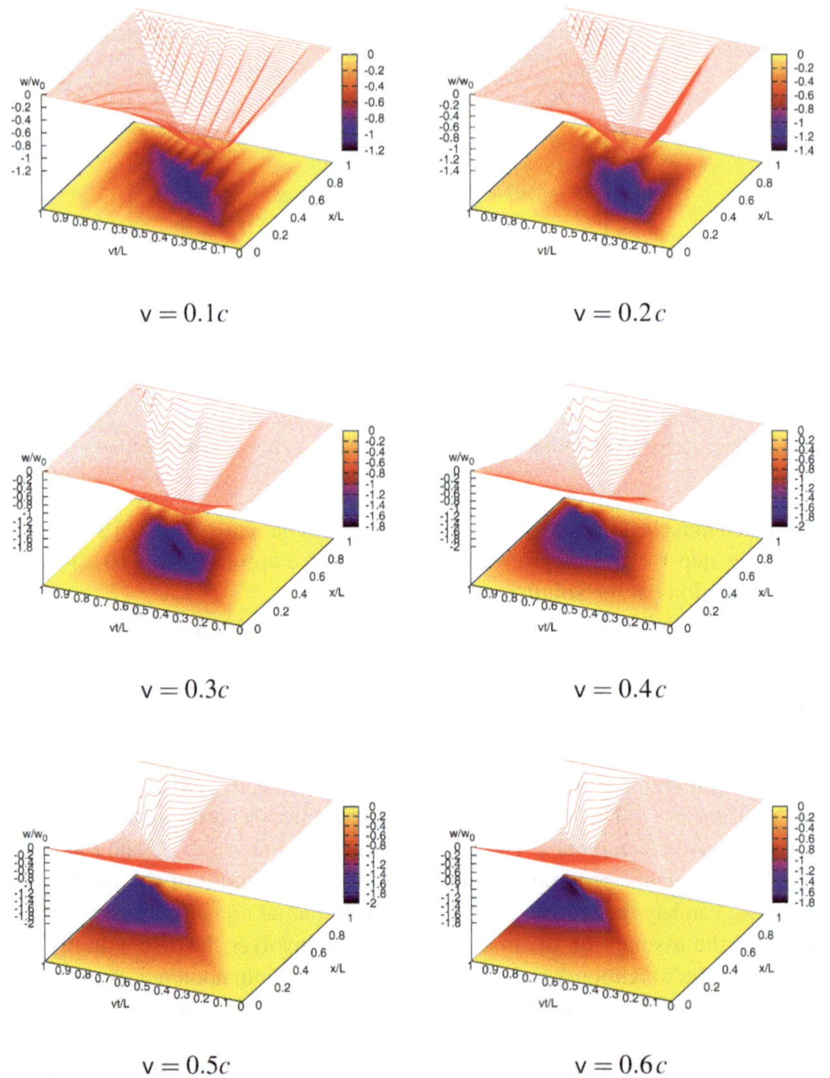

Fig. 3.5 Simulation of the string motion under the mass moving at $v = 0.1\,c$, $0.2c$, $0.3c$, $0.4c$, $0.5c$, and $0.6c$.

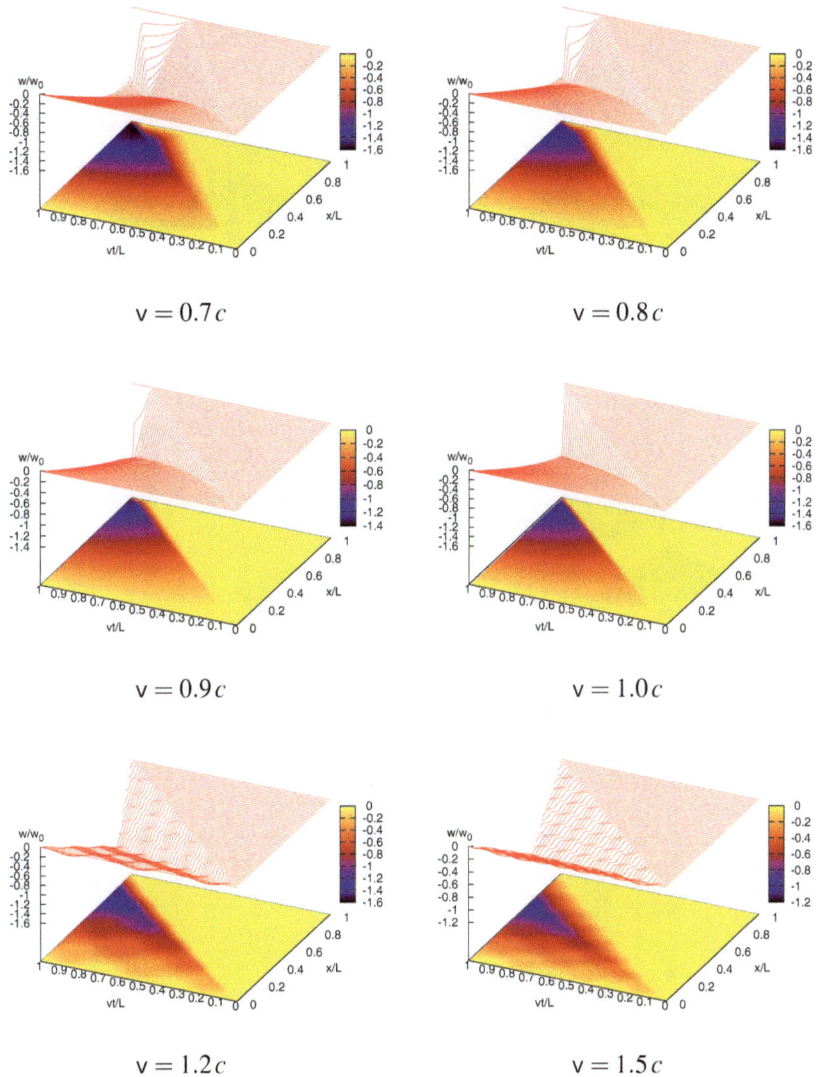

Fig. 3.6 Simulation of the string motion under the mass moving at $v = 0.7c$, $0.8c$, $0.9c$, $1.0c$, $1.2c$, and $1.5c$.

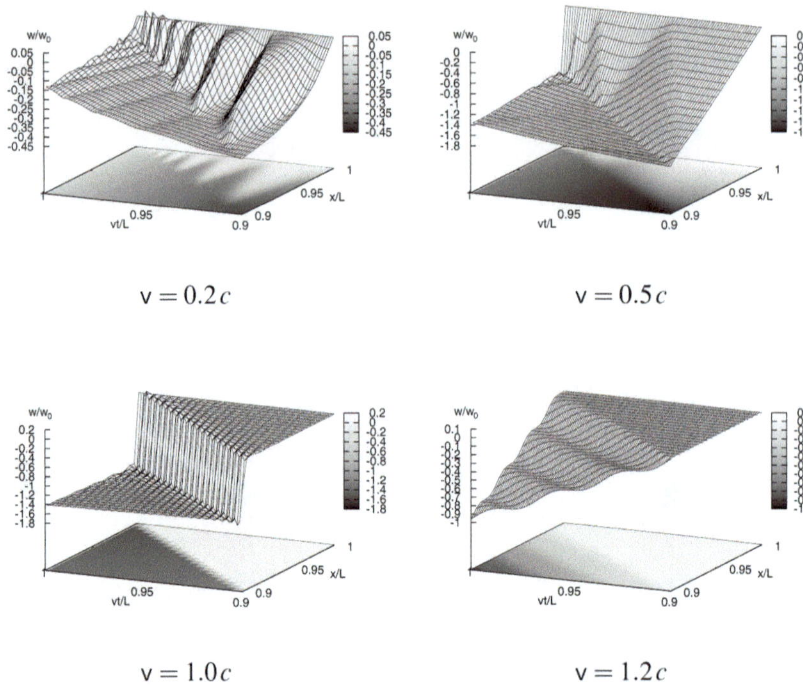

Fig. 3.7 Last stage of the mass motion at $v = 0.2c$, $0.5c$, $1.0c$ and $1.2c$.

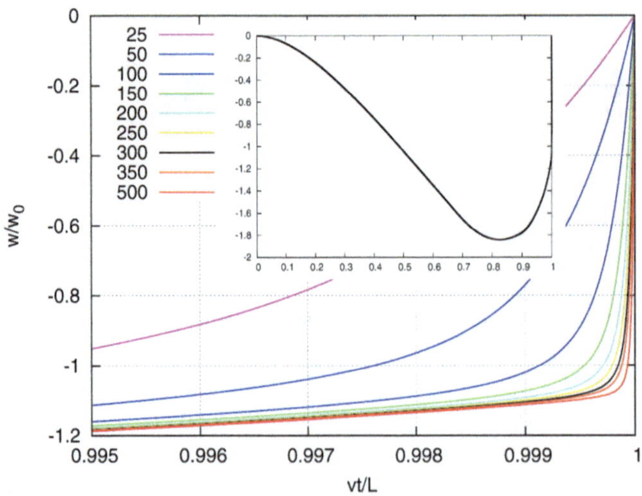

Fig. 3.8 The convergence of the mass trajectory travelling with $v = 0.5c$ near the end point, for various numbers of terms (25, 50, ..., 1000) in Eq. (3.105).

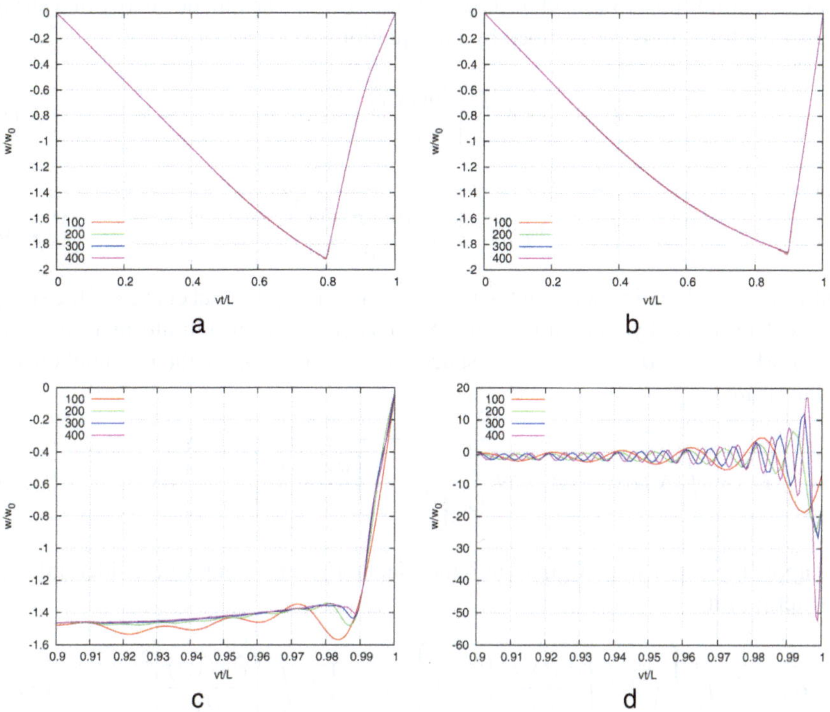

Fig. 3.9 The string at time: (a) $t = 0.80 l/\mathrm{v}$, (b) $0.90\ l/\mathrm{v}$, (c) $0.99\ l/\mathrm{v}$, and (d) $1.00\ l/\mathrm{v}$, with various number of terms.

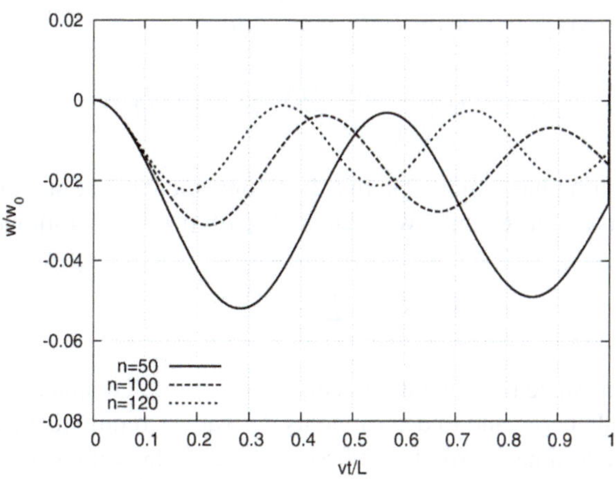

Fig. 3.10 Convergence of displacements under the mass at the speed $\mathrm{v} = 1.05c$.

boundary conditions (3.2) and initial conditions (3.3). The kinetic energy of a string and a travelling mass is described by the equation

$$E_k = \frac{1}{2}\rho A \int_0^l \left[\frac{\partial w(x,t)}{\partial t}\right]^2 \mathrm{d}x + E_{km} \ , \tag{3.18}$$

where

$$E_{km} = \frac{1}{2} m \left[\frac{\mathrm{d}w(vt,t)}{\mathrm{d}t}\right]^2 \tag{3.19}$$

contributes the kinetic energy of the moving mass. The potential energy of the string can be determined by computing of the δx to δs change of its infinitesimal segment. The work $N(\delta x - \delta s)$ integrated in space allow us to compute the potential energy of the string

$$E_p = \int_0^l N(\delta s - \delta x) = N \int_0^l \left\{ \sqrt{1 + \left[\frac{\partial w(x,t)}{\partial x}\right]^2} - 1 \right\} \mathrm{d}x \ . \tag{3.20}$$

We apply the expansion of (3.20) into the Maclaurin series and we consider only the first term of it

$$E_p = N \int_0^l \left\{ \sqrt{1 + \left[\frac{\partial w(x,t)}{\partial x}\right]^2} - 1 \right\} \mathrm{d}x \approx \frac{1}{2} N \int_0^l \left[\frac{\partial w(x,t)}{\partial x}\right]^2 \mathrm{d}x \ . \tag{3.21}$$

If we neglect next terms of the series, we assume higher powers of $\partial w(x,t)/\partial x$ to be nearly equal to zero. The equation (3.21) can be applied to the problem of small displacements of the string only. Finally the potential energy of the system, *i.e.* the string and the moving constant force P gains a form

$$E_p = \frac{1}{2} N \int_0^l \left[\frac{\partial w(x,t)}{\partial x}\right]^2 \mathrm{d}x - P w(vt,t) \ . \tag{3.22}$$

The examined string has a finite length l. It is convenient to use standing waves for description of its displacements. We assume the solution in the following form:

$$w(x,t) = \sum_{i=1}^{\infty} U_i(x)\xi_i(t) \ . \tag{3.23}$$

$\xi_i(t)$ are the generalized coordinate functions. In order to compute both the kinetic and the potential energy and to determine its derivatives required, we express them by generalized coordinates. We derive first the Eqn. (3.23) with respect to t

$$\frac{\partial w(x,t)}{\partial t} = \sum_{i=1}^{\infty} U_i(x)\dot{\xi}_i(t) \ , \tag{3.24}$$

and with respect to spatial variable x

$$\frac{\partial w(x,t)}{\partial x} = \sum_{i=1}^{\infty} U_i'(x)\xi_i(t) \ . \tag{3.25}$$

The displacement of the string in the contact point with a travelling mass is given by the equation

$$w(\mathrm{v}t,t) = \sum_{i=1}^{\infty} U_i(\mathrm{v}t)\xi_i(t) \ . \tag{3.26}$$

The transverse velocity of the moving mass is expressed by a composite derivative. It expresses the load travelling along the string

$$\frac{\mathrm{d}w(\mathrm{v}t,t)}{\mathrm{d}t} = \mathrm{v}\sum_{i=1}^{\infty} U_i'(x)\xi_i(t)\Bigg|_{x=\mathrm{v}t} + \sum_{i=1}^{\infty} U_i(x)\dot{\xi}_i(t)\Bigg|_{x=\mathrm{v}t} \ . \tag{3.27}$$

According to the above equation the velocity $\mathrm{d}w(\mathrm{v}t,t)/\mathrm{d}t$ is expressed as a function of both generalized coordinates and the derivative of generalized coordinates with respect to time

$$\frac{\mathrm{d}w(\mathrm{v}t,t)}{\mathrm{d}t} = f\left(\xi_i,\dot{\xi}_i\right) \ . \tag{3.28}$$

After rearrangement of the equation (3.18), with respect to (3.24), the total energy is given the by the following form

$$E_k = \frac{1}{2}\rho A \sum_{i,j=1}^{\infty} \dot{\xi}_i(t)\dot{\xi}_j(t) \int_0^l U_i(x)U_j(x)\mathrm{d}x + \frac{1}{2}m\left[\frac{\mathrm{d}w(\mathrm{v}t,t)}{\mathrm{d}t}\right]^2 \ . \tag{3.29}$$

We assume orthogonal functions which fulfill boundary conditions (3.2)

$$U_i(x) = \sin\frac{i\pi x}{l} \ . \tag{3.30}$$

The orthogonality of functions $U_i(x)$ allows us to write

$$\int_0^l U_i(x)U_j(x)\mathrm{d}x = \begin{cases} \frac{1}{2}l & \text{if } i=j, \\ 0 & \text{if } i \neq j \ . \end{cases} \tag{3.31}$$

The kinetic energy of the system (3.29) according to (3.30) and (3.31) is described by the relation

$$E_k = \frac{1}{4}\rho A l\,\dot{\xi}_i^2(t) + \frac{1}{2}m\left[\frac{\mathrm{d}w(\mathrm{v}t,t)}{\mathrm{d}t}\right]^2 \ . \tag{3.32}$$

In the case of the potential energy the Eqn. (3.25), with respect to the Eqn. (3.22) integrated by parts, has the following form:

$$E_p = \frac{1}{2}N\xi_i(t)\xi_j(t)\int_0^l U_i'(x)U_j'(x)\mathrm{d}x - Pw(\mathrm{v}t,t) = \\ = -\frac{1}{2}N\xi_i(t)\xi_j(t)\int_0^l U_i''(x)U_j(x)\mathrm{d}x - PU_i(\mathrm{v}t)\xi_i(t)\ . \tag{3.33}$$

We can derive the function (3.30)

$$U_i''(x) = -\frac{i^2\pi^2}{l^2}U_i(x)\ . \tag{3.34}$$

The equation (3.33) with respect to (3.34) can be written in the form

$$E_p = \frac{1}{2}N\frac{i^2\pi^2}{l^2}\xi_i(t)\xi_j(t)\int_0^l U_i(x)U_j(x)\mathrm{d}x - PU_i(\mathrm{v}t)\xi_i(t)\ . \tag{3.35}$$

Finally the potential energy of the string, with respect to (3.31) can be described by the equation

$$E_p = \frac{1}{4}Nl\frac{i^2\pi^2}{l^2}\xi_i^2(t) - P\xi_i(t)\sin\frac{i\pi\mathrm{v}t}{l}\ . \tag{3.36}$$

Now, when we have kinetic and potential energy described in generalized coordinates and the derivative of generalized coordinates with respect to time, we can formulate the Lagrange equation, which general form is given by the equation

$$\frac{\mathrm{d}}{\mathrm{d}t}\left(\frac{\partial E_k}{\partial \dot{\xi}_i}\right) - \frac{\partial E_k}{\partial \xi_i} + \frac{\partial E_p}{\partial \xi_i} = 0\,. \tag{3.37}$$

In order to obtain the Lagrange equation describing our problem, we must compute the required terms.

From (3.19) and (3.28) we have the derivatives of the kinetic energy of the travelling mass E_{km} with respect to ξ_i and $\dot{\xi}_i$

$$\frac{\partial E_{km}}{\partial \xi_i} = m\frac{\mathrm{d}w(\mathrm{v}t,t)}{\mathrm{d}t}\frac{\mathrm{d}}{\mathrm{d}\xi_i}\left(\frac{\mathrm{d}w(\mathrm{v}t,t)}{\mathrm{d}t}\right)\,, \tag{3.38}$$

$$\frac{\partial E_{km}}{\partial \dot{\xi}_i} = m\frac{\mathrm{d}w(\mathrm{v}t,t)}{\mathrm{d}t}\frac{\mathrm{d}}{\mathrm{d}\dot{\xi}_i}\left(\frac{\mathrm{d}w(\mathrm{v}t,t)}{\mathrm{d}t}\right)\,. \tag{3.39}$$

We compute the derivatives of the kinetic energy for the whole system (3.32) with respect to ξ_i and $\dot{\xi}_i$, taking into account (3.38) and (3.39):

$$\frac{\partial E_k}{\partial \xi_i} = \frac{\partial E_{km}}{\partial \xi_i} = \\ = m\left[\mathrm{v}^2\sum_{j=1}^{\infty}\frac{ij\pi^2}{l^2}\cos\frac{i\pi\mathrm{v}t}{l}\cos\frac{j\pi\mathrm{v}t}{l}\xi_j(t) + \mathrm{v}\sum_{j=1}^{\infty}\frac{i\pi}{l}\cos\frac{i\pi\mathrm{v}t}{l}\sin\frac{j\pi\mathrm{v}t}{l}\dot{\xi}_j(t)\right]\,, \tag{3.40}$$

$$\frac{\partial E_k}{\partial \dot{\xi}_i} = \frac{1}{2}\rho A l\,\dot{\xi}_i(t) + \frac{\partial E_{km}}{\partial \dot{\xi}_i} = \frac{1}{2}\rho A l\,\dot{\xi}_i(t) + \\ + m\left[\mathrm{v}\sum_{j=1}^{\infty}\frac{j\pi}{l}\sin\frac{i\pi \mathrm{v}t}{l}\cos\frac{j\pi \mathrm{v}t}{l}\xi_j(t) + \sum_{j=1}^{\infty}\sin\frac{i\pi \mathrm{v}t}{l}\sin\frac{j\pi \mathrm{v}t}{l}\dot{\xi}_j(t)\right]. \tag{3.41}$$

The derivative of the potential energy (3.36) with respect to generalized coordinates ξ_i is

$$\frac{\partial E_p}{\partial \xi_i} = \frac{1}{2}Nl\frac{i^2\pi^2}{l^2}\xi_i(t) - P\sin\frac{i\pi \mathrm{v}t}{l}. \tag{3.42}$$

The derivative of (3.41) with respect to t is

$$\frac{\mathrm{d}}{\mathrm{d}t}\left(\frac{\partial E_k}{\partial \dot{\xi}_i}\right) = \frac{1}{2}\rho A l\,\ddot{\xi}_i(t) + m\left\{\sum_{j=1}^{\infty}\frac{j\pi \mathrm{v}}{l}\frac{\mathrm{d}}{\mathrm{d}t}\left[\sin\frac{i\pi \mathrm{v}t}{l}\cos\frac{j\pi \mathrm{v}t}{l}\xi_j(t)\right] + \right. \\ \left. + \sum_{j=1}^{\infty}\frac{\mathrm{d}}{\mathrm{d}t}\left[\sin\frac{i\pi \mathrm{v}t}{l}\sin\frac{j\pi \mathrm{v}t}{l}\dot{\xi}_j(t)\right]\right\}. \tag{3.43}$$

Finally, the Lagrange equation (3.37) in the case of our problem of an inertial string subjected to a moving inertial force has the following form:

$$\frac{1}{2}\rho A l\,\ddot{\xi}_i(t) + m\sum_{j=1}^{\infty}\frac{j\pi \mathrm{v}}{l}\frac{\mathrm{d}}{\mathrm{d}t}\left[\sin\frac{i\pi \mathrm{v}t}{l}\cos\frac{j\pi \mathrm{v}t}{l}\right]\xi_j(t) + \\ + m\sum_{j=1}^{\infty}\frac{j\pi \mathrm{v}}{l}\sin\frac{i\pi \mathrm{v}t}{l}\cos\frac{j\pi \mathrm{v}t}{l}\dot{\xi}_j(t) + m\sum_{j=1}^{\infty}\frac{\mathrm{d}}{\mathrm{d}t}\left[\sin\frac{i\pi \mathrm{v}t}{l}\sin\frac{j\pi \mathrm{v}t}{l}\right]\dot{\xi}_j(t) + \\ + m\sum_{j=1}^{\infty}\sin\frac{i\pi \mathrm{v}t}{l}\sin\frac{j\pi \mathrm{v}t}{l}\ddot{\xi}_j(t) + \frac{1}{2}Nl\frac{i^2\pi^2}{l^2}\xi_i(t) - \\ - m\left[\mathrm{v}^2\sum_{j=1}^{\infty}\frac{ij\pi^2}{l^2}\cos\frac{i\pi \mathrm{v}t}{l}\cos\frac{j\pi \mathrm{v}t}{l}\xi_j(t) + \mathrm{v}\sum_{j=1}^{\infty}\frac{i\pi}{l}\cos\frac{i\pi \mathrm{v}t}{l}\sin\frac{j\pi \mathrm{v}t}{l}\dot{\xi}_j(t)\right] = \\ = P\sin\frac{i\pi \mathrm{v}t}{l}, \tag{3.44}$$

where

$$\frac{\mathrm{d}}{\mathrm{d}t}\left[\sin\frac{i\pi \mathrm{v}t}{l}\cos\frac{j\pi \mathrm{v}t}{l}\right] = \frac{i\pi \mathrm{v}}{l}\cos\frac{i\pi \mathrm{v}t}{l}\cos\frac{j\pi \mathrm{v}t}{l} - \frac{j\pi \mathrm{v}}{l}\sin\frac{i\pi \mathrm{v}t}{l}\sin\frac{j\pi \mathrm{v}t}{l}, \tag{3.45}$$

$$\frac{\mathrm{d}}{\mathrm{d}t}\left[\sin\frac{i\pi \mathrm{v}t}{l}\sin\frac{j\pi \mathrm{v}t}{l}\right] = \frac{i\pi \mathrm{v}}{l}\cos\frac{i\pi \mathrm{v}t}{l}\sin\frac{j\pi \mathrm{v}t}{l} + \frac{j\pi \mathrm{v}}{l}\sin\frac{i\pi \mathrm{v}t}{l}\cos\frac{j\pi \mathrm{v}t}{l}. \tag{3.46}$$

We have a differential equation with variable coefficients. Finally (3.44) can be written in the following form

$$\ddot{\xi}_i(t)+\frac{2m}{\rho Al}\sum_{j=1}^{\infty}\ddot{\xi}_j(t)\sin\frac{i\pi vt}{l}\sin\frac{j\pi vt}{l}+\frac{4m}{\rho Al}\sum_{j=1}^{\infty}\frac{j\pi v}{l}\dot{\xi}_j(t)\sin\frac{i\pi vt}{l}\cos\frac{j\pi vt}{l}+$$
$$+\frac{N}{\rho A}\frac{i^2\pi^2}{l^2}\xi_i(t)-\frac{2m}{\rho Al}\sum_{j=1}^{\infty}\frac{j^2\pi^2v^2}{l^2}\xi_j(t)\sin\frac{i\pi vt}{l}\sin\frac{j\pi vt}{l}=\frac{2P}{\rho Al}\sin\frac{i\pi vt}{l}\,. \tag{3.47}$$

These two methods lead us to the identical differential equation with variable coefficients (3.47) as in (3.10)–(3.11).

3.2 Bernoulli–Euler Beam

The problem of bridge spans under a moving inertial load has existed since the beginning of the development of railways. The turning point in the literature was established by two historical publications [70, 124]. These analytical papers were elaborated with significant mathematical simplifications. The authors considered a complex acceleration of the moving mass. Its geometrical interpretation was presented by Renaudot [118]. Although the number of publications on the moving mass problem has exceeded a thousand items, still we do not have its detailed and fully analytical solutions. The approach given by Smith [127] seems to be a positive exception. He considered, however, the massless string only. There exist numerous review papers [106, 146, 43, 133] which discuss the problems treated in hundreds of other publications. For a long time the mainstream of works treated the problem in an analytical-numerical way [122, 138, 48, 47] or strictly numerically [52, 120, 24].

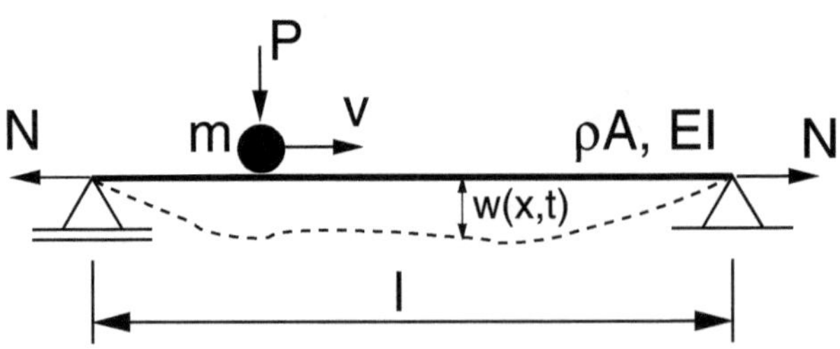

Fig. 3.11 A string-beam under a moving mass.

Together with the increasing velocity of trains, the influence of the wave phenomenon is rising as well. Dynamic effects are generated by the load of train current collectors, travelling through the power supply cable of the overhead contact line. In this chapter we consider a cable as a string-beam model, since it has a certain flexible stiffness. The Bernoulli–Euler beam with an additional tensile effect comprises this phenomenon (Figure 3.11). The differential equation of the motion of a string-beam is derived from the Lagrange equation of the second kind.

3.2.1 Fourier Solution

Let us consider a simply supported Bernoulli–Euler beam (B–E) with additional tensile effect, of length l under a moving point mass m travelling with constant speed v. EI, N, and ρA are, respectively, the beam stiffness, tensile force, and the string-beam mass density. The differential equation of motion describing the string-beam B–E under a moving mass point is

$$EI\frac{\partial^4 w(x,t)}{\partial x^4} - N\frac{\partial^2 w(x,t)}{\partial x^2} + \rho A\frac{\partial^2 w(x,t)}{\partial t^2} = \delta(x-\mathrm{v}t)P - \delta(x-\mathrm{v}t)m\frac{\mathrm{d}^2 w(\mathrm{v}t,t)}{\mathrm{d}t^2} \,. \tag{3.48}$$

We assume a simply supported beam, and therefore assume the following boundary conditions

$$w(0,t) = 0\,, \quad w(l,t) = 0\,, \quad \left.\frac{\partial^2 w(x,t)}{\partial x^2}\right|_{x=0} = 0\,, \quad \left.\frac{\partial^2 w(x,t)}{\partial x^2}\right|_{x=l} = 0\,. \tag{3.49}$$

In addition, we assume zero initial conditions

$$w(x,0) = 0\,, \quad \left.\frac{\partial w(x,t)}{\partial t}\right|_{t=0} = 0\,. \tag{3.50}$$

Using the sine Fourier transform in a finite interval $< 0,l >$, we can write

$$V(j,t) = \int_0^l w(x,t)\sin\frac{j\pi x}{l}\,\mathrm{d}x\,, \tag{3.51}$$

where

$$w(x,t) = \frac{2}{l}\sum_{j=1}^{\infty} V(j,t)\sin\frac{j\pi x}{l}\,, \tag{3.52}$$

which fulfills the boundary conditions. Using equation (3.52), we can determine the vertical acceleration of the moving mass at the point $x = \mathrm{v}t$

$$\begin{aligned}\frac{\mathrm{d}^2 w(\mathrm{v}t,t)}{\mathrm{d}t^2} = \frac{2}{l}\sum_{k=1}^{\infty}\Bigg[&\ddot{V}(k,t)\sin\frac{k\pi\mathrm{v}t}{l} + \frac{2k\pi\mathrm{v}}{l}\dot{V}(k,t)\cos\frac{k\pi\mathrm{v}t}{l} - \\ &- \frac{k^2\pi^2\mathrm{v}^2}{l^2}V(k,t)\sin\frac{k\pi\mathrm{v}t}{l}\Bigg]\,.\end{aligned} \tag{3.53}$$

The Fourier transformation allows us to transform a partial differential equation (3.48) into a system of ordinary differential equations as follows

$$\begin{aligned}EI\frac{j^4\pi^4}{l^4}V(j,t) + N\frac{j^2\pi^2}{l^2}V(j,t) + \rho A\ddot{V}(j,t) = P\sin\frac{j\pi\mathrm{v}t}{l} - \\ - m\frac{\mathrm{d}^2 w(\mathrm{v}t,t)}{\mathrm{d}t^2}\int_0^l \delta(x-\mathrm{v}t)\sin\frac{j\pi x}{l}\,\mathrm{d}x\,,\end{aligned} \tag{3.54}$$

where

$$\int_0^l \delta(x - \mathrm{v}t)\sin\frac{j\pi x}{l}\,\mathrm{d}x = \sin\frac{j\pi \mathrm{v}t}{l}\,. \tag{3.55}$$

Equation (3.54) for the vertical acceleration of the moving mass (3.53) can be written in the following form:

$$\begin{aligned}
EI\frac{j^4\pi^4}{l^4}V(j,t)+N\frac{j^2\pi^2}{l^2}V(j,t)+\rho A\ddot{V}(j,t) &= P\sin\frac{j\pi \mathrm{v}t}{l}-\\
&-\frac{2m}{l}\sum_{k=1}^{\infty}\ddot{V}(k,t)\sin\frac{k\pi \mathrm{v}t}{l}\sin\frac{j\pi \mathrm{v}t}{l}-\\
&-\frac{2m}{l}\sum_{k=1}^{\infty}\frac{2k\pi \mathrm{v}}{l}\dot{V}(k,t)\cos\frac{k\pi \mathrm{v}t}{l}\sin\frac{j\pi \mathrm{v}t}{l}+\\
&+\frac{2m}{l}\sum_{k=1}^{\infty}\frac{k^2\pi^2\mathrm{v}^2}{l^2}V(k,t)\sin\frac{k\pi \mathrm{v}t}{l}\sin\frac{j\pi \mathrm{v}t}{l}\,.
\end{aligned} \tag{3.56}$$

After re-arranging the above equation, we obtain the simpler form

$$\begin{aligned}
\ddot{V}(j,t)+\alpha\sum_{k=1}^{\infty}\ddot{V}(k,t)\sin\omega_k t\sin\omega_j t+2\alpha\sum_{k=1}^{\infty}\omega_k\dot{V}(k,t)\cos\omega_k t\sin\omega_j t+\\
+\Omega^2V(j,t)-\alpha\sum_{k=1}^{\infty}\omega_k^2V(k,t)\sin\omega_k t\sin\omega_j t=\frac{P}{\rho A}\sin\omega_j t\,,
\end{aligned} \tag{3.57}$$

using the following notation

$$\omega_k=\frac{k\pi \mathrm{v}}{l}\,,\qquad \omega_j=\frac{j\pi \mathrm{v}}{l}\,,\qquad \Omega^2=\frac{EI}{\rho A}\frac{j^4\pi^4}{l^4}+\frac{N}{\rho A}\frac{j^2\pi^2}{l^2}\,,\qquad \alpha=\frac{2m}{\rho A l}\,. \tag{3.58}$$

As in the case of a string, no full analytical solution of the system of equations (3.57) is yet known. In a further step, we integrate the system of differential equations numerically. For this purpose, we write (3.57) in matrix form:

$$\mathbf{M}\begin{bmatrix}\ddot{V}(1,t)\\ \ddot{V}(2,t)\\ \vdots\\ \ddot{V}(n,t)\end{bmatrix}+\mathbf{C}\begin{bmatrix}\dot{V}(1,t)\\ \dot{V}(2,t)\\ \vdots\\ \dot{V}(n,t)\end{bmatrix}+\mathbf{K}\begin{bmatrix}V(1,t)\\ V(2,t)\\ \vdots\\ V(n,t)\end{bmatrix}=\mathbf{P}\,, \tag{3.59}$$

or in short

$$\mathbf{M}\ddot{\mathbf{V}}+\mathbf{C}\dot{\mathbf{V}}+\mathbf{K}\mathbf{V}=\mathbf{P}\,. \tag{3.60}$$

The matrices of inertia, damping and stiffness, and the vectors on the right sides take the following forms

$$\mathbf{M} = \begin{bmatrix} 1 & 0 & \cdots & 0 \\ 0 & 1 & \cdots & 0 \\ \vdots & \vdots & \ddots & \vdots \\ 0 & 0 & \cdots & 1 \end{bmatrix} + \tag{3.61}$$

$$+\alpha \begin{bmatrix} \sin\frac{1\pi vt}{l}\sin\frac{1\pi vt}{l} & \sin\frac{1\pi vt}{l}\sin\frac{2\pi vt}{l} & \cdots & \sin\frac{1\pi vt}{l}\sin\frac{n\pi vt}{l} \\ \sin\frac{2\pi vt}{l}\sin\frac{1\pi vt}{l} & \sin\frac{2\pi vt}{l}\sin\frac{2\pi vt}{l} & \cdots & \sin\frac{2\pi vt}{l}\sin\frac{n\pi vt}{l} \\ \vdots & \vdots & \ddots & \vdots \\ \sin\frac{n\pi vt}{l}\sin\frac{1\pi vt}{l} & \sin\frac{n\pi vt}{l}\sin\frac{2\pi vt}{l} & \cdots & \sin\frac{n\pi vt}{l}\sin\frac{n\pi vt}{l} \end{bmatrix}, \tag{3.62}$$

$$\mathbf{C} = 2\alpha \begin{bmatrix} \frac{1\pi v}{l}\sin\frac{1\pi vt}{l}\cos\frac{1\pi vt}{l} & \frac{2\pi v}{l}\sin\frac{1\pi vt}{l}\cos\frac{2\pi vt}{l} & \cdots & \frac{n\pi v}{l}\sin\frac{1\pi vt}{l}\cos\frac{n\pi vt}{l} \\ \frac{1\pi v}{l}\sin\frac{2\pi vt}{l}\cos\frac{1\pi vt}{l} & \frac{2\pi v}{l}\sin\frac{2\pi vt}{l}\cos\frac{2\pi vt}{l} & \cdots & \frac{n\pi v}{l}\sin\frac{2\pi vt}{l}\cos\frac{n\pi vt}{l} \\ \vdots & \vdots & \ddots & \vdots \\ \frac{1\pi v}{l}\sin\frac{n\pi vt}{l}\cos\frac{1\pi vt}{l} & \frac{2\pi v}{l}\sin\frac{n\pi vt}{l}\cos\frac{2\pi vt}{l} & \cdots & \frac{n\pi v}{l}\sin\frac{n\pi vt}{l}\cos\frac{n\pi vt}{l} \end{bmatrix}, \tag{3.63}$$

$$\mathbf{K} = \begin{bmatrix} \frac{1^4\pi^4}{l^4}\frac{EI}{\rho A} + \frac{1^2\pi^2}{l^2}\frac{N}{\rho A} & 0 & \cdots & 0 \\ 0 & \frac{2^4\pi^4}{l^4}\frac{EI}{\rho A} + \frac{2^2\pi^2}{l^2}\frac{N}{\rho A} & \cdots & 0 \\ \vdots & \vdots & \ddots & \vdots \\ 0 & 0 & \cdots & \frac{n^4\pi^4}{l^4}\frac{EI}{\rho A} + \frac{n^2\pi^2}{l^2}\frac{N}{\rho A} \end{bmatrix} - \tag{3.64}$$

$$-\alpha \begin{bmatrix} \frac{1^2\pi^2 v^2}{l^2}\sin\frac{1\pi vt}{l}\sin\frac{1\pi vt}{l} & \frac{2^2\pi^2 v^2}{l^2}\sin\frac{1\pi vt}{l}\sin\frac{2\pi vt}{l} & \cdots & \frac{n^2\pi^2 v^2}{l^2}\sin\frac{1\pi vt}{l}\sin\frac{n\pi vt}{l} \\ \frac{1^2\pi^2 v^2}{l^2}\sin\frac{2\pi vt}{l}\sin\frac{1\pi vt}{l} & \frac{2^2\pi^2 v^2}{l^2}\sin\frac{2\pi vt}{l}\sin\frac{2\pi vt}{l} & \cdots & \frac{n^2\pi^2 v^2}{l^2}\sin\frac{2\pi vt}{l}\sin\frac{n\pi vt}{l} \\ \vdots & \vdots & \ddots & \vdots \\ \frac{1^2\pi^2 v^2}{l^2}\sin\frac{n\pi vt}{l}\sin\frac{1\pi vt}{l} & \frac{2^2\pi^2 v^2}{l^2}\sin\frac{n\pi vt}{l}\sin\frac{2\pi vt}{l} & \cdots & \frac{n^2\pi^2 v^2}{l^2}\sin\frac{n\pi vt}{l}\sin\frac{n\pi vt}{l} \end{bmatrix},$$

$$\mathbf{P} = \frac{P}{\rho A}\begin{bmatrix} \sin\frac{1\pi vt}{l} \\ \sin\frac{2\pi vt}{l} \\ \vdots \\ \sin\frac{n\pi vt}{l} \end{bmatrix}. \tag{3.65}$$

All the matrices are time–dependent. The inertia matrix **M** is symmetric, while the remaining damping matrix **C** and stiffness matrix **K** are unsymmetrical. Calculating numerically at each time step the vector **V** and substituting into the series (3.52) describes the displacement at any point on the beam. Figures 3.12 and 3.13 represent the solution obtained. The critical velocity in the Euler beam is equal to $v_{cr} = \pi/l \cdot \sqrt{EI/\rho/A} = 0.314$.

3.2.2 The Lagrange Equation of the Second Kind

The kinetic energy of a string-beam and a travelling mass is described by

$$E_k = \frac{1}{2}\rho A \int_0^l \left[\frac{\partial w(x,t)}{\partial t}\right]^2 \mathrm{d}x + \frac{1}{2} m \left[\frac{\mathrm{d}w(\mathrm{v}t,t)}{\mathrm{d}t}\right]^2 . \tag{3.66}$$

The potential energy of a string-beam and a moving force is

$$E_p = \frac{1}{2} N \int_0^l \left[\frac{\partial w(x,t)}{\partial x}\right]^2 \mathrm{d}x + \frac{1}{2} EI \int_0^l \left[\frac{\partial^2 w(x,t)}{\partial x^2}\right]^2 \mathrm{d}x - P w(\mathrm{v}t,t) . \tag{3.67}$$

In order to separate variables, the displacement can be written in the form of an infinite series and then the integrals in space x in equations (3.66) and (3.67) can be computed

$$w(x,t) = \sum_{i=1}^{\infty} U_i(x)\xi_i(t) . \tag{3.68}$$

According to (3.68) the displacement under a moving load has the following form:

$$w(\mathrm{v}t,t) = \sum_{i=1}^{\infty} U_i(\mathrm{v}t)\xi_i(t) . \tag{3.69}$$

The velocity of the displacement is determined by the chain rule

$$\frac{\mathrm{d}w(\mathrm{v}t,t)}{\mathrm{d}t} = \mathrm{v}\sum_{i=1}^{\infty} U_i'(x)\xi_i(t)\Big|_{x=\mathrm{v}t} + \sum_{i=1}^{\infty} U_i(x)\dot{\xi}_i(t)\Big|_{x=\mathrm{v}t} . \tag{3.70}$$

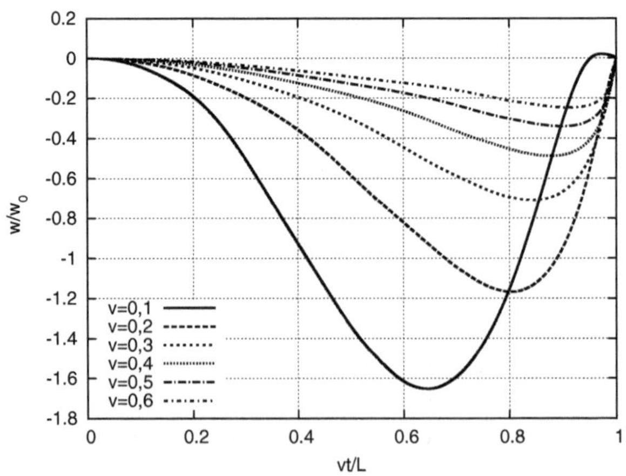

Fig. 3.12 Trajectories of the moving mass travelling along Bernoulli–Euler beam at different speeds v (for $N = 0$).

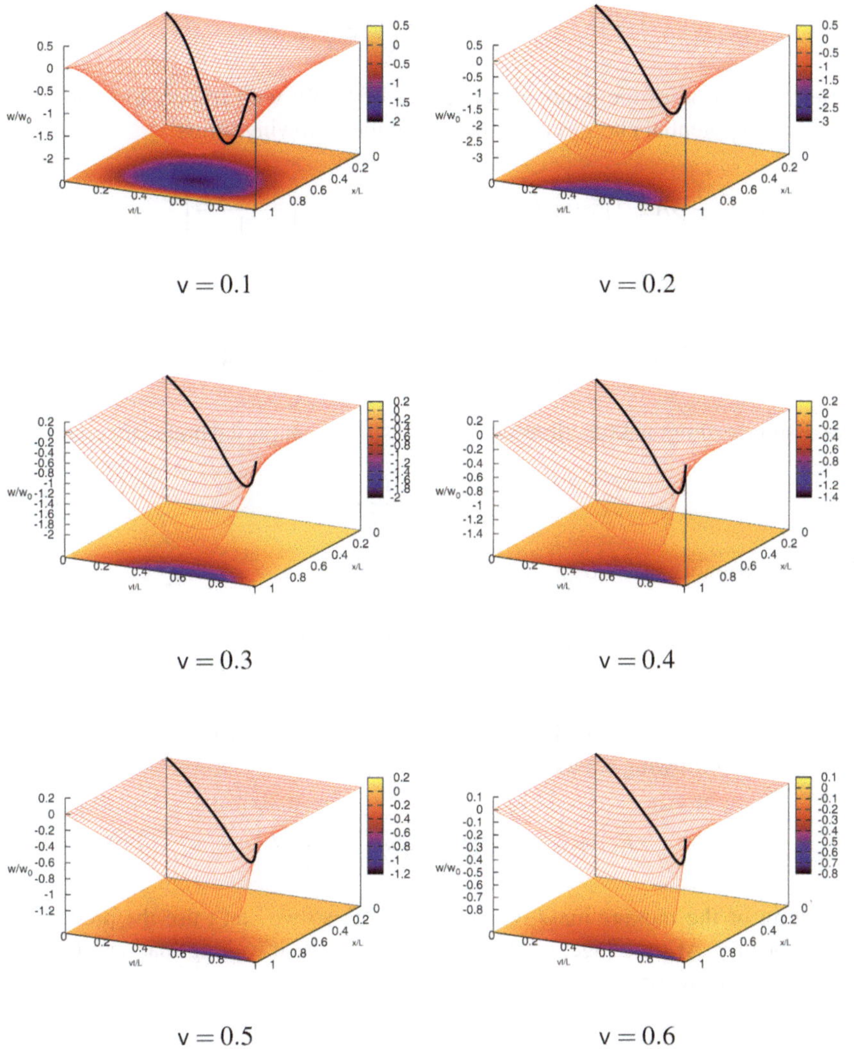

Fig. 3.13 Vibrations of Bernoulli–Euler beam under inertial load in the velocity v =0.1, 0.2, 0.3, 0.4, 0.5 and 0.6.

It is a function of general coordinates as well as their velocities

$$\frac{\mathrm{d}w(vt,t)}{\mathrm{d}t} = f\left(\xi_i, \dot{\xi}_i\right) . \tag{3.71}$$

After calculation of the required derivatives of (3.68) with respect to t and x, the kinetic and potential energy can be written in the following forms

$$E_k = \frac{1}{2}\rho A\,\dot{\xi}_i(t)\dot{\xi}_j(t)\int_0^l U_i(x)U_j(x)\mathrm{d}x + \frac{1}{2}m\left[\frac{\mathrm{d}w(vt,t)}{\mathrm{d}t}\right]^2 , \tag{3.72}$$

$$\begin{aligned} E_p = &\frac{1}{2}N\xi_i(t)\xi_j(t)\int_0^l U_i'(x)U_j'(x)\mathrm{d}x+ \\ &+\frac{1}{2}EI\xi_i(t)\xi_j(t)\int_0^l U_i''(x)U_j''(x)\mathrm{d}x - P\sum_{i=1}^{\infty}U_i(vt)\xi_i(t) . \end{aligned} \tag{3.73}$$

We assume orthogonal functions which fulfil simply supported boundary conditions

$$U_i(x) = \sin\frac{i\pi x}{l} . \tag{3.74}$$

The orthogonality of the $U_i(x)$ allows us to write

$$\int_0^l U_i(x)U_j(x)\mathrm{d}x = \begin{cases} \frac{1}{2}l & \text{if } i=j, \\ 0 & \text{if } i \neq j. \end{cases} \tag{3.75}$$

With respect to (3.68), (3.74) and (3.75) the kinetic energy of the hole system is given by the following equation

$$E_k = \frac{1}{4}\rho Al\,\dot{\xi}_i^2(t) + \frac{1}{2}m\left[\frac{\mathrm{d}w(vt,t)}{\mathrm{d}t}\right]^2 . \tag{3.76}$$

The term of the moving mass is not an integral, so we can't use the property of orthogonality. At this stage, the kinetic energy of the travelling load is left in the original form. According to (3.74) we have

$$U_i''(x) = -\frac{i^2\pi^2}{l^2}U_i(x) . \tag{3.77}$$

After integration by parts and taking into account (3.77), the potential energy (3.73) can be written in the form

$$\begin{aligned} E_p = &\frac{1}{2}N\frac{i^2\pi^2}{l^2}\xi_i(t)\xi_j(t)\int_0^l U_i(x)U_j(x)\mathrm{d}x+ \\ &+\frac{1}{2}EI\frac{i^2j^2\pi^4}{l^4}\xi_i(t)\xi_j(t)\int_0^l U_i(x)U_j(x)\mathrm{d}x - PU_i(vt)\xi_i(t) . \end{aligned} \tag{3.78}$$

Finally, the orthogonality of (3.75) allows us to write

$$E_p = \frac{1}{4}Nl\frac{i^2\pi^2}{l^2}\xi_i^2(t) + \frac{1}{4}EIl\frac{i^4\pi^4}{l^4}\xi_i^2(t) - P\xi_i(t)\sin\frac{i\pi vt}{l} . \tag{3.79}$$

The equation of motion of the string-beam under a moving inertial load is obtained from the Lagrange equation of the second kind, of which the general form is given by equation (3.37). This method results in the differential equation with variable coefficients

$$\begin{aligned}\rho A\ddot{\xi}_i(t)+&\frac{2m}{l}\sum_{j=1}^{\infty}\ddot{\xi}_j(t)\sin\frac{i\pi vt}{l}\sin\frac{j\pi vt}{l}+\frac{4m}{l}\sum_{j=1}^{\infty}\frac{j\pi v}{l}\dot{\xi}_j(t)\sin\frac{i\pi vt}{l}\cos\frac{j\pi vt}{l}+\\ &+N\frac{i^2\pi^2}{l^2}\xi_i(t) + EI\frac{i^4\pi^4}{l^4}\xi_i(t) - \frac{2m}{l}\sum_{j=1}^{\infty}\frac{j^2\pi^2v^2}{l^2}\xi_j(t)\sin\frac{i\pi vt}{l}\sin\frac{j\pi vt}{l} =\\ &= \frac{2P}{l}\sin\frac{i\pi vt}{l} .\end{aligned} \tag{3.80}$$

The system of equations (3.80) can not be easily solved and we must integrate it numerically. We use here the matrix notation

$$\mathbf{M}\begin{bmatrix}\ddot{\xi}_1(t)\\ \ddot{\xi}_2(t)\\ \vdots\\ \ddot{\xi}_n(t)\end{bmatrix} + \mathbf{C}\begin{bmatrix}\dot{\xi}_1(t)\\ \dot{\xi}_2(t)\\ \vdots\\ \dot{\xi}_n(t)\end{bmatrix} + \mathbf{K}\begin{bmatrix}\xi_1(t)\\ \xi_2(t)\\ \vdots\\ \xi_n(t)\end{bmatrix} = \mathbf{P} \tag{3.81}$$

which results in the short form $\mathbf{M}\ddot{\xi} + \mathbf{C}\dot{\xi} + \mathbf{K}\xi = \mathbf{P}$, where $\mathbf{M}$, $\mathbf{C}$ and $\mathbf{K}$ are square matrices for $i = j = 1, 2, \ldots, n$.

Then we calculate the value of the general coordinates $\xi_i(t)$ for each i to n. Finally we can compute the displacements of the string-beam $w(x,t)$

$$w(x,t) = \sum_{i=1}^{\infty}\xi_i(t)\sin\frac{i\pi x}{l} . \tag{3.82}$$

The displacements given in the example below are dimensionless. They were calculated in relation to the static deflection w_0 of the string-beam loaded at the midpoint by the point force P: $w_0 = w_{0s}w_{0b}/(w_{0s}+w_{0b})$. w_{0s} and w_{0b} are static deflections in the case of a string and a beam, respectively.

Example Let us assume the following data: $E = 1$, $I = 0.01$, $N = 1$, $\rho = 1$, $A = 1$, $l = 1$, $P = -1$ and $m = 1$. We solve the problem for different speeds v of the moving load. The mass trajectory is depicted in Figure 3.14. The simulation of the string-beam motion is depicted in Figure 3.15.

Let us look at Figure 3.16. The bending stiffness smooths the load trajectory. In the case of lower bending stiffness, the acceleration near the end support increases significantly. It tends to infinity in the limit case of a pure string. Although the

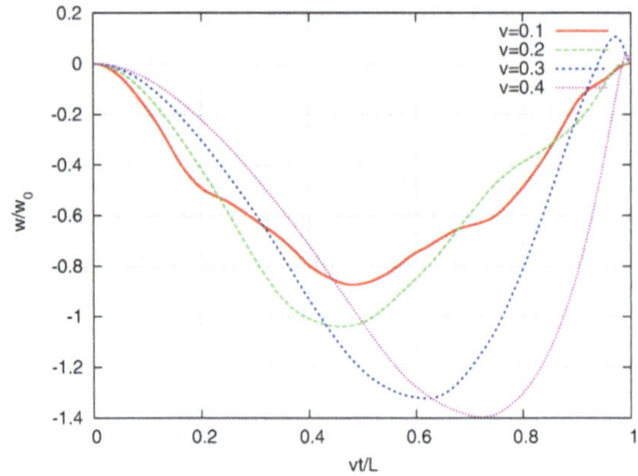

Fig. 3.14 Mass trajectory for different speeds v.

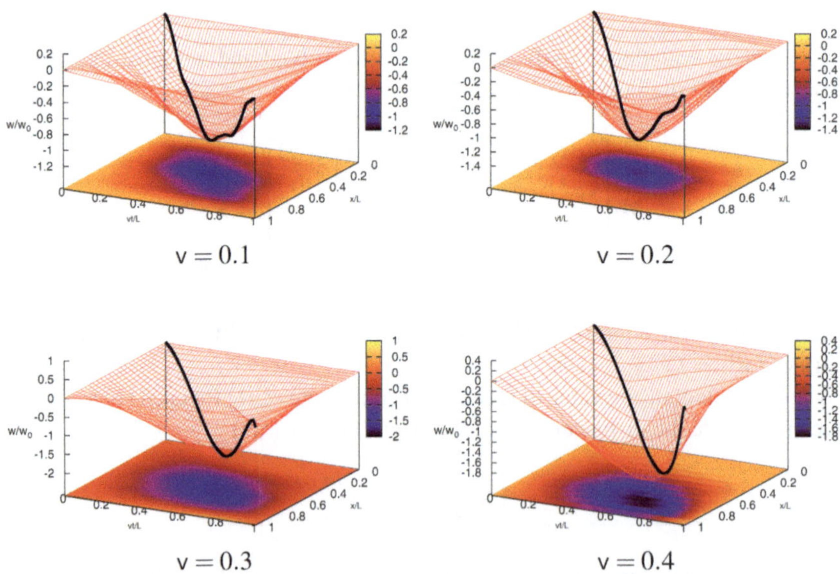

Fig. 3.15 Simulation of the string-beam motion under a mass moving at speeds v =0.1, 0.2, 0.3 and 0.4.

plots exhibit high jumps in all cases, we can not say whether the mathematical discontinuity occurs only in the case of a lack of bending or whether it can be demonstrated in the cases of mixed tensile-bending deformation. The question is: whether the introduced string effects result in discontinuity or whether bending effects eliminate the discontinuity? Further mathematical analysis could clarify this.

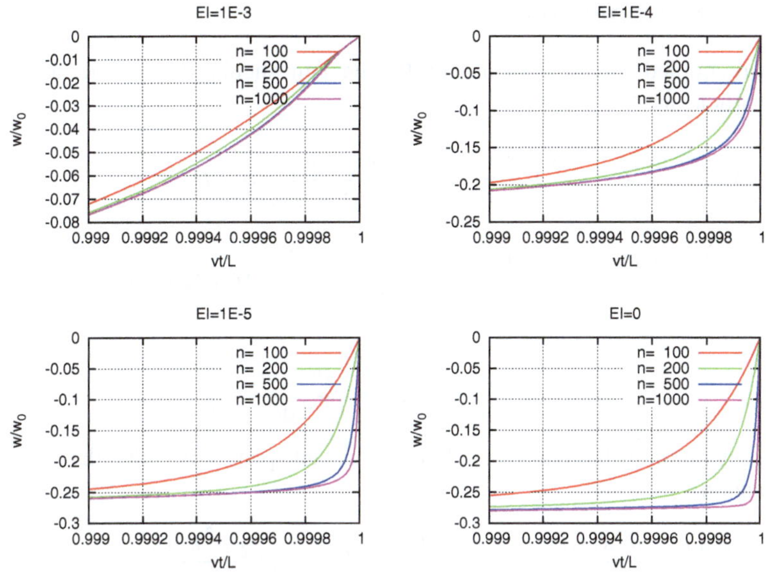

Fig. 3.16 Influence of bending on the discontinuity of the mass trajectory.

3.2.3 Conclusions

The above presented the Bernoulli–Euler beam with an additional tensile effect under a moving inertial load. The proposed semi-analytical approach can be applied in through the entire range of speeds and for all points of the string-beam span. The accuracy of the solution (3.82) depends on the number of terms taken in the infinite series (Figure 3.16). The examined series of displacements is convergent, so we may limit the number of terms in our example to $n = 130$.

If we reduce the flexible stiffness of the system, we observe a discontinuity near the end support. This was broadly presented and proved in [48]. The discontinuity had appeared also in the case of the Timoshenko beam [47]. In the matrix form (3.81), we can use classical numeric methods for the integration of the final differential equation of the motion, for example the Newmark method.

3.3 Timoshenko Beam

In this chapter we derive the solution of the mass particle travelling on a Timoshenko beam. The problem is complex since the product of the Dirac delta function with the vertical acceleration commonly used in the literature to problems with a moving point mass contributes certain discontinuities to the governing differential equation. The Lagrange equation of the second kind allowed us to solve the problem and to prove the correctness of the results. They are identical with the direct transformation of the differential equations of motion.

3.3.1 Fourier Solution

The two coupled equations of motion for the Timoshenko beam under a moving mass particle are

$$\rho A\frac{\partial^2 w(x,t)}{\partial t^2}-\frac{GA}{k}\left(\frac{\partial^2 w(x,t)}{\partial x^2}-\frac{\partial\psi(x,t)}{\partial x}\right)=q(x,t)\,,$$
$$\rho I\frac{\partial^2\psi(x,t)}{\partial t^2}-EI\frac{\partial^2\psi(x,t)}{\partial x^2}-\frac{GA}{k}\left(\frac{\partial w(x,t)}{\partial x}-\psi(x,t)\right)=0\;, \tag{3.83}$$

where

$$q(x,t)=\delta(x-\mathrm{v}t)m\left(g-\frac{\mathrm{d}^2w(\mathrm{v}t,t)}{\mathrm{d}t^2}\right)\;. \tag{3.84}$$

The acceleration of the mass particle at a constant speed v is called the Renaudot formulation

$$\frac{\mathrm{d}^2w(\mathrm{v}t,t)}{\mathrm{d}t^2}=\left.\frac{\partial^2 w(x,t)}{\partial t^2}\right|_{x=\mathrm{v}t}+2\mathrm{v}\left.\frac{\partial^2 w(x,t)}{\partial x\partial t}\right|_{x=\mathrm{v}t}+\mathrm{v}^2\left.\frac{\partial^2 w(x,t)}{\partial x^2}\right|_{x=\mathrm{v}t}\;. \tag{3.85}$$

The solution by Fourier transformation starts from the partial differential equation (3.83) and by direct mathematical transformation is reduced to a second order matrix ordinary differential equation. In the case of a simply supported Timoshenko beam we apply the sine transform to w and the cosine transform to ψ. The last stage is performed numerically and for this reason we call this solution semi-analytical. This solution has one disadvantage. The formulation (3.84) contains the term $\delta(x-\mathrm{v}t)m\mathrm{d}^2w(\mathrm{v}t,t)/\mathrm{d}t^2$ which defines the inertial force of the mass particle in space x and time t. The Dirac delta contributes discontinuities to the formulation. Although the solution can be defined, we can not prove that it verifies the equation of the problem. That is why we intend to apply another method which avoids this discontinuous formulation. In this way we can eliminate a weak point of the investigation.

The problem of a moving mass is important since in the case of a string and simply supported Timoshenko beam, the results exhibit a discontinuity in the mass trajectory at the end support. This phenomenon in the case of a string was presented and discussed for the first time in our previous paper [48], and in the case of the massless string it was mathematically proved. It can also be noticed in engineering practice. In railway traction systems cables are broken just before the end support. Also road plates are destroyed at their ends.

3.3.2 The Lagrange Equation

The energetic description of the issue removes a weak point of the analysis. The kinetic energy of a moving mass particle m travelling with a constant speed v is described by the equation

$$E_{km} = \frac{1}{2} m \left(\frac{\mathrm{d}w(\mathrm{v}t,t)}{\mathrm{d}t} \right)^2 + \frac{1}{2} m \mathrm{v}^2 \, . \tag{3.86}$$

The effect of the moving force of gravity mg can be written as the potential energy

$$E_{pm} = mg\, w(\mathrm{v}t,t) \, . \tag{3.87}$$

A moving mass is always in a pure rigid contact with the beam. The displacement of the point of a beam in contact with the mass particle is described by the same relation as a travelling point mass motion.

Let us consider a simply supported Timoshenko beam with constant cross-sectional area A, mass density ρ, and moment of area I. The examined beam has a finite length l. The kinetic and potential energy of the beam take the form

$$E_k = \frac{1}{2} \rho A \int_0^l \left(\frac{\partial w(x,t)}{\partial t} \right)^2 \mathrm{d}x + \frac{1}{2} \rho I \int_0^l \left(\frac{\partial \psi(x,t)}{\partial t} \right)^2 \mathrm{d}x \, , \tag{3.88}$$

$$E_p = \frac{1}{2} EI \int_0^l \left(\frac{\partial \psi(x,t)}{\partial x} \right)^2 \mathrm{d}x + \frac{1}{2} \frac{GA}{k} \int_0^l \left(\frac{\partial w(x,t)}{\partial x} - \psi(x,t) \right)^2 \mathrm{d}x \, . \tag{3.89}$$

Here, E is Young's modulus, G is the shear modulus, and k is the shear coefficient, which depends on the shape of cross-section of the beam.

We impose boundary conditions

$$w(x,t)|_{x=0} = 0 \, , \quad w(x,t)|_{x=l} = 0 \, , \quad \left. \frac{\partial \psi(x,t)}{\partial x} \right|_{x=0} = 0 \, , \quad \left. \frac{\partial \psi(x,t)}{\partial x} \right|_{x=l} = 0 \, , \tag{3.90}$$

and initial conditions

$$w(x,t)|_{t=0} = 0 \, , \quad \left. \frac{\partial w(x,t)}{\partial t} \right|_{t=0} = 0 \, ,$$

$$\psi(x,t)|_{t=0} = 0 \, , \quad \left. \frac{\partial \psi(x,t)}{\partial t} \right|_{t=0} = 0 \, . \tag{3.91}$$

We assume a general solution in the following form

$$w(x,t) = \sum_{j=1}^{n} X_{1j}(x) \xi_j(t) \, , \quad \psi(x,t) = \sum_{j=1}^{n} X_{2j}(x) \gamma_j(t) \, , \tag{3.92}$$

where $X_{1j}(x)$ and $X_{2j}(x)$ are orthogonal functions which fulfil the boundary conditions (3.90):

$$X_{1j}(x) = \sin \frac{j\pi x}{l} \, , \quad X_{2j}(x) = \cos \frac{j\pi x}{l} \, . \tag{3.93}$$

The displacement of the beam at the contact point with a travelling mass is

$$w(vt,t) = \sum_{j=1}^{n} \xi_j(t) \sin\frac{j\pi vt}{l} \ . \tag{3.94}$$

According to the rules of differentiation, we then obtain the following formula

$$\frac{\mathrm{d}w(vt,t)}{\mathrm{d}t} = \sum_{j=1}^{n} \dot{\xi}_j(t) \sin\frac{j\pi vt}{l} + \sum_{j=1}^{n} \xi_j(t)\frac{j\pi v}{l}\cos\frac{j\pi vt}{l} \ . \tag{3.95}$$

The kinetic energy of the moving inertial point (3.86) is expressed as a function of both the generalized coordinates and the derivatives of the generalized coordinates with respect to time

$$E_{km} = f\left(\xi,\dot{\xi}\right) \ . \tag{3.96}$$

The required derivation of the above quantity has important consequences. According to the formula for kinetic and potential energy of the beam we obtain two coupled equations of motion. These two equations can be reduced to two uncoupled equations with ξ and γ. Let us focus our attention on the displacement case

$$\begin{aligned}
&\ddddot{\xi}_j(t) + \beta\sum_{k=1}^{n} f_1(j,k,t)\ddddot{\xi}_k(t) + 2\beta\sum_{k=1}^{n}\left(\omega_j f_2(j,k,t) + 2f_3(j,k,t)\right)\dddot{\xi}_k(t) + \\
&+\frac{A}{I}c_1^2\ddot{\xi}_j(t) + \frac{\omega_j^2}{v^2}\left(c_1^2+c_2^2\right)\ddot{\xi}_j(t) + \sum_{k=1}^{n}\left[g(j)f_1(j,k,t) + 6\beta\left(\omega_j\omega_k f_4(j,k,t) - \right.\right.\\
&\left.\left.-\omega_k^2 f_1(j,k,t)\right)\right]\ddot{\xi}_k(t) + 2\sum_{k=1}^{n}\left[g(j)\omega_k f_3(j,k,t) - \beta\left(3\omega_j\omega_k^2 f_2(j,k,t) + \right.\right.\\
&\left.\left.+2\omega_k^3 f_3(j,k,t)\right)\right]\dot{\xi}_k(t) + \frac{\omega_j^4}{v^4}c_1^2c_2^2\xi_j(t) - \sum_{k=1}^{n}\left[g(j)\omega_k^2 f_1(j,k,t) + \right.\\
&\left.+\beta\left(2\omega_j\omega_k^3 f_4(j,k,t) - \omega_k^4 f_1(j,k,t)\right)\right]\xi_k(t) = \frac{mg}{\rho A\beta}g(j)\sin\omega_j t \ ,
\end{aligned} \tag{3.97}$$

where

$$c_1 = \sqrt{\frac{G}{k\rho}}\ ,\quad c_2 = \sqrt{\frac{E}{\rho}}\ ,\quad \beta = \frac{2m}{\rho A l}\ ,\quad \omega_k = \frac{k\pi v}{l}\ ,\quad \omega_j = \frac{j\pi v}{l}\ , \tag{3.98}$$

$$\begin{aligned}
f_1(j,k,t) &= \sin\omega_j t\sin\omega_k t\ ,\\
f_2(j,k,t) &= \cos\omega_j t\sin\omega_k t\ ,\\
f_3(j,k,t) &= \sin\omega_j t\cos\omega_k t\ ,\\
f_4(j,k,t) &= \cos\omega_j t\cos\omega_k t\ ,
\end{aligned} \tag{3.99}$$

$$g(j) = \beta\left[\frac{A}{I}c_1^2 + \omega_j^2\left(\frac{c_2^2}{v^2} - 1\right)\right] \ . \tag{3.100}$$

The coefficients c_1 and c_2 are the shear and bending wave velocity in a Timoshenko beam, respectively.

Lagrangian methods lead us to the system of differential equations (3.97) with variable coefficients. This system of equations can not be easily solved in an analytical way and we must integrate it numerically. We perform this integration with the use of the Runge–Kutta method. Equations (3.97) can be written in the short form

$$\mathbf{G}\ddddot{\xi} + \mathbf{U}\dddot{\xi} + \mathbf{M}\ddot{\xi} + \mathbf{C}\dot{\xi} + \mathbf{K}\xi = \mathbf{P}\,. \tag{3.101}$$

The matrices **G**, **U**, **M**, **C**, **K** and the vector **P** are given as follows in Table 3.1.

Formula (3.101) constitutes a system of ordinary differential equations of the fourth order with respect to time, so we need two additional initial conditions [54]:

$$\left.\frac{\partial^2 w(x,t)}{\partial t^2}\right|_{t=0} = \left.\frac{1}{\rho A}q(x,t)\right|_{t=0}, \quad \left.\frac{\partial^3 w(x,t)}{\partial t^3}\right|_{t=0} = \left.\frac{1}{\rho A}\frac{\partial q(x,t)}{\partial t}\right|_{t=0}, \tag{3.102}$$

where $q(x,t)$ is given by equation (3.84). By the sine Fourier transformation over a finite range of the initial conditions (3.91) and (3.102), we can write initial subvectors for displacements in the following form

$$\xi_j(t)\big|_{t=0} = 0\,, \quad \dot{\xi}_j(t)\Big|_{t=0} = 0\,, \tag{3.103}$$

$$\ddot{\xi}_j(t)\Big|_{t=0} = 0\,, \quad \dddot{\xi}_j(t)\Big|_{t=0} = \frac{mg\omega_j}{\rho A}\,. \tag{3.104}$$

Finally, the displacements of an arbitrary point of the beam can be determined from the following relation (see equation 3.92):

$$w(x,t) = \sum_{i=1}^{n} \xi_i(t) \sin\left(\frac{i\pi x}{l}\right)\,. \tag{3.105}$$

3.3.3 Examples

We use dimensionless data $l = 1$, $\rho = 1$, $A = 1$, $I = 0.01$, $E = 1$, $G = 0.4$ and $k = 1$. These data result in a shear wave speed of $c_1 = 0.63$ and a bending wave speed of $c_2 = 1.0$ (equation 3.98). Results of the semi-analytical solution are depicted in Figure 3.17. The displacements are related to the amplitude of the quasi-static displacement of the beam mid-point w_0. A more detailed presentation of the Timoshenko beam motion is given in Figure 3.18. Both types of waves are noticeable. We emphasize the sharp edge of the wave and the reflections from the support and from the moving mass point. The velocity $\mathrm{v} = 0.5\,c_2$ is characteristic in our example since the discontinuity of the mass trajectory is quite visible. Further tests will be performed with this velocity. The convergence is slow and we examined it in relation to the number of terms (Figure 3.19) taken in the equation (3.105). The plot with a small number of terms is smooth in the neighbourhood of the support. Taking

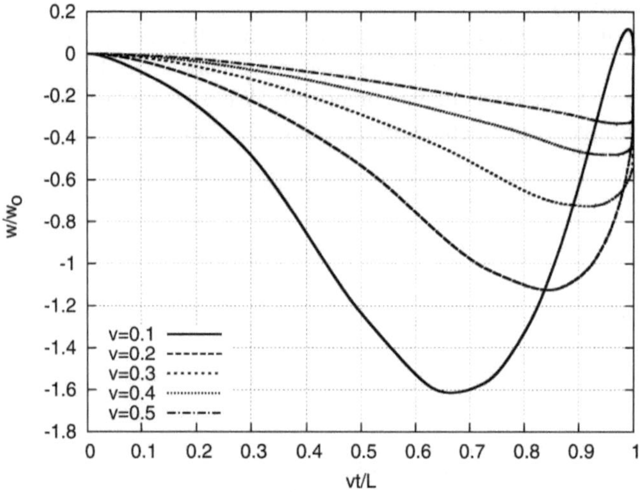

Fig. 3.17 Semi-analytical solution of the mass trajectory moving along the Timoshenko beam at various velocities ($c_1 = 0.63, c_2 = 1.00$).

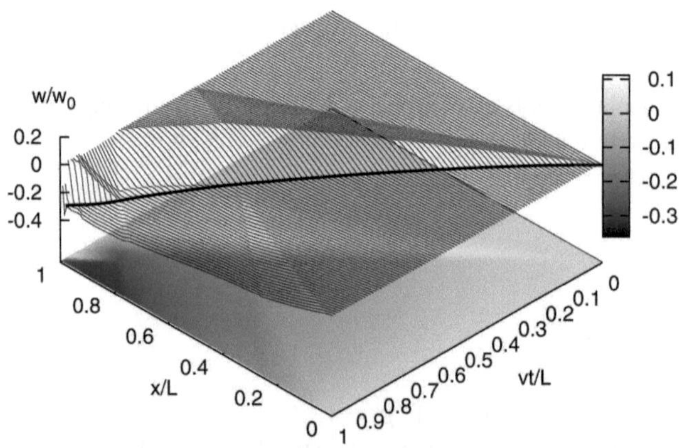

Fig. 3.18 Deflection in time of the Timoshenko beam under a load moving at the speed $v = 0.5c_2$.

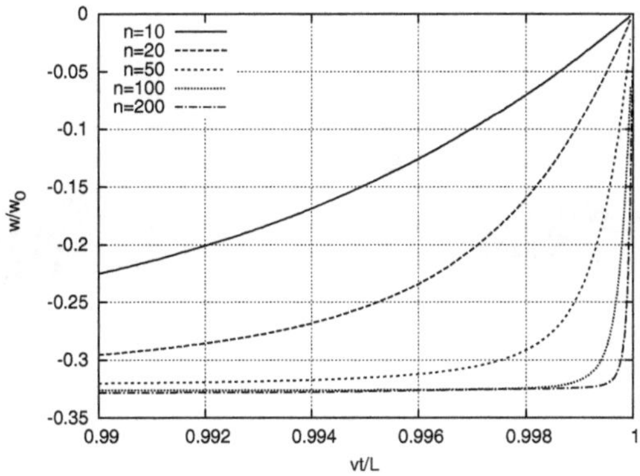

Fig. 3.19 The convergence of the mass trajectory travelling with $v = 0.5c_2$ near the end point, for various numbers of terms (10, 20, ..., 200) in Eq. (3.105).

an increasing number of terms sharpens the plot of the last 1% of the trajectory. This can be compared with the same phenomenon obtained for a string [48]. Examples show the same type of discontinuity of the solution in the case of the Timoshenko beam. Although we can not mathematically establish the presence of this feature in the case of an inertial Timoshenko beam, we can say that for practical purposes, the differential equation of the Timoshenko beam motion under the assumption of small displacements involves a discontinuity of the structure in the neighbourhood of the support. Such a phenomenon is observed in real structures (a track or a bridge plate) in the form of impacts of high magnitude. Figure 3.20 depicts the deflection of the Timoshenko beam in time and reflections of the transverse wave c_1 and longitudinal wave c_2 at subcritical and critical speed.

3.3.4 Conclusions and Discussion

The solution of the problem discussed here can not be simply applied to complex problems, for example strings, beams, or three dimensional bodies, subjected to a system of masses or composed of segments with variable rigidity. In such cases discrete methods should be applied. However, it enables us to exhibit the qualitative features and validate numerical solutions. The existing numerical approaches fail in the case of inertial loads. Although the solutions converge in some cases, the error in the case of mass motion, near the critical speed, is significant.

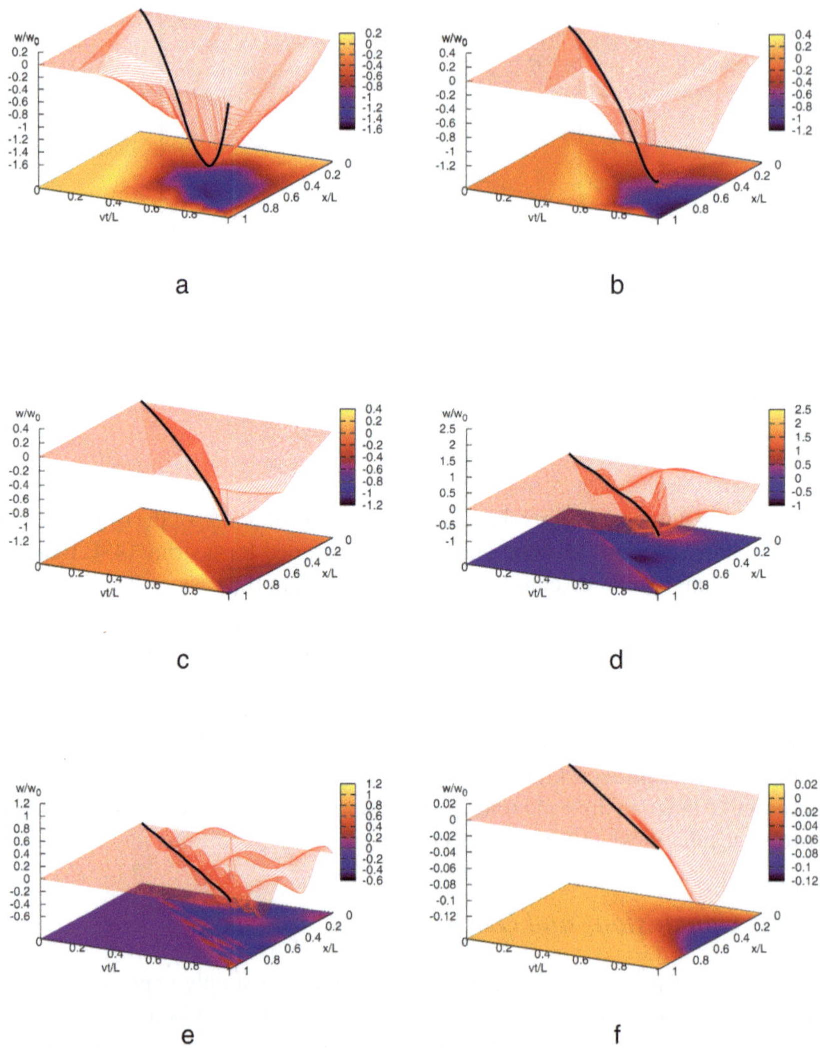

Fig. 3.20 Simulation of the Timoshenko beam under the moving inertial point at the speed: (a) v = 0.2, (b) v = 0.4, (c) v = 0.6, (d) v = 0.7, (e) v = 0.9 and (f) v = 1.0 ($c_1 = 0.63$, $c_2 = 1.00$).

Table 3.1 Matrices **G**, **U**, **M**, **C**, **K** and the vector **P** (3.101)

$$\mathbf{G} = \begin{bmatrix} 1 & 0 & \cdots & 0 \\ 0 & 1 & \cdots & 0 \\ \vdots & \vdots & \ddots & \vdots \\ 0 & 0 & \cdots & 1 \end{bmatrix} + \beta \begin{bmatrix} f_1(1,1,t) & f_1(1,2,t) & \cdots & f_1(1,n,t) \\ f_1(2,1,t) & f_1(2,2,t) & \cdots & f_1(2,n,t) \\ \vdots & \vdots & \ddots & \vdots \\ f_1(n,1,t) & f_1(n,2,t) & \cdots & f_1(n,n,t) \end{bmatrix}$$

$$\mathbf{U} = 2\beta \begin{bmatrix} \omega_1 f_2(1,1,t) & \omega_1 f_2(1,2,t) & \cdots & \omega_1 f_2(1,n,t) \\ \omega_2 f_2(2,1,t) & \omega_2 f_2(2,2,t) & \cdots & \omega_2 f_2(2,n,t) \\ \vdots & \vdots & \ddots & \vdots \\ \omega_n f_2(n,1,t) & \omega_n f_2(n,2,t) & \cdots & \omega_n f_2(n,n,t) \end{bmatrix} + 4\beta \begin{bmatrix} f_3(1,1,t) & f_3(1,2,t) & \cdots & f_3(1,n,t) \\ f_3(2,1,t) & f_3(2,2,t) & \cdots & f_3(2,n,t) \\ \vdots & \vdots & \ddots & \vdots \\ f_3(n,1,t) & f_3(n,2,t) & \cdots & f_3(n,n,t) \end{bmatrix}$$

$$\mathbf{M} = \frac{A}{I} c_1^2 \begin{bmatrix} 1 & 0 & \cdots & 0 \\ 0 & 1 & \cdots & 0 \\ \vdots & \vdots & \ddots & \vdots \\ 0 & 0 & \cdots & 1 \end{bmatrix} + \frac{c_1^2 + c_2^2}{v^2} \begin{bmatrix} \omega_1 & 0 & \cdots & 0 \\ 0 & \omega_2 & \cdots & 0 \\ \vdots & \vdots & \ddots & \vdots \\ 0 & 0 & \cdots & \omega_n \end{bmatrix} + \begin{bmatrix} g(1) f_1(1,1,t) & g(1) f_1(1,2,t) & \cdots & g(1) f_1(1,n,t) \\ g(2) f_1(2,1,t) & g(2) f_1(2,2,t) & \cdots & g(2) f_1(2,n,t) \\ \vdots & \vdots & \ddots & \vdots \\ g(n) f_1(n,1,t) & g(n) f_1(n,2,t) & \cdots & g(n) f_1(n,n,t) \end{bmatrix} +$$

Table 3.1 *(continued)*

$$+6\beta\begin{bmatrix}\omega_1\omega_1 f_4(1,1,t) & \omega_1\omega_2 f_4(1,2,t) & \cdots & \omega_1\omega_n f_4(1,n,t)\\ \omega_2\omega_1 f_4(2,1,t) & \omega_2\omega_2 f_4(2,2,t) & \cdots & \omega_2\omega_n f_4(2,n,t)\\ \vdots & \vdots & \ddots & \vdots\\ \omega_n\omega_1 f_4(2,n,t) & \omega_n\omega_2 f_4(n,2,t) & \cdots & \omega_n\omega_n f_4(n,n,t)\end{bmatrix}-\begin{bmatrix}\omega_1^2 f_1(1,1,t) & \omega_2^2 f_1(1,2,t) & \cdots & \omega_n^2 f_1(1,n,t)\\ \omega_1^2 f_1(2,1,t) & \omega_2^2 f_1(2,2,t) & \cdots & \omega_n^2 f_1(2,n,t)\\ \vdots & \vdots & \ddots & \vdots\\ \omega_1^2 f_1(n,1,t) & \omega_2^2 f_1(n,2,t) & \cdots & \omega_n^2 f_1(n,n,t)\end{bmatrix}$$

$$\mathbf{C}=2\begin{bmatrix}g(1)\omega_1 f_3(1,1,t) & g(1)\omega_2 f_3(1,2,t) & \cdots & g(1)\omega_n f_3(1,n,t)\\ g(2)\omega_1 f_3(2,1,t) & g(2)\omega_2 f_3(2,2,t) & \cdots & g(2)\omega_n f_3(2,n,t)\\ \vdots & \vdots & \ddots & \vdots\\ g(n)\omega_1 f_3(n,1,t) & g(n)\omega_2 f_3(n,2,t) & \cdots & g(n)\omega_n f_3(n,n,t)\end{bmatrix}-$$

$$-6\beta\begin{bmatrix}\omega_1\omega_1^2 f_2(1,1,t) & \omega_1\omega_2^2 f_2(1,2,t) & \cdots & \omega_1\omega_n^2 f_2(1,n,t)\\ \omega_2\omega_1^2 f_2(2,1,t) & \omega_2\omega_2^2 f_2(2,2,t) & \cdots & \omega_2\omega_n^2 f_2(2,n,t)\\ \vdots & \vdots & \ddots & \vdots\\ \omega_n\omega_1^2 f_2(n,1,t) & \omega_n\omega_2^2 f_2(n,2,t) & \cdots & \omega_n\omega_n^2 f_2(n,n,t)\end{bmatrix}+4\begin{bmatrix}\omega_1^3 f_3(1,1,t) & \omega_2^3 f_3(1,2,t) & \cdots & \omega_n^3 f_3(1,n,t)\\ \omega_1^3 f_3(2,1,t) & \omega_2^3 f_3(2,2,t) & \cdots & \omega_n^3 f_3(2,n,t)\\ \vdots & \vdots & \ddots & \vdots\\ \omega_1^3 f_3(n,1,t) & \omega_2^3 f_3(n,2,t) & \cdots & \omega_n^3 f_3(n,n,t)\end{bmatrix}$$

Table 3.1 (*continued*)

$$\mathbf{K} = \frac{c_1^2 c_2^2}{v^4} \begin{bmatrix} \omega_1^4 & 0 & \cdots & 0 \\ 0 & \omega_2^4 & \cdots & 0 \\ \vdots & \vdots & \ddots & \vdots \\ 0 & 0 & \cdots & \omega_n^4 \end{bmatrix} - \begin{bmatrix} g(1)\omega_1^2 f_1(1,1,t) & g(1)\omega_2^2 f_1(1,2,t) & \cdots & g(1)\omega_n^2 f_1(1,n,t) \\ g(2)\omega_1^2 f_1(2,1,t) & g(2)\omega_2^2 f_1(2,2,t) & \cdots & g(2)\omega_n^2 f_1(2,n,t) \\ \vdots & \vdots & \ddots & \vdots \\ g(n)\omega_1^2 f_1(n,1,t) & g(n)\omega_2^2 f_1(n,2,t) & \cdots & g(n)\omega_n^2 f_1(n,n,t) \end{bmatrix} +$$

$$+2\beta \begin{bmatrix} \omega_1\omega_1^3 f_4(1,1,t) & \omega_1\omega_2^3 f_4(1,2,t) & \cdots & \omega_1\omega_n^3 f_4(1,n,t) \\ \omega_2\omega_1^3 f_4(2,1,t) & \omega_2\omega_2^3 f_4(2,2,t) & \cdots & \omega_2\omega_n^3 f_4(2,n,t) \\ \vdots & \vdots & \ddots & \vdots \\ \omega_n\omega_1^3 f_4(n,1,t) & \omega_n\omega_2^3 f_4(n,2,t) & \cdots & \omega_n\omega_n^3 f_4(n,n,t) \end{bmatrix} - \begin{bmatrix} \omega_1^4 f_1(1,1,t) & \omega_2^4 f_1(1,2,t) & \cdots & \omega_n^4 f_1(1,n,t) \\ \omega_1^4 f_1(2,1,t) & \omega_2^4 f_1(2,2,t) & \cdots & \omega_n^4 f_1(2,n,t) \\ \vdots & \vdots & \ddots & \vdots \\ \omega_1^4 f_1(n,1,t) & \omega_2^4 f_1(n,2,t) & \cdots & \omega_n^4 f_1(n,n,t) \end{bmatrix}$$

$$\mathbf{P} = \frac{mg}{\rho A\beta} \begin{bmatrix} g(1)\sin\frac{1\pi vt}{l} \\ g(2)\sin\frac{2\pi vt}{l} \\ \vdots \\ g(n)\sin\frac{n\pi vt}{l} \end{bmatrix}$$

3.4 Bernoulli–Euler Beam vs. Timoshenko Beam

The Bernoulli–Euler beam does not exhibit the discussed discontinuity of the solution. Figure 3.21 compares of the trajectories of the moving inertial point travelling along the Euler beam and the Timoshenko beam.

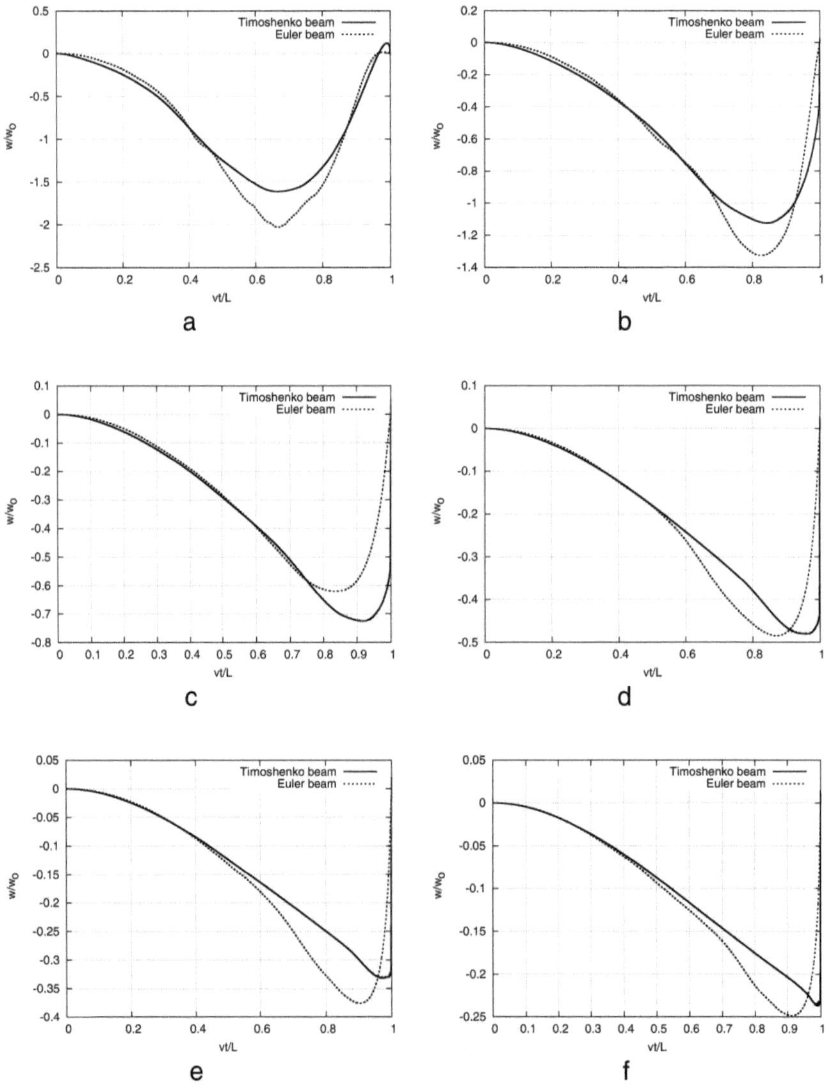

Fig. 3.21 Trajectory of a mass travelling at different speeds: (a) $v = 0.1$, (b) $v = 0.2$, (c) $v = 0.3$, (d) $v = 0.4$, (e) $v = 0.5$, and (f) $v = 0.6$ ($c_1 = 0.63$, $c_2 = 1.00$).

Although a discontinuity of the trajectory is a characteristic feature of the differential equations used in modelling the Timoshenko beam, high jumps of the travelling contact point are observed also in the case of the Bernoulli–Euler beam (Figure 3.21 e) and f)). The final stage of the load trajectory's exhibiting a discontinuity is a feature which is important in practice. This is especially true for short spans with high cross sections and, what is important, in the case of higher travelling speed, regardless of the type of beam.

An identical problem occurs for plates. Further chapters will present semi-analytical solutions. The numerical results given in Chapter 7.4.2, especially Figure 7.26, illustrate the same effect. In the case of plates, the sharpness of the trajectory is reduced by the influence of the stiffness of the plate in the direction perpendicular to the load path. The effect can vary with the proportion of the plate edges.

3.5 Plate

Let us consider a simply supported rectangular plate subjected to an inertial moving load. We employ the notations used in Figure 3.22. Thus, l_x and l_y are the dimensions of the plate. We, then, consider a moving point mass m with accompanying force P travelling with speed v based on function f. The plate thickness is denoted by h.

In order to simplify the writing of the equations, we write the plate displacement at the point with coordinates (x,y) as $w = w(x,y,t)$ and the functions describing the position of a moving load $f_x = f_x(t)$ and $f_y = f_y(t)$. The equation of motion of a Kirchhoff plate under a moving inertial load can be written in the following form

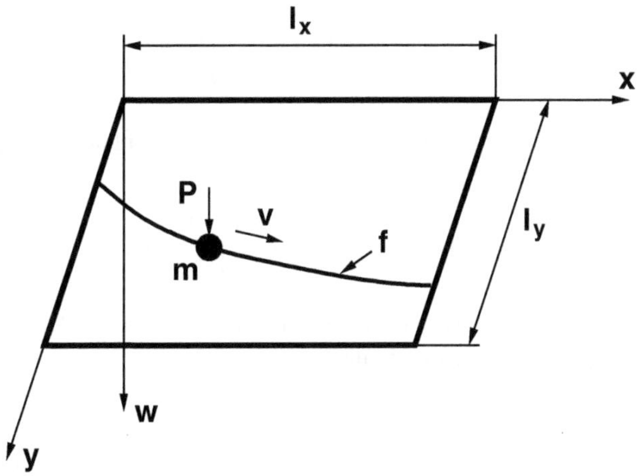

Fig. 3.22 The scheme of the Kirchhoff plate under an inertial load.

$$D\left[\frac{\partial^4 w}{\partial x^4}+2\frac{\partial^4 w}{\partial x^2\partial y^2}+\frac{\partial^4 w}{\partial y^4}\right]+\mu\frac{\partial^2 w}{\partial t^2}= \\ =\delta(x-f_x)\delta(y-f_y)P-\delta(x-f_x)\delta(y-f_y)m\frac{\mathrm{d}^2 w(f_x,f_y,t)}{\mathrm{d}t^2}\,, \tag{3.106}$$

where D is the plate stiffness, described by

$$D=\frac{Eh^3}{12\,(1-\nu^2)}\,. \tag{3.107}$$

Here, μ is the mass per unit area, E is the Young modulus for the plate, and ν is the Poisson ratio. As mentioned at the beginning, the considered plate is simply supported at the edges, and thus the boundary conditions take the following form

$$\begin{aligned}
&w(0,y,t)=0, &&\left.\left(\frac{\partial^2 w(x,y,t)}{\partial x^2}+\nu\frac{\partial^2 w(x,y,t)}{\partial y^2}\right)\right|_{x=0}=0\,,\\
&w(l_x,y,t)=0, &&\left.\left(\frac{\partial^2 w(x,y,t)}{\partial x^2}+\nu\frac{\partial^2 w(x,y,t)}{\partial y^2}\right)\right|_{x=l_x}=0\,,\\
&w(x,0,t)=0, &&\left.\left(\frac{\partial^2 w(x,y,t)}{\partial y^2}+\nu\frac{\partial^2 w(x,y,t)}{\partial x^2}\right)\right|_{y=0}=0\,,\\
&w(x,l_y,t)=0, &&\left.\left(\frac{\partial^2 w(x,y,t)}{\partial y^2}+\nu\frac{\partial^2 w(x,y,t)}{\partial x^2}\right)\right|_{y=l_y}=0\,.
\end{aligned} \tag{3.108}$$

The displacements and bending moments at the edges of plates are zero. We also assume zero initial conditions:

$$w(x,y,0)=0\,,\qquad \left.\frac{\partial w(x,y,t)}{\partial t}\right|_{t=0}=0\,. \tag{3.109}$$

The inertia of the moving mass can be described by the material derivative. Writing $\mathrm{d}^2 w/\mathrm{d}t^2=\mathrm{d}^2 w(f_x,f_y,t)/\mathrm{d}t^2$, we obtain

$$\frac{\mathrm{d}^2 w}{\mathrm{d}t^2}=\frac{\partial^2 w}{\partial t^2}+2\dot{f}_x\frac{\partial^2 w}{\partial x\partial t}+2\dot{f}_y\frac{\partial^2 w}{\partial y\partial t}+2\dot{f}_x\dot{f}_y\frac{\partial^2 w}{\partial x\partial y}+\dot{f}_x^2\frac{\partial^2 w}{\partial x^2}+\dot{f}_y^2\frac{\partial^2 w}{\partial y^2}+ \\ +\ddot{f}_x\frac{\partial w}{\partial x}+\ddot{f}_y\frac{\partial w}{\partial y}\,. \tag{3.110}$$

This is a general form of the equation describing the inertia of a moving mass, taking into account the changes of velocity both in the direction of x and y.

In order to solve the partial differential equation of motion (3.106), we use the Fourier method and separation of variables. In the case of a simply supported plate, we use the following integral transformation

$$V(j,k,t) = \int_0^{l_x}\int_0^{l_y} w(x,y,t)\sin\frac{j\pi x}{l_x}\sin\frac{k\pi y}{l_y}\,\mathrm{d}x\,\mathrm{d}y\,, \tag{3.111}$$

$$w(x,y,t) = \frac{4}{l_x l_y}\sum_{j=1}^{\infty}\sum_{k=1}^{\infty} V(j,k,t)\sin\frac{j\pi x}{l_x}\sin\frac{k\pi y}{l_y}\,, \tag{3.112}$$

which satisfies the imposed boundary conditions (3.108). As a result of the Fourier transformation of equation (3.106), we obtain an infinite system of ordinary differential equations:

$$D\left(\frac{j^2\pi^2}{l_x^2}+\frac{k^2\pi^2}{l_y^2}\right)^2 V(j,k,t)+\rho A\ddot{V}(j,k,t) = \left(P-m\frac{\mathrm{d}^2w}{\mathrm{d}t^2}\right)\sin\frac{j\pi f_x}{l_x}\sin\frac{k\pi f_y}{l_y}\,. \tag{3.113}$$

Limit the task to a finite system of equations of dimension n. In order to solve equations (3.113), we need to represent the vertical acceleration of the moving mass by the series (3.112). Writing $V_{pq} = V(p,q,t)$, and using (3.112), the equation (3.110) can be re-written:

$$\begin{aligned}\frac{\mathrm{d}^2w}{\mathrm{d}t^2} = \frac{4}{l_x l_y}\sum_{p=1}^{\infty}\sum_{q=1}^{\infty}\Bigg\{&\ddot{V}_{pq}\sin\frac{p\pi f_x}{l_x}\sin\frac{q\pi f_y}{l_y}+2\dot{f}_x\dot{V}_{pq}\frac{p\pi}{l_x}\cos\frac{p\pi f_x}{l_x}\sin\frac{q\pi f_y}{l_y}+\\ &+2\,\dot{f}_y\dot{V}_{pq}\sin\frac{p\pi f_x}{l_x}\frac{q\pi}{l_y}\cos\frac{q\pi f_y}{l_y}+2\dot{f}_x\dot{f}_y V_{pq}\frac{p\pi}{l_x}\cos\frac{p\pi f_x}{l_x}\frac{q\pi}{l_y}\cos\frac{q\pi f_y}{l_y}-\\ &-\dot{f}_x^2 V_{pq}\frac{p^2\pi^2}{l_x^2}\sin\frac{p\pi f_x}{l_x}\sin\frac{q\pi f_y}{l_y}-\dot{f}_y^2 V_{pq}\sin\frac{p\pi f_x}{l_x}\frac{q^2\pi^2}{l_y^2}\sin\frac{q\pi f_y}{l_y}+\\ &+\ddot{f}_x V_{pq}\frac{p\pi}{l_x}\cos\frac{p\pi f_x}{l_x}\sin\frac{q\pi f_y}{l_y}+\ddot{f}_y V_{pq}\sin\frac{p\pi f_x}{l_x}\frac{q\pi}{l_y}\cos\frac{q\pi f_y}{l_y}\Bigg\}\,.\end{aligned} \tag{3.114}$$

Using (3.114), the system of equations (3.113) can be written in matrix form:

$$\mathbf{M}\ddot{\mathbf{V}}_{pq}+\mathbf{C}\dot{\mathbf{V}}_{pq}+\mathbf{K}\mathbf{V}_{pq} = \mathbf{P}\,, \tag{3.115}$$

where **M**, **C** and **K** are, respectively, the matrices of inertia, damping, and stiffness. Each matrix is full and time-dependent, thus the matrix equation (3.115) is a system of differential equations with variable coefficients. Because there are no methods for solving the system of equations (3.115) in an analytical way, we have to solve it numerically. As a result, we obtain the vector $\mathbf{V}_{pq}$. Finally, the displacement at a given point of the plate is described by a series (3.112).

Let us consider a point mass travelling parallel to the edge of a plate defined by the axis x, with mass m and accompanying force P travelling with speed v along the

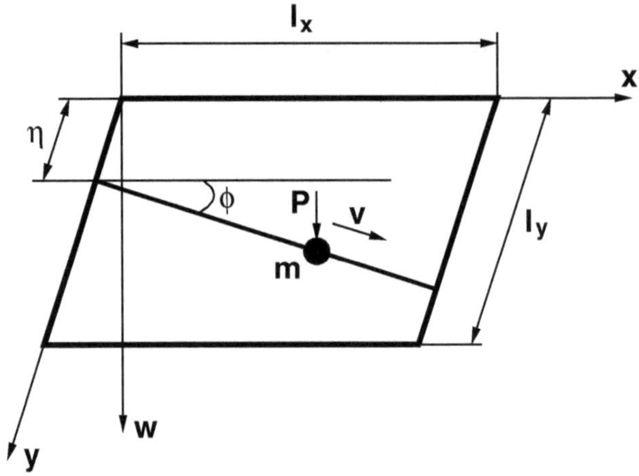

Fig. 3.23 The mass travelling along a straight line over the plate.

line defined by η and ϕ (Figure 3.23). This passage is described by the following parameters

$$f_x(t) = \mathrm{v}t\,, \quad f_y(t) = \eta\,, \quad \phi = 0\,. \tag{3.116}$$

The vertical acceleration of an inertial moving point (3.114) takes the simple form, written as follows

$$\frac{\mathrm{d}^2 w}{\mathrm{d}t^2} = \frac{4}{l_x l_y} \sum_{p=1}^{\infty} \sum_{q=1}^{\infty} \left\{ \ddot{V}_{pq} \sin\frac{p\pi \mathrm{v}t}{l_x} \sin\frac{q\pi\eta}{l_y} + 2\dot{V}_{pq} \frac{p\pi \mathrm{v}}{l_x} \cos\frac{p\pi \mathrm{v}t}{l_x} \sin\frac{q\pi\eta}{l_y} - \right.$$
$$\left. - V_{pq} \frac{p^2\pi^2\mathrm{v}^2}{l_x^2} \sin\frac{p\pi \mathrm{v}t}{l_x} \sin\frac{q\pi\eta}{l_y} \right\} . \tag{3.117}$$

3.6 The Renaudot Approach vs. The Yakushev Approach

In most papers devoted to dynamics of structures subjected to moving loads, the load term $q(x,t)$ is written in the following form

$$q(x,t) = \delta(x - \mathrm{v}t) m \left(g - \frac{\mathrm{d}^2 w}{\mathrm{d}t^2} \right) . \tag{3.118}$$

In most analytical and numerical solutions, the authors do not care about the correct solution process. In practice, neither the Renaudot nor the Yakushev approach have been much employed, since engineering computations are simplified in practice. However, an extensive discussion took place in the literature on the correctness of these approaches [83]. We can reasonably conclude that both formulations describe different mechanical problems. Let us look at the details.

3.6.1 The Renaudot Approach

Modelling a moving load with constant force of gravity is a significant simplification of reality. Nevertheless, its relatively simple mathematical apparatus is often used in both analytical solutions and numerical calculations. These problems yield to complete analytical solutions in the form of a series. In the case of a computer simulation of a moving force with a fixed value, the task leads to modifying the right hand side vector of the dynamic layout, step by step. However, literature reports say that with increasing speed of the travelling load, the inertial effects of the moving mass start to dominate, which changes the dynamic response of the entire system. Omission of of these effects leads to qualitative and quantitative deviations in the resulting solution. So we should take into account the inertia of moving masses, especially for higher speeds of the travelling load. Taking into account the inertia of the moving point mass leads to a component expressing the inertial d'Alembert force [133]. For $m = const$ we obtain

$$\delta(x-\mathrm{v}t)\frac{\mathrm{d}}{\mathrm{d}t}\left[m\frac{\mathrm{d}w(\mathrm{v}t,t)}{\mathrm{d}t}\right] = \delta(x-\mathrm{v}t)m\frac{\mathrm{d}^2w(\mathrm{v}t,t)}{\mathrm{d}t^2}\,. \tag{3.119}$$

The second derivative of the vertical acceleration of the point of the structure w under a moving mass can be written using a material derivative. Assuming a constant velocity v of the travelling mass, we obtain the Renaudot formula [118]

$$\frac{\mathrm{d}^2w(\mathrm{v}t,t)}{\mathrm{d}t^2} = \left.\frac{\partial^2 w(x,t)}{\partial t^2}\right|_{x=\mathrm{v}t} + 2\mathrm{v}\left.\frac{\partial^2 w(x,t)}{\partial x\partial t}\right|_{x=\mathrm{v}t} + \mathrm{v}^2\left.\frac{\partial^2 w(x,t)}{\partial x^2}\right|_{x=\mathrm{v}t}\,. \tag{3.120}$$

The reasoning given in the paper [118] is as follows According to Figure 3.24, at time t, the moving load is at the point N. After a time $\mathrm{d}t$, the moving load has shifted to the point N', if the deflection of the beam line AB remains unchanged. In fact, after an elapse of time $\mathrm{d}t$, the line deflection of the beam passes to a new position, and the moving load is at N''. For this reason, determining the vertical acceleration of the moving point mass or continuous load requires taking the full derivative of the function of w.

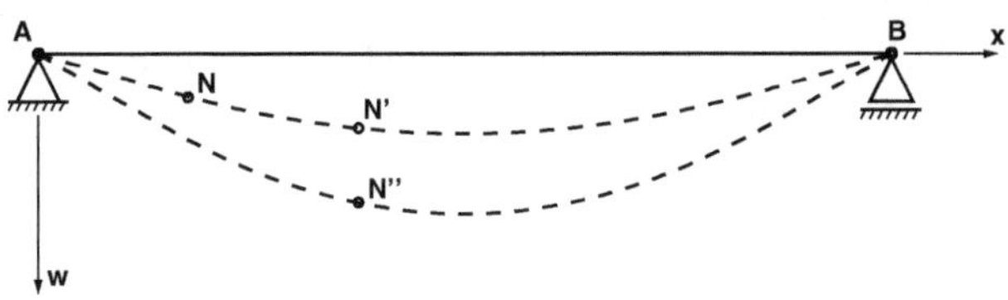

Fig. 3.24 Renaudot's reasoning to determine the acceleration of the moving mass.

3.6.2 The Yakushev Approach

In 1974, Yakushev published the paper [146], in which he proposed a new model for the inertial moving load. The Renaudot approach is there reformulated and refined in order to more accurately take into account the effect of a moving inertial load on the dynamic deflection of the beam. The foundation of that approach is to describe the effect of inertia in the following form:

$$\frac{\mathrm{d}}{\mathrm{d}t}\left[\delta(x-\mathrm{v}t)m\frac{\mathrm{d}w(\mathrm{v}t,t)}{\mathrm{d}t}\right]=-\delta'(x-\mathrm{v}t)m\mathrm{v}\frac{\mathrm{d}w(\mathrm{v}t,t)}{\mathrm{d}t}+\delta(x-\mathrm{v}t)m\frac{\mathrm{d}^2w(\mathrm{v}t,t)}{\mathrm{d}t^2}\,. \tag{3.121}$$

In this approach, the Dirac δ describes the location of a moving load and directly affects the momentum of the beam at the point $x=\mathrm{v}t$. This leads to changes in the description of the load compared to the Renaudot assumptions. In the Yakushev approach, the inertia of a point mass is described by not only the members of the second derivative of (3.120), but also a member of the first derivative

$$\frac{\mathrm{d}w(\mathrm{v}t,t)}{\mathrm{d}t}=\left.\frac{\partial w(x,t)}{\partial t}\right|_{x=\mathrm{v}t}+\mathrm{v}\left.\frac{\partial w(x,t)}{\partial x}\right|_{x=\mathrm{v}t}\,. \tag{3.122}$$

Equation (3.121) has a definite physical interpretation. Almost ten years later, in 1993, Sześniak [134] reviewed and analysed Yakushev's approach. In 1995, Langer and Klasztorny [83] discussed the correctness of the approach. They concluded that the Renaudot approach is the only correct formulation.

Two years later the differential equation term describing the moving inertial particle was also treated in the paper [41]. The effect of a moving mass is determined by Newton's second law:

$$F=\frac{\mathrm{d}p}{\mathrm{d}t}=\frac{\mathrm{d}m}{\mathrm{d}t}\dot{u}+m\frac{\mathrm{d}\dot{u}}{\mathrm{d}t}\,. \tag{3.123}$$

The resultant force F, acting on an inertial particle m, is defined as equal to the first derivative of impulse p, equivalent to the product of the velocity $\dot{u}$ by the variation of mass $\mathrm{d}m/\mathrm{d}t$ augmented by the product of the mass m and the acceleration $\ddot{u}=\mathrm{d}\dot{u}/\mathrm{d}t$.

Unfortunately, the authors of the last paper make no reference to any of the numerous papers devoted to the Yakushev approach. The authors illustrate their computations with several figures. They consider four types of load. Type I (3.124) describes the static problem. Type II (3.125) corresponds to the dynamic problem, however, it does not contain the mass of the moving load. Such a mass is considered in type III (3.126). Type IV (3.127), in turn, includes the damping resulting from the time dependence of the masses.

$$\mathbf{Ku} = \mathbf{F_t(t)} \tag{3.124}$$

$$\mathbf{M\ddot{u}} + \mathbf{Ku} = \mathbf{F_t(t)} \tag{3.125}$$

$$(\mathbf{M} + \mathbf{M_R})\mathbf{\ddot{u}} + \mathbf{Ku} = \mathbf{F_t(t)} \tag{3.126}$$

$$(\mathbf{M} + \mathbf{M_R})\mathbf{\ddot{u}} + \mathbf{\dot{M}_R\dot{u}} + \mathbf{Ku} = \mathbf{F_t(t)} \tag{3.127}$$

The computations were carried out with the following data:
– mass $M_1 = 0.135$ kg,
– forces $F_1 = 0.147$ N,
– number of elements = 20,
– time step $\Delta t = 0.00215$ s,
– length of beam = 1000 mm,
– height of beam = 1 mm,
– width of beam = 20 mm,
– mass density = $7.85{\cdot}10^6$ kg mm^{-2},
– Young's modulus = 204 833 N mm^{-2},
– speed = 2320 mm s^{-1}.
Figure 3.25 depicts the results given in [41] compared to accurate analytical results. We can compare the results obtained by Delgado with our results. We depict only the moving mass case and compare it with the Bernoulli–Euler semi-analytical solution (Figure 3.26).

The Renaudot approach and the Yakushev approach are independent. Each version is based on certain assumptions in a certain way theoretically justified. Any demonstration of the superiority of one variant of the theory over the other, leading to a more accurate description of physical phenomena, requires experimental verification. The difference between both approaches is depicted in Figure 3.27.

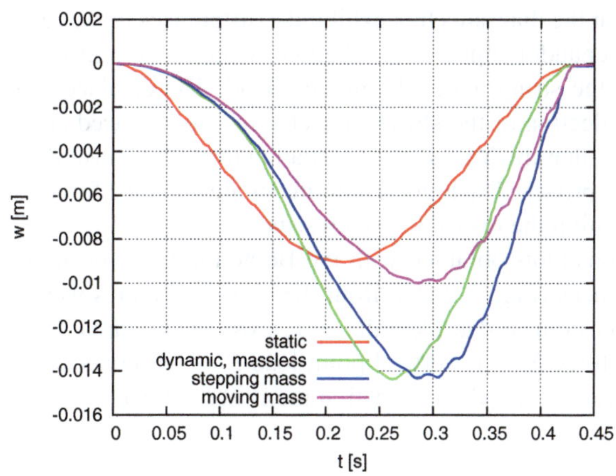

Fig. 3.25 Comparison of four load types based on [41].

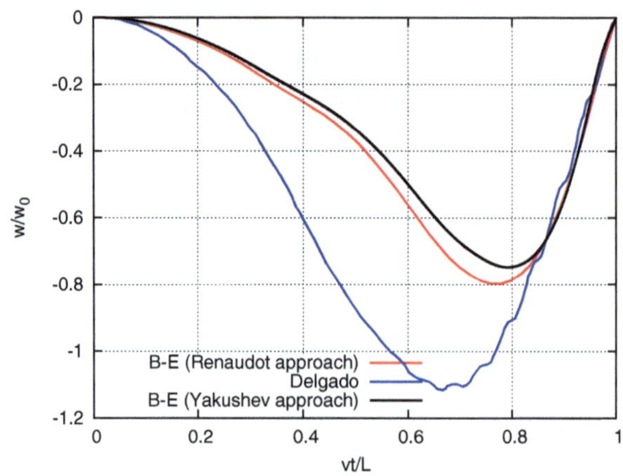

Fig. 3.26 Comparison of the displacement under the moving mass.

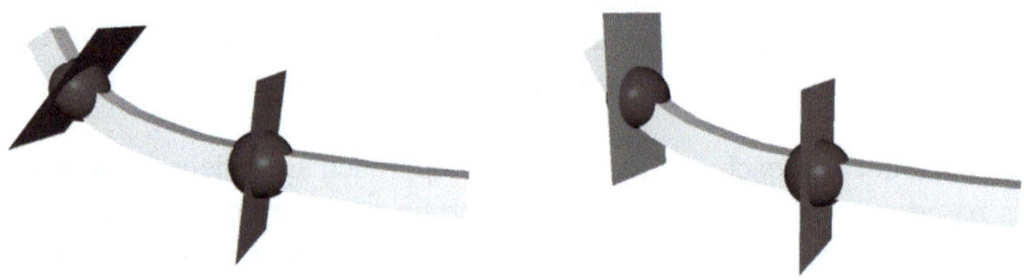

Fig. 3.27 Yakushev (a) and Renaudot (b) model of the moving mass problem.

Let us look at the diagrams that exhibit the differences in the results of the Yakushev and the Renaudot approaches. The first two figures (3.28) depict vertical displacements of the string under the moving load. The displacements were related to the static deflection of the string subjected to a load placed at the middle of the span. We notice that the Yakushev formulation results in lower displacements than the Renaudot approach. The highest amplitudes are obtained for the simple massless force, which exhibit the lowest inertial resistance.

The analogous plots obtained with the Bernoulli–Euler beam also exhibits significant differences (Figure 3.29). Both the above examples were solved with the following dimensionless data: length $l = 1$, cross section area $A = 1$, mass density $\rho = 1$, tensile force $N = 1$, vertical point load $P = -1$, concentrated mass $m = 1$, Young modulus $E = 1$, and cross section moment $I = 0.01$.

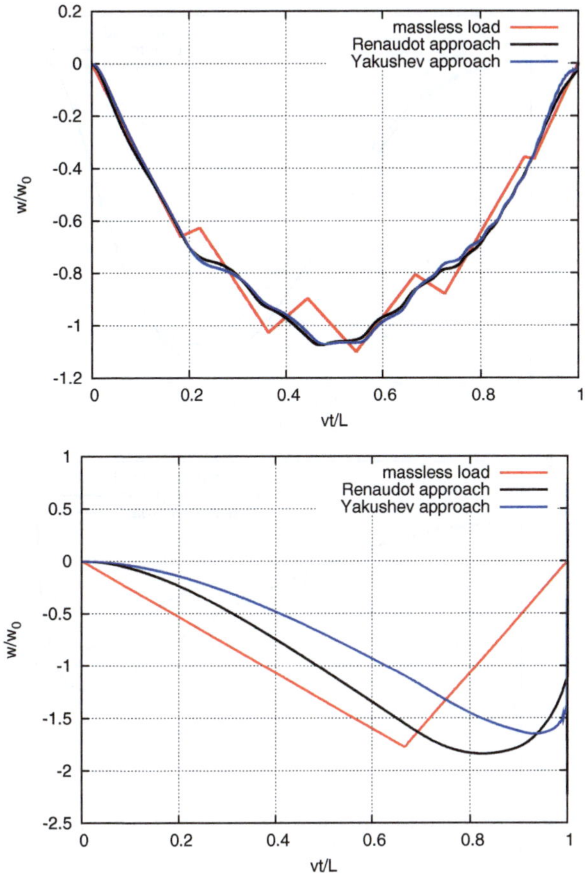

a

b

Fig. 3.28 Vertical displacements of the string under the load for the massless load, the Yakushev approach and the Renaudot approach, in the case of the speed of the moving load: $0.1\,c$ (upper), $0.5\,c$ (lower), $c = \sqrt{N/\rho/A}$.

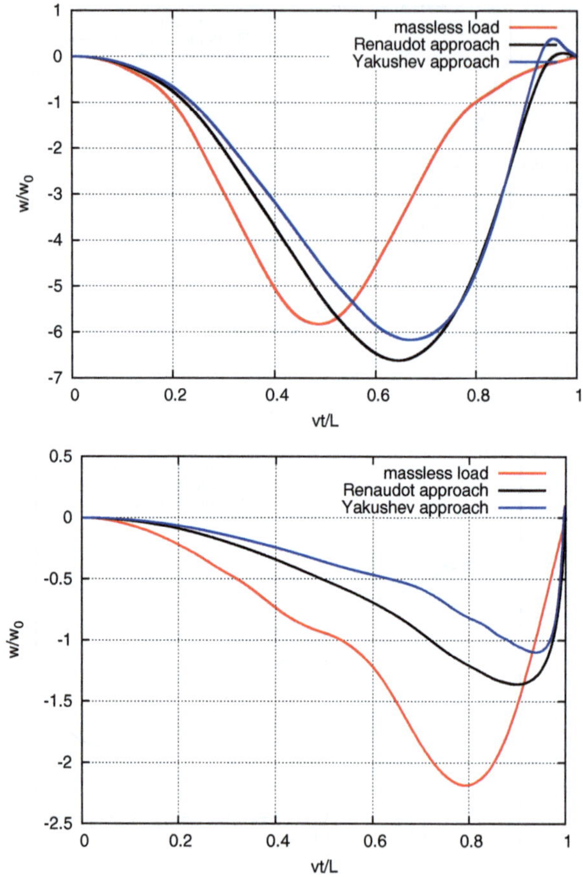

a

b

Fig. 3.29 Vertical displacements of the Bernoulli–Euler beam under a massless load, the Yakushev approach, and the Renaudot approach, for speeds of the moving load: $0.1\,c$ (upper), $0.5\,c$ (lower), $v_{cr} = \pi/l \cdot \sqrt{EI/\rho/A} = 0.314$.

Chapter 4
Review of Numerical Methods of Solution

In this chapter we will discuss the numerical approaches to the moving load problem given in the literature. Most of them concern beam deflection. Unfortunately, comparison with exact analytical or semi-analytical results are rarely given. In most cases the authors compare their results with curves published by other researchers. The authors compute examples using different data and boundary conditions. They usually emphasize the agreement of their results with other computational methods. Unfortunately, results which coincide with an approximate method are not necessarily accurate as well. We should relate the results to analytical solutions or at least to solutions which fulfill the governing differential equations with possibly the lowest error. In this chapter we will compare the curves presented in these publications with semi-analytical results.

First we will consider a string, although this type of a structure is not frequently studied in the literature. Then we will describe the Bernoulli–Euler beam and the Timoshenko beam. The approaches in the literature deal with a moving non-inertial force and an inertial force. Some of them are devoted to a system with a point load or a distributed load.

Some published papers, even ones extensively cited by other authors, do not give an objective measure of the error. The authors claim that a slight visual coincidence with other curves proves its correctness. Moreover, they expect that differences in the results justify the advantages of the published approach and should convince one of its correctness.

First of all, we must warn against using an ad-hoc inertia lumping in neighbouring nodes of the mesh. The mass distribution proportional to the distances to the nodal points fails, and the results can not be accepted. In the case of a beam we are dealing with a parabolic differential equation in space. This fact results in smooth and infinitely fast bending of the entire structure. The local deformation is strongly influenced by other parts of the beam. The influence of a concentrated load of a different type is lower than in the case of a string or a membrane.

The development of computer methods has led to a series of works on numerical calculations, especially using the finite element method (FEM). This method is much more comprehensive than the analytical or semi-analytical methods. Papers

C.I. Bajer and B. Dyniewicz: Numerical Analysis of Vibrations of Structures, LNACM 65, pp. 77–93.
springerlink.com

discussing moving loads with constant or periodic amplitude [91, 145] are simple and rely on modifying the vector on the right hand side, step by step. The resulting work is presented in papers devoted to modelling the motion of a vehicle as a group of oscillators [37, 65, 89, 92, 135]. These problems require the coincidence of the displacements and forces of two subsystems: the main structure and the moving oscillator. For balancing the respective quantities in both systems a simple iterative procedure is applied. This method also involves a modification of the right-hand side vector. At the first stage the structure is loaded by dynamic forces at the contact points corresponding with the oscillators. As a result, the nodal displacements of a discrete structure are obtained. This allows us to determine the vertical displacement of a beam or a plate at the contact points with the oscillators. Displacements assumed as boundary conditions force the motion of the oscillator. This iterative procedure results in force–displacement equilibrium in a single time step. Unfortunately, the convergence of such a scheme is limited to a certain range of parameters, such as the travelling velocity, stiffness of the structure, inertia, and especially—the time step. Otherwise the iterative procedure must be more complex and time consuming.

The insertion of the inertia of the moving load effect requires not only a modification of the right-hand side vector, but also selected parts of the global inertia, damping, and stiffness matrices of the system, in every time step. The first study discussing the influence of the inertia of the moving mass was reported in [150]. An inertial load moving at a constant speed on the Euler beam was considered. Further works [38, 52, 120] are also related to beams or plates in which the nodal displacements and angles are interpolated by cubic polynomials. In these papers the derived matrices are not general. They are not suitable for use for the string or Timoshenko beam in which the nodal displacements and angles are interpolated by a linear function independently. In the literature, you can also find examples of the discrete element method for moving loads [100, 149]. This consists of replacing a beam by a system of rigid rods, connected among themselves on the basis of the compatibility of the rotation of adjacent elements.

The acceleration of a mass particle in the space-time domain is described by the Renaudot formula [118]. The different parts of the equation describe the lateral acceleration, Coriolis acceleration and centrifugal acceleration. The interpolation of the nodal displacements by a third order polynomial allows us to derive the matrices responsible for the travelling mass particle. Unfortunately, the Euler beam equation is not a wave equation. The study of a wave phenomena is possible by using a more complex model of the Timoshenko beam in which the vibration equation takes into account the influence of lateral forces and rotatory inertia on the deflection line of the beam. The angle formed by the axis of the deformed beam is composed of the pure bending angle and the angle corresponding to the deformation of the pure shear. Independent interpolation of displacements and rotation angles of the Timoshenko beam causes serious problems. Linear interpolation of nodal shape features renders impossible the designation of the centrifugal acceleration of a moving mass particle. In the previous works [24, 25, 26], we presented a method for determining the matrices responsible for the description of the moving mass by the space-time finite element method with the use of a linear interpolation. We developed unique finite elements carrying a moving mass particle. It was not yet the general solution for practitioners. In engineering practice real structures possess a characteristic critical speed (of the load) in a range of about 200 km/h (in the case of railway tracks) and this can vary depending on structural details and environmental conditions.

4.1 Oscillator

The main advantage of computations based on the idea of only the spring elements being in contact with the main structure is the simplicity of the algorithms involved. Let us discuss the main features of this method.

First of all we can avoid the need to include the matrices resulting from time derivation of the displacement in the follower point. Incorporation of such matrices may cause serious problems. The matrices corresponding to the transverse acceleration and centrifugal acceleration are symmetric. Unfortunately the matrix describing the Coriolis acceleration is unsymmetrical. This fact may complicate the procedure, especially if we intend to use existing code which was prepared for symmetric matrices. Moreover we can not decompose the system of equations by using the diagonal inertia matrix and explicit time integration procedure. The main advantage of the massless spring's being in contact with the structure are having constant matrices in the system of algebraic equations. The right hand side vector is modified. However, this modification must be performed in an iterative way at each time step and convergence in the case of a complex structure can be poor or even lost in some circumstances.

The use of the oscillator instead of a rigid contact of the mass with a structure simplifies the computer codes and allows of performing computations in two separate threads, which enables a simple mutual exchange of displacements and forces. An example of such computations is presented in Algorithm 1.

4.1.1 String Vibrations under a Moving Oscillator

The simplest way of modelling a moving inertial load is an oscillator placed at the travelling contact point. In the case of a sufficiently rigid spring, we can consider the system as mass placed at a point. Unfortunately, the rigidity can not be assumed to be arbitrarily high. A low value does not ensure sufficient accuracy, while high values do not assure convergence. The algorithm to solve this has two stages. In the first one, we compute the point contact force on the base of the displacements of both ends of the spring of the oscillator. Then the string is subjected to this force, proportionally distributed to two neighbouring nodal points. In the second stage, the string displacements are computed. The displacements of the two subjected points allow of interpolating the vertical displacement at the intermediate points of the contact with the oscillator. This displacement, in turn, is imposed on the oscillator and determines the nodal contact force. Iteratively, we can equilibrate the forces in the two separate systems: the string and the oscillator (Algorithm 1). The

Algorithm 1. Solution of one time step performed for two structures in contact with mutual exchange of information

- Define the initial state at $t = t_i$:
  - external load applied to oscillator (substructure A),
  - displacements of the main structure (substructure B),
  - boundary condition at the contact point (c) of the oscillator (A), obtained with the interpolation formula, based on displacements of the substructure (B),
- execute one step of the calculations of the motion of the oscillator,
- determine the reaction (R) at point (c) of the oscillator,
- distribute the action force ($-R$) at the contact point (c) among the neighbouring nodes in the mesh by using a simple interpolation formula,
- solve one step of the motion of the structure (B) subjected to the forces resulting from the action ($-R$),
- select transverse displacements of the structure (B) in nodes neighbouring to the contact point (c),
- interpolate the displacements at point (c) based on the selected displacements at the neighbouring nodes in the mesh of the structure (B),
- use these displacements as the boundary conditions imposed on the contact points (c) of the oscillator (A),
- modify the external load to which is subjected the oscillator (A), if required,
- repeat the loop starting from the second point of this table until convergence is achieved,
- go to the next time step t_{i+1} and return to the beginning of this table.

---

convergence of the iterations will allow us to proceed to the next time step. However, wise choice of the solution methods applied to both stages of time integration is essential. Below we will present the results of a simple test. A moving oscillator forces the string motion.

In the first approach, we use the central difference method for the time integration of the equation of motion of the oscillator and the Newmark method for the remaining part, i.e. the string. We write: m – mass of the oscillator, k – oscillator stiffness, y_c – forced contact displacement of the bottom point of the oscillator, equal to $y_i^{(2)}$, $y_i^{(j)}$ – the jth component of the displacement vector at time t_i, ie. $y_i^{(1)}$ is the displacement of the point which is in contact and $y_i^{(2)}$ is the displacement of the free end of the spring of the oscillator. We use the following formula for the oscillator:

$$\mathbf{y}_{i+1} = \begin{Bmatrix} y_{i+1}^{(1)} \\ y_{i+1}^{(2)} \end{Bmatrix} = \begin{Bmatrix} q_0 h^2/m + (2 - kh^2/m) y_i^{(1)} + kh^2/m \cdot y_i^{(2)} - y_{i-1}^{(1)} \\ y_c \end{Bmatrix} . \tag{4.1}$$

The contact force q is proportional to the spring elongation

$$q = k\left(y_{i+1}^{(1)} - y_{i+1}^{(2)}\right) \ . \tag{4.2}$$

Simulations allowed us to verify the iterative procedure and its efficiency. We have two ways to fit the displacements of a contact point of the oscillator with the respective points of the string or beam. The first one is direct coincidence of the displacements of both points. In the second approach, we compare the velocities and we expect that if the time sequences of the velocities of both contact points are equal, the resulting contact of the two bodies is ensured. Still the influence of the spring rigidity on results may be severe. We performed this test with the hypothesis of the agreement of the displacements. The spatial domain of the string was divided into 100 finite elements. Dimensionless data was assumed: $l = 1, N = 1, \rho A = 1$. An extremely short time step was used ($h = 10^{-6}$). With the velocity $\mathrm{v} = 0.1$, this requires a large number, 10^7, of time steps. The spring stiffness $k = 10^2 - 10^4$ was assumed. The stiffness $k = 10^4$ resulted in divergence for a longer time step, $h = 10^{-4}$. The results are depicted in Figure 4.1. The displacements are scaled by the maximum static deflection w_0. We must emphasize that both in the above case and in the next example, the computations were performed at the limit of stability and convergence because of the values of the time step and spring stiffness. We notice that the accepted set of both parameters, i.e. time step and spring stiffness, gives procedures which are in practice ineffective.

Figure 4.2 shows the displacements of both ends of the oscillator spring. The mutual displacements are proportional to the contact force. We see that the contact force oscillates and is influenced by the spatial partition of the structure.

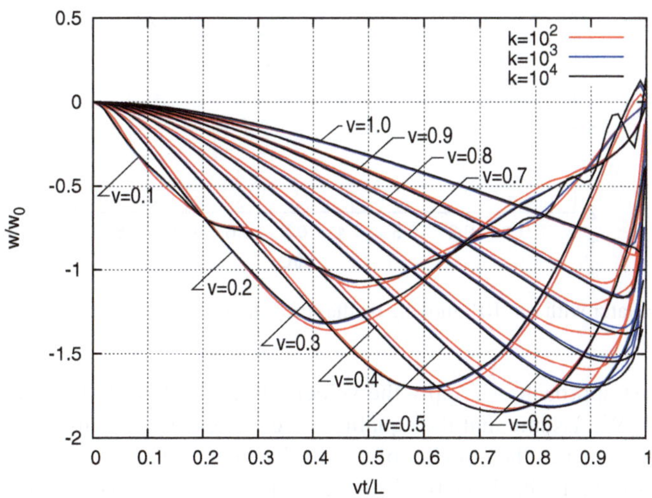

Fig. 4.1 Oscillator with $k = 10^2$, $k = 10^3$, and $k = 10^4$, moving on the string at various speeds.

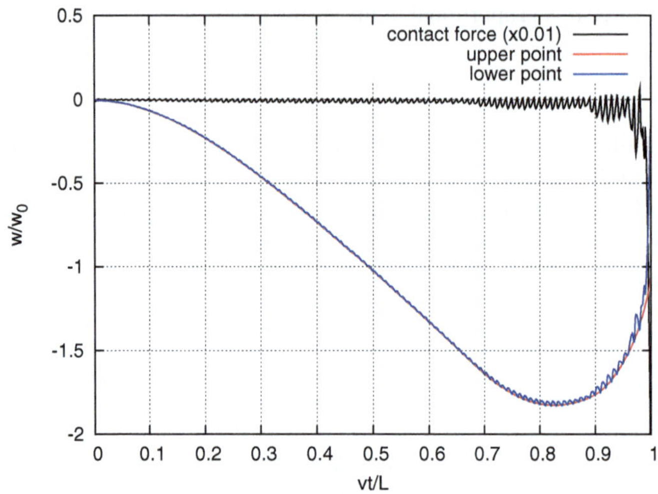

Fig. 4.2 Displacements of the upper and lower point of the spring of the oscillator at the speed $\mathrm{v} = 0.5\,c$, $k = 10^3$, $h = 10^{-5}$.

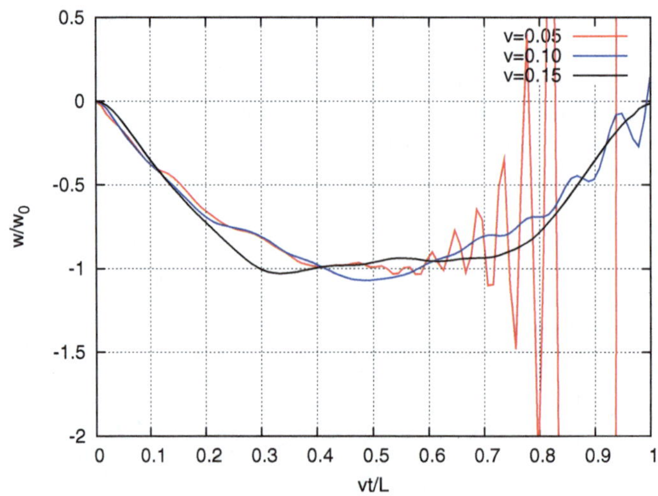

Fig. 4.3 Oscillator with $k = 10^4$, moving on the string at the speeds $\mathrm{v} = 0.05, 0.10$, and $0.15\,c$.

In the next approach, we apply the velocity formulas both to the oscillator and to the string. We assume that the equality of velocities of both substructures at the contact points ensures the equality of the displacements. The stages of the solution method are similar to the first case. Unfortunately, the convergence is unsatisfactory. First, we can not assume a sufficiently high rigidity of the spring. Decreasing the time step does not improve the situation. Neither can we complete the computations for a higher velocity range. Figure 4.3 shows that sufficiently accurate results are obtained for the low speed of $\mathsf{v} \leq 0.1\,c$.

4.1.2 Beam Vibrations under a Moving Oscillator

Filho [52] considers a simple supported beam subjected to a moving oscillator composed of two masses: one supported with a Kelvin–Voigt spring–damper system m_1 and one placed at the contact point m_2 (Figure 4.4). The oscillator moves at a variable speed $\mathsf{v}(t)$ and constant horizontal acceleration a_0. The dynamic equilibrium equation of the mass m_1 is of the form

$$m_1\ddot{y}+c(\dot{y}-\dot{w})+k(y-w)=0\ . \tag{4.3}$$

The dynamic equilibrium of the mass m_2 and the structure was written in the matrix form

$$\mathbf{m}\ddot{\mathbf{d}}+\mathbf{c}\dot{\mathbf{d}}+\mathbf{k}\mathbf{d}=\mathbf{N}^T f_0\ . \tag{4.4}$$

In the above equations, **m**, **c**, and **k** are the inertia, damping, and stiffness matrices of the structure, respectively, **d** is the nodal displacement vector, and f_0 is the point load. The matrix **N** has zeros in all places except those corresponding to the degrees of freedom of the element with the positioned point force. **N** in fact is the interpolation matrix which distributes the point load between both element's nodes.

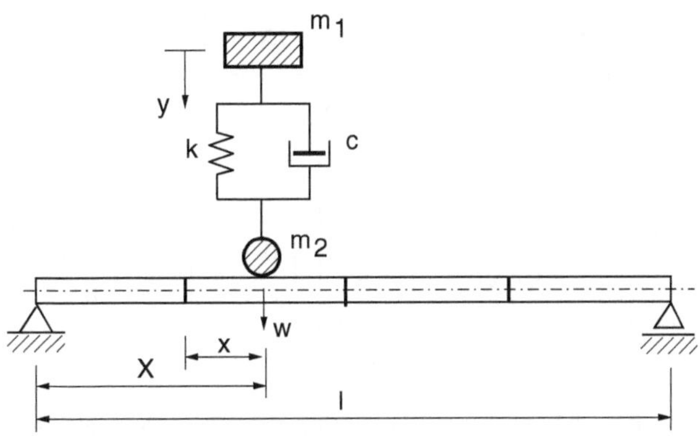

Fig. 4.4 Scheme of the discretized beam [52].

The force f_0 acting at the loaded element is composed of the gravity load of both masses, the inertial force of m_2, the damping force, and the spring force of the oscillator:

$$f_0=(m_1+m_2)g-m_2\ddot{w}+c(\dot{y}-\dot{w})+k(y-w)\ . \tag{4.5}$$

Here, the deflection w is determined at the moving contact point $x=\mathsf{v}t$. Its time derivatives are given by

$$\frac{\mathrm{d}w}{\mathrm{d}t} = \frac{\partial w}{\partial t} + \mathrm{v}\frac{\partial w}{\partial x}, \tag{4.6}$$

$$\frac{\mathrm{d}^2 w}{\mathrm{d}t^2} = \frac{\partial^2 w}{\partial t^2} + 2\mathrm{v}\frac{\partial^2 w}{\partial x \partial t} + \mathrm{v}^2\frac{\partial^2 w}{\partial x^2} + \frac{\mathrm{d}\mathrm{v}}{\mathrm{d}t}\frac{\partial w}{\partial x} . \tag{4.7}$$

The deflection w is interpolated by nodal displacements

$$w(x,t) = \mathbf{N}(\mathbf{x})\mathbf{d}(t) . \tag{4.8}$$

The respective derivatives are computed

$$\frac{\partial^2 w}{\partial x^2} = \mathbf{N}''\mathbf{d}, \quad \frac{\partial^2 w}{\partial x \partial t} = \mathbf{N}'\dot{\mathbf{d}}, \quad \frac{\partial w}{\partial x} = \mathbf{N}'\mathbf{d}, \quad \frac{\partial^2 w}{\partial t^2} = \mathbf{N}\ddot{\mathbf{d}} , \tag{4.9}$$

and the distance is determined by integration $x = \mathrm{v}_0 t + a_0 t^2/2$. In such a case, $\dot{x} = \mathrm{v}_0 + a_0 t$ and $\ddot{x} = a_0$.

When we substitute the above formulas into (4.6) and (4.7), we obtain

$$\begin{aligned} \dot{w} &= (\mathrm{v}_0 + a_0 t)\mathbf{N}' + \mathbf{N}\dot{d} , \\ \ddot{w} &= (\mathrm{v}_0 + a_0 t)^2\mathbf{N}''\mathbf{d} + 2(\mathrm{v}_0 + a_0 t)\mathbf{N}'\dot{\mathbf{d}} + a_0\mathbf{N}'\mathbf{d} + \mathbf{N}\ddot{\mathbf{d}} . \end{aligned} \tag{4.10}$$

We put (4.10) into (4.8) and into the equilibrium equation (4.3). Finally we have

$$\left[\begin{array}{c|c} \mathbf{M}+ & \\ +m_2\mathbf{N}^T\mathbf{N} & \mathbf{0} \\ \hline \mathbf{0}^T & m_1 \end{array}\right]\left\{\begin{array}{c}\ddot{\mathbf{d}}\\ \ddot{\mathbf{y}}\end{array}\right\} + \left[\begin{array}{c|c} \mathbf{C} + c\mathbf{N}^T\mathbf{N}+ & \\ +2m_2(\mathrm{v}_0 + a_0 t)\mathbf{N}^T\mathbf{N}' & -c\mathbf{N} \\ \hline -c\mathbf{N}^T & c \end{array}\right]\left\{\begin{array}{c}\dot{\mathbf{d}}\\ \dot{\mathbf{y}}\end{array}\right\} +$$

$$+\left[\begin{array}{c|c} \mathbf{K} + m_2(\mathrm{v}_0 t + a_0 t^2)^2\mathbf{N}^T\mathbf{N}''+ & \\ +m_2 a_0\mathbf{N}^T\mathbf{N}' + k\mathbf{N}^T\mathbf{N}+ & \\ +c(\mathrm{v}_0 + a_0 t)\mathbf{N}^T\mathbf{N}' & -k\mathbf{N}^T \\ \hline -c(\mathrm{v}_0 + a_0 t)\mathbf{N}' - k\mathbf{N} & k \end{array}\right]\left\{\begin{array}{c}\mathbf{d}\\ \mathbf{y}\end{array}\right\} = \left[\begin{array}{c} (m_1 + m_2)g\mathbf{N}^T \\ \mathbf{0} \end{array}\right] . \tag{4.11}$$

The matrix equation (4.11) is a linear equation with variable coefficients depending on t. In the particular case of $k = 0$, $c = 0$, and $f_0 = (m_1 + m_2)g$, we have a problem with a non-inertial force.

4.2 Inertial Load

In this section we will discuss numerical approaches to inertial loads given in the literature. Most of them concern beams. Unfortunately, in the literature, comparisons with exact analytical results are rarely given. In most cases, the authors compare their results with other published curves obtained from other approximate methods.

Let us look at the simple ad-hoc mass insertion: the mass is added proportionally to the distance from the element joints. Such a procedure can be performed by finite element analysis software if we can only push the load through the spatial element. A string vibration analysis enables plotting curves only in the case of a low string inertia and a low speed of the motion.

Plots for the inertial load in the case of the mass moving over the string differ from the analytical ones. Figure 4.5 shows the results at various speeds for the mass $m = 10$. Figure 4.6 presents the results for $m = 1$ and $\mathrm{v} = 0.1$, for $m = 1000$ and $\mathrm{v} = 0.2$, and for low values of both parameters: $m = 0.01$ and $\mathrm{v} = 0.1$. We notice that only in the case of low mass, i.e. almost a non-inertial load and at the low speed $\mathrm{v} = 0.1$ do we have agreement with the semi-analytical results.

4.2.1 A Bernoulli–Euler Beam Subjected to an Inertial Load

The solution procedure is relatively simple in the case of the Bernoulli–Euler beam with third order shape functions.

Consider Figure 4.7. Results in the literature [2] were compared with our results obtained with the space-time finite element method. The displacements of a free and of a cantilever beam subjected to a moving inertial point load are depicted. The velocity v of the moving load was equal to 0.27 of the critical speed $\mathrm{v}_{cr} = \pi/l \cdot \sqrt{EI/\rho/A}$. The following data was assumed: $l = 7.62$ m, $E = 20.68 \cdot 10^{10}$ Pa, $I = 4.58 \cdot 10^{-5}$ m^2, and $\mathrm{v} = 50.8$ m/s. Notice that the curve exhibiting the influence of the mass does not coincide with the numerical results (Figure 4.7a). On the other hand, the massless load in the same example has a perfect coincidence (Figure 4.7b).

Although Figure 4.7 was plotted for relatively high speed, other examples in the literature present results at low speeds around 0.01—0.05 of the wave speed. Unfortunately, in practice, we have a significantly higher ratio of the travelling velocity to the wave velocity. Such occurs in track dynamics, with a ballast that significantly increases the inertia and reduces the wave speeds. In such cases, the ratio can reach 0.1—0.5. Moreover, the existence of a point mass of a relatively high magnitude, compared with the beam mass, instead of a spring–mass system, increases the differences between the results obtained with the various methods at high speeds. That is why a simple, efficient, and accurate numerical tool is essential for engineering practice.

We next show results of numerical analyses of the Bernoulli–Euler beam taken from the literature. We have taken data from Sadiku [122] and Stanisic [128]. They are as follows: $l = 6$ m, $\mathrm{v} = 6$ m/s, $EI/\rho/A = 275.4408$ m^4/s^2, $m/\rho/A/l$=0.2, $g =$ 9.81 m/s^2.

We present a plot with displacements under a mass moving along a simply supported Bernoulli–Euler beam (Figure 4.8). The results given by Stanisic practically coincide with the precise finite element solution. The data for creating the figure was taken from a digitalized article's plot and may not be accurate. This could explain the slight differences visible between both curves.

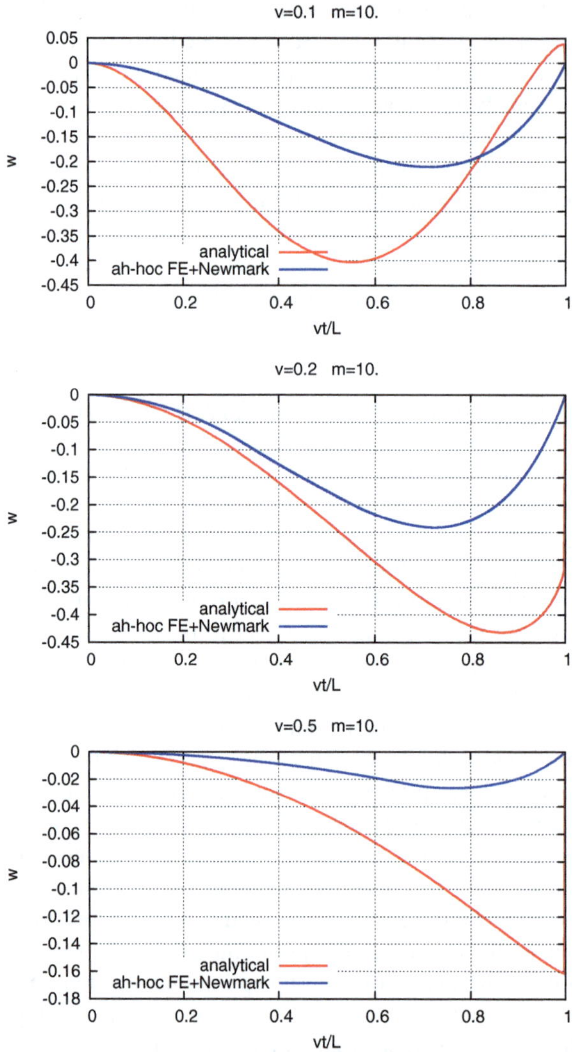

Fig. 4.5 Comparison of the ad-hoc finite element solution with semi-analytical results for a ratio of the moving mass to the string mass equal to 10, and at velocities $0.1\,c$, $0.2\,c$, and $0.5\,c$ $(c = 1)$.

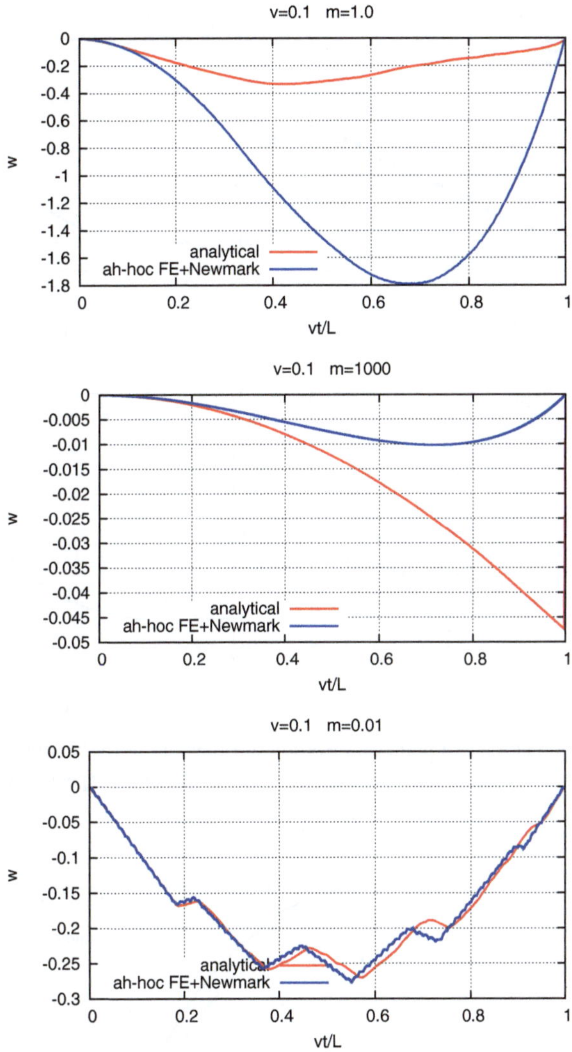

Fig. 4.6 Comparison of the ad-hoc finite element solution with semi-analytical results at various masses and velocities in the case of a string.

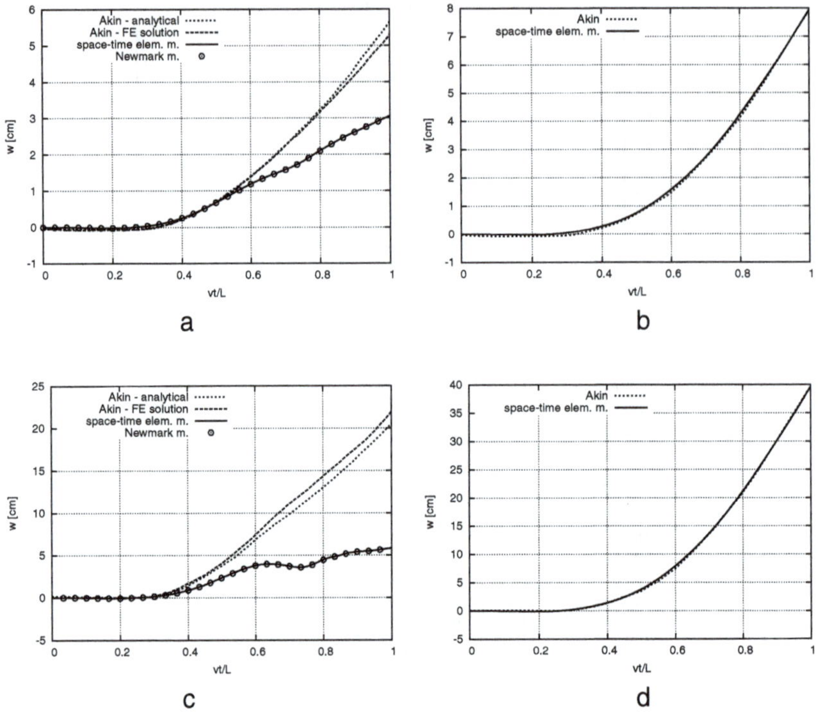

Fig. 4.7 Deflection at the end point of a Bernoulli–Euler cantilever beam subjected to (a) an inertial and (b) a non-inertial load for mass $m = 525.35$ kg and, respectively, (c) and (d) for $m = 2626.75$ kg [2].

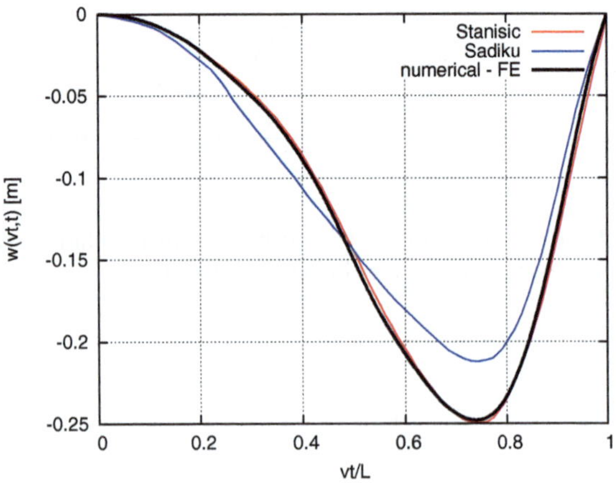

Fig. 4.8 Deflection under the moving mass.

The same data was used in Lee [85]. The clamped–clamped beam has been solved. Also, these results were assumed for comparison (Figures 4.9 and 4.10).

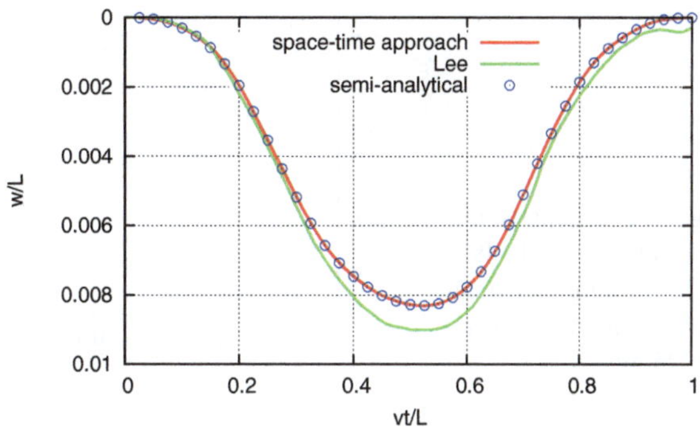

Fig. 4.9 The dimensionless deflection under a moving load for a clamped–clamped Euler beam at speed 0.5v.

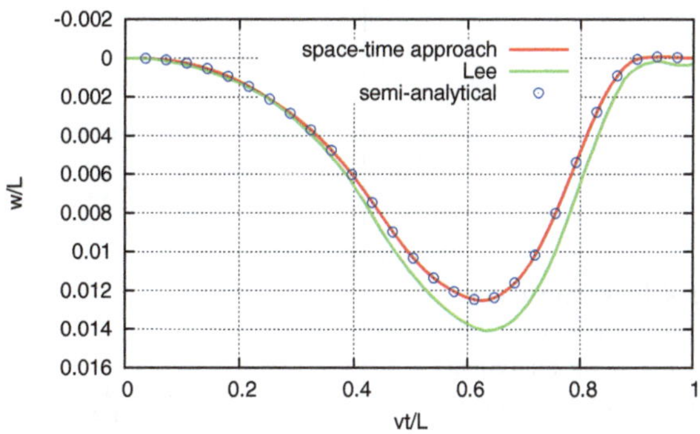

Fig. 4.10 The dimensionless deflection under a moving load for a clamped–clamped Euler beam at speed 1.2v.

In further chapters we will see that a correct derivation of the term describing the displacement of the moving point is essential. These results differ and do not coincide with the analytical results. What is more, the analytical curves given in the literature are also doubtful. In this book we will show both the correct analytical or semi-analytical and numerical solutions which coincide precisely with them.

4.2.2 A Timoshenko Beam Subjected to an Inertial Load

Mackertish in [95] presents the dynamic response of a simply supported Timoshenko beam excited by a moving mass that moves at a constant speed. The solution was obtained analytically and then the final results were computed numerically. The shear deformation and rotatory inertia effects were included.

The following data were assumed: $l = 50$ m, $\rho = 2400$ kg/m^3, $E = 3.36 \cdot 10^{10}$ Pa, $G = 1.40 \cdot 10^{10}$ Pa, $A = 2$ m^2, $I = 1.042$ m^4, and the shape factor $k = 1.2$. The moving mass m_0 =50,000 kg was moving with speeds v = 25 m/s, 50 m/s, and

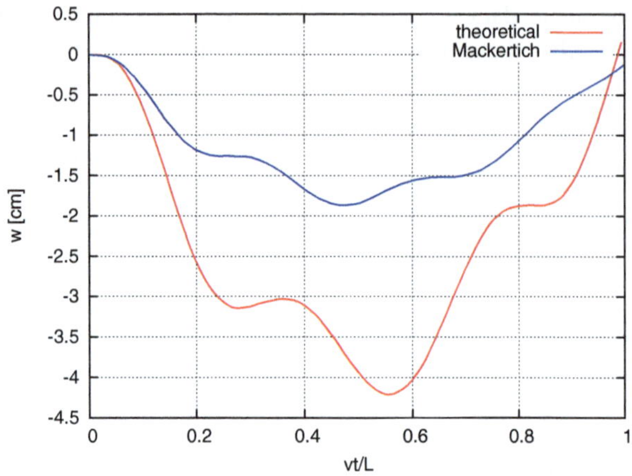

Fig. 4.11 Comparison of results given in [95] with semi-analytical results at the speed v = 25 m/s.

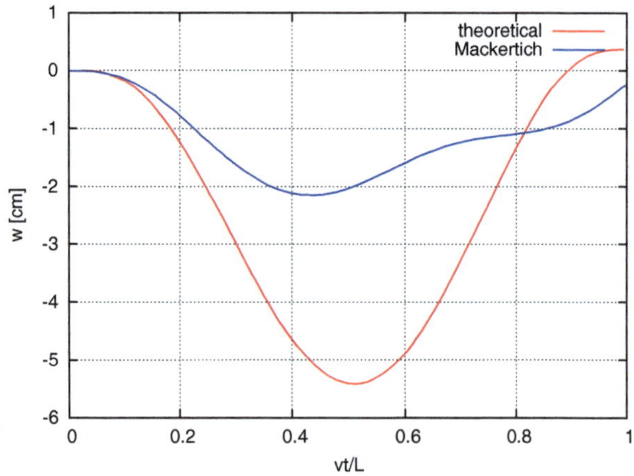

Fig. 4.12 Comparison of results given in [95] with semi-analytical results at the speed v = 50 m/s.

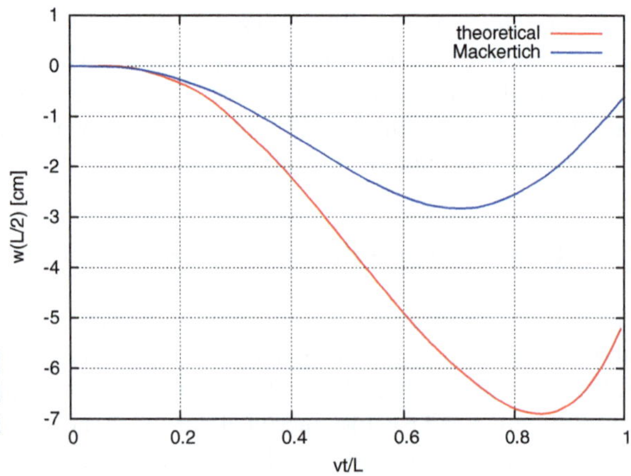

Fig. 4.13 Comparison of results given in [95] with semi-analytical results at the speed v = 100 m/s.

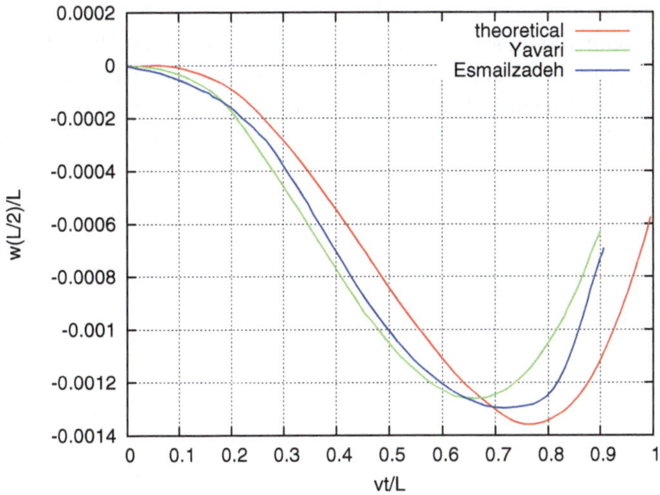

Fig. 4.14 Deflection in the middle of the beam [148] compared with semi-analytical results — rigid beam.

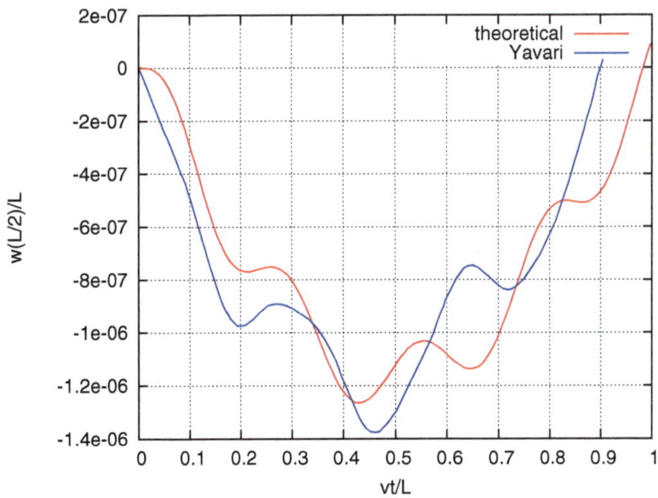

Fig. 4.15 Deflection in the middle of the beam [148] compared with semi-analytical results — soft beam.

100 m/s. The shear wave speed was $c_1 = 2200$ m/s and the bending wave speed was $c_2 = 3740$ m/s. All three Figures 4.11, 4.12, and 4.13 exhibit significant underestimation of the displacements in comparison with the semi-analytical results. The amplitudes were underestimated by more than a factor of two.

Yavari, Noufi, and Mofid in 2002 [148] solved the problem with a discrete method. They called the method the discrete element technique. The continuous flexible beam elements were replaced by a system of rigid bars and flexible joints. The authors compared their results with those obtained by Esmailzadeh and Gorashi [50]. We compare the displacements in the middle of the span with the theoretical, semi-analytical results. The following data were assumed: $l = 435.2$ cm, $\rho = 15.267$ g/cm^3, $E = 2.02 \cdot 10^{11}$ Pa, $G = 0.77 \cdot 10^{11}$ Pa, $A = 13.1$ cm^2, $I = 57.1$ cm^4,

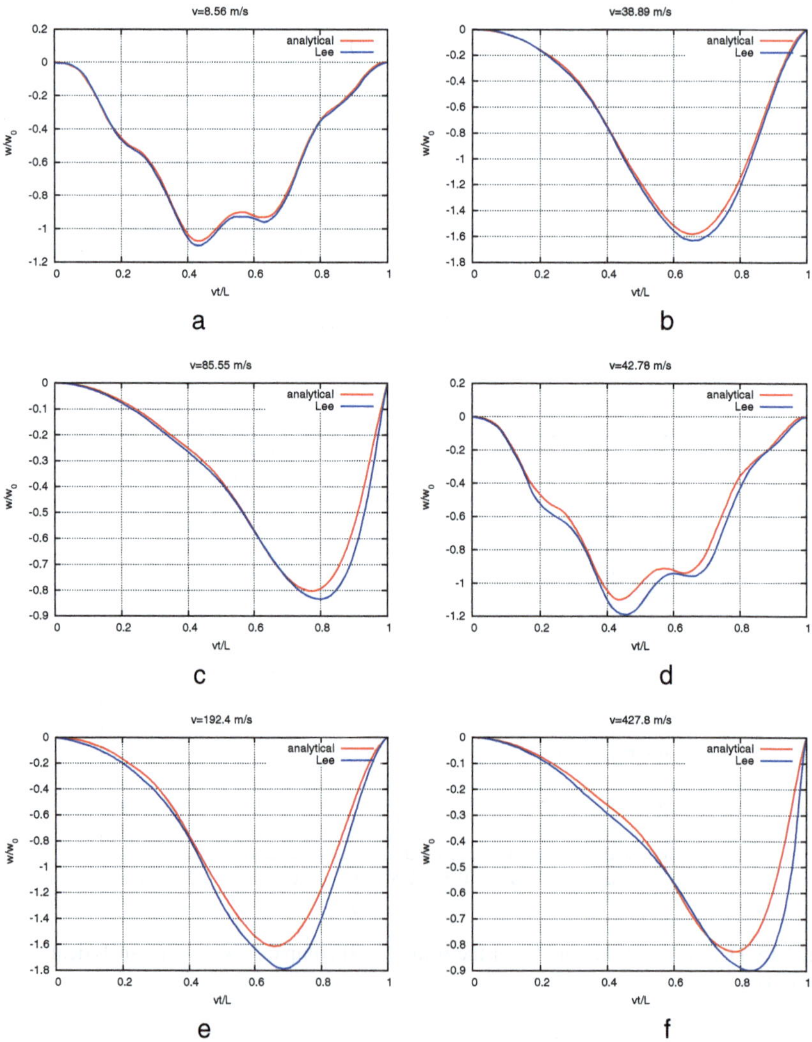

Fig. 4.16 Displacements under a moving mass: a) – $A = 2.865$ cm^2, $I = 0.653$ cm^4, v $= 8.56$ m/s, b) – $A = 2.865$ cm^2, $I = 0.653$ cm^4, v $= 38.89$ m/s, c) – $A = 2.865$ cm^2, $I = 0.653$ cm^4, v $= 85.55$ m/s, d) – $A = 71.62$ cm^2, $I = 408.18$ cm^4, v $= 42.78$ m/s, e) – $A = 71.62$ cm^2, $I = 408.18$ cm^4, v $= 192.4$ m/s, f) – $A = 71.62$ cm^2, $I = 408.18$ cm^4, v $= 427.8$ m/s.

and the shape factor $k = 1.43$. The mass of the span was $M = 87.04$ kg and the moving mass was $m_0 = 21.83$ kg. The speed of the motion was $\mathrm{v} = 27.49$ m/s. We can also give the shearing and bending wave speed: $c_1 = 1878$ m/s, $c_2 = 3637$ m/s. In Figure 4.14, we compare the results from the literature. We notice that the moving mass is relatively small and the speed v, related to the wave speed, is low. In such a case the influence of the moving mass on the beam dynamics is slight. Even in this case, both results are underestimated in comparison to the semi-analytical results. These last we assume to be accurate.

The second example presented in the same paper was computed for the following data: $l = 435.2$ cm, $\rho = 0.0988$ g/cm^3, $E = 2.02 \cdot 10^{11}$ Pa, $G = 0.77 \cdot 10^{11}$ Pa, $A = 2025$ cm^2, $I = 3.42 \cdot 10^5$ cm^4, and the shape factor $k = 1.18$. The mass of the span was $M = 200$ kg and the moving mass was $m_0 = 87.04$ kg. The speed of the motion was $\mathrm{v} = 50$ m/s. The shearing and bending wave speeds are equal to, respectively, c_1=2570 m/s and $c_2 = 4522$ m/s. In Figure 4.15, the mid-span displacements are depicted and compared with the semi-analytical results. Notice that although the maximum amplitude does not differ significantly, the shapes of the curves do not fit.

Lee in [86] expressed the bending and shear energy as a sine and cosine series. The Lagrangian formulation was used. He applied the following data to his examples: $l = 1$ m, $E = 2.07 \cdot 10^{11}$ Pa, $G = 0.776 \cdot 10^{11}$ Pa, and the shape factor $k = 1.11$. The mass density of the span was $\rho = 7700$ kg/m^3. In the first three tests (Figure 4.16a, b, c) the cross sectional area was $A = 2.865$ cm^2 and the cross sectional inertia moment was $I = 0.653$ cm^4. The velocity was increased in each of the three cases and was equal to 8.56, 38.89, and 85.55 m/s, respectively.

Next tests were performed for the following data: $A = 71.62$ cm^2 and $I = 408.18$ cm^4. The velocities were, respectively, 42.78, 192.4, and 427.8 m/s. The moving mass was 0.2 of the mass of the beam in each case. The results are presented in Figure 4.16d, e, f.

Chapter 5
Classical Numerical Methods of Time Integration

The development of electronics and the dissemination of computer technology has led to the development of methods for computational mechanics. First, previously published methods were implemented. Then, more effective solutions were sought. New methods were created incomparably faster. With the increasing computational power of computers, new and more complex issues were studied: the problems of geometric and material non-linearities in the dynamics of structures and problems with complex geometry.

Let us try to outline the group of methods used in the dynamics of structures. It is essential to look at the computational methods of the statics of structures, because most of the methods for the simulation of the dynamics can be reduced to a series of static tasks. Good knowledge of the former allows of better understanding the results of computer simulations of problems of structural dynamics.

The traditional methods of computational dynamics are described in many academic books. Classical methods such as the central difference method and the Newmark method, are widely used. Others, despite undeniable advantages over traditional methods, did not gain suitable popularity. In this chapter, in addition to brief descriptions of the central difference method and the Newmark method, we will present the advantages of the Newmark–Bossak method, the Park–Hausner method, and the Adams method. In the next chapter we will focus only on the space-time element method.

Figure 5.1 shows a systematic classification of computational methods. Approximate numerical methods will be the area of our interest. They are used in engineering practice. A sufficient accuracy allows us to solve complex tasks. The efforts of researchers and users are moving towards understanding the properties of computer methods in order to apply them to new categories of problems, such as in our case a moving mass problem.

The computational methods of dynamics can be divided into two main groups: explicit and implicit. They differ in their arguments. All of them are determined in a previous time step in the case of explicit methods. In the case of implicit methods, the derivatives of the unknown quantities can be defined in the final time level. This idea is shown in Figure 5.2. Explicit methods are performed with relatively

C.I. Bajer and B. Dyniewicz: Numerical Analysis of Vibrations of Structures, LNACM 65, pp. 95–122.
springerlink.com

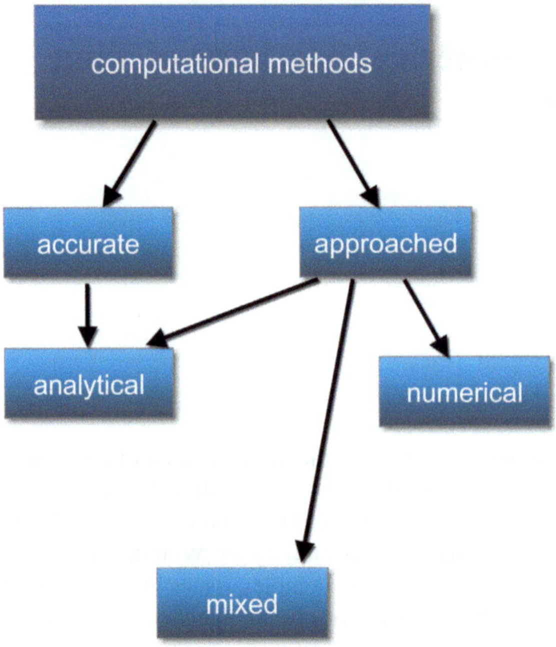

Fig. 5.1 Classification of computational mechanics methods.

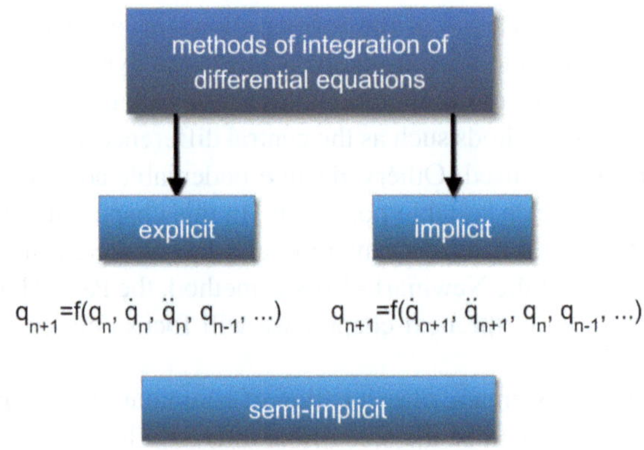

Fig. 5.2 Calculation methods of dynamics.

low costs, since they can be reduced to the solution of simple systems of algebraic equations. Implicit methods require supplementary equations to compute the right-hand side derivatives of the unknowns in the succeeding time level. The cost is thus increased. However, such methods have a higher accuracy in solving the reference equation $\ddot{q}+q=0$. Moreover they are have no limitations imposed by stability criteria.

5.1 Integration of the First Order Differential Equations

We should mention here the methods of integrating differential equations of the first order. The second order equations can simply be reduced to a first order equation by the simple substitution of a new variable for the first derivative of the unknown. In this way we double the number of unknowns. In the same way, we can lower the level of each higher order differential equation. Methods developed for solving first order equations, for example of parabolic type, can be used for solving the hyperbolic type. Although the computational cost of a single time step is higher, this way of reduction of the problem to a first order matrix equation is attractive because of the availability of numerical procedures and extensive descriptions of the equations of the first order.

First, the second-order Runge–Kutta method will be described. It is also called the midpoint method or a modified method of trapezia. Stability analysis shows that the scheme of integration applied to the equations of motion, is divergent for $\Delta t > 0$. Although the pace of discrepancy is small, a vibration analysis of the issues with this method is not suitable. The fourth order Runge–Kutta method is definitely better, although the analysis of its convergence in the problem of vibration is difficult to implement for the general case.

In order not to groundlessly reject the second order Runge–Kutta method as unsuitable for our task, we will briefly present an analysis of its accuracy and stability. The method is usually written in the form of the following algorithm:

$$\begin{aligned} y_{i+1} &= y_i + hf(\bar{y}_{i+1/2}, t_{i+1/2}), \\ \bar{y}_{i+1/2} &= y_i + h/2\, f(y_i, t_i), \quad t_{i+1/2} = t_i + h/2\,. \end{aligned} \tag{5.1}$$

The scheme (5.1) is obtained by the development of the derivative $\dot{\mathbf{y}}$ in the neighbourhood of t_i in a Taylor series and truncating it after the first term. The function $\mathbf{y}$ is developed in a Taylor series, but now including the second term.

From both expansions, the member $\ddot{\mathbf{y}}$ is eliminated with the use of the definition of the tasks, we then get

$$\mathbf{y}_{i+1} = \mathbf{y}_i + \frac{h}{2}\left[\mathbf{f}(\mathbf{y}_i, t_i) + \mathbf{f}(\mathbf{y}_{i+1}, t_{i+1})\right] \,. \tag{5.2}$$

In the case of a system of ordinary differential equations, the right-hand side of the equation $\dot{\mathbf{y}}(t) = \mathbf{f}(\mathbf{y}, t)$, $\mathbf{y}(0) = \mathbf{y}_0$ can be re-written

$$\mathbf{f}(\mathbf{y}, t) = \mathbf{A}\mathbf{y}(t) \,, \tag{5.3}$$

where $\mathbf{A}$ is independent of time. Then the Runge–Kutta method takes the form

$$\mathbf{y}_{i+1} = \mathbf{y}_i + h\mathbf{f}\left(\mathbf{y}_i + \frac{h}{2}\mathbf{f}(\mathbf{y}_i, t_i), t_{i+1/2}\right) = \mathbf{y}_i + h\mathbf{A}\left(\mathbf{y}_i + \frac{h}{2}\mathbf{A}\mathbf{y}_i\right) \,. \tag{5.4}$$

Finally, we have

$$\mathbf{y}_{i+1} = \left(\mathbf{I} + h\mathbf{A} + \frac{h^2}{2}\mathbf{A}^2\right)\mathbf{y}_i \,. \tag{5.5}$$

In the general case, instead of $\dot{\mathbf{y}}(t_i) = (\mathbf{y}_{i+1} - \mathbf{y}_i)/h$, we use the approximation

$$\dot{\mathbf{y}}(t_i + \alpha h) = \frac{\mathbf{y}_{i+1} - \mathbf{y}_i}{h} \,. \tag{5.6}$$

Then the primary equation reduces to

$$\mathbf{y}_{i+1} = \mathbf{y}_i + h\mathbf{f}(\mathbf{y}_{i+\alpha}, t_{i+\alpha}) \,. \tag{5.7}$$

The value of the function $\mathbf{f}$ at intermediate points of time is interpolated:

$$\mathbf{f}(\mathbf{y}_{i+\alpha}, t_{i+\alpha}) = (1-\alpha)\mathbf{f}(\mathbf{y}_i, t_i) + \alpha\mathbf{f}(\mathbf{y}_{i+1}, t_{i+1}) = \mathbf{y}_i + h(1-\alpha)\mathbf{A}\mathbf{y}_i + h\alpha\mathbf{y}_{i+1} \,. \tag{5.8}$$

At $\alpha = 1/2$, we get the scheme (5.1). After reordering (5.8), we obtain

$$(\mathbf{I} - h\alpha\mathbf{A})\mathbf{y}_{i+1} = (\mathbf{I} + h(1-\alpha)\mathbf{A})\mathbf{y}_i \,. \tag{5.9}$$

The matrix coefficients of the above system of equations with $\alpha > 0$ must be nonsingular. Then we obtain the solution which corresponds to the transition matrix

$$\mathbf{y}_{i+1} = (\mathbf{I} - h\alpha\mathbf{A})^{-1}(\mathbf{I} + h(1-\alpha)\mathbf{A})\mathbf{y}_i \,. \tag{5.10}$$

Modal analysis allows us to separate variables, and the differential equations uncouple. Although the variables in the generalized vector $\mathbf{y}(t)$ are replaced by the appropriate ones from the revised database, one can ask whether the stability of this system of differential equations is reduced to the problem of the stability of the n single differential equations.

Positively defined matrix $\mathbf{A}$ is diagonalized by using the substitution $\mathbf{y} = \boldsymbol{\Phi}\mathbf{z}$. The matrix $\boldsymbol{\Phi}$ is composed of eigenvectors of the matrix $\mathbf{A}$. Then we have

$$\dot{\mathbf{z}} = \boldsymbol{\Phi}^{-1}\mathbf{A}\boldsymbol{\Phi}\mathbf{z} \,, \;\; \Lambda = \boldsymbol{\Phi}^{-1}\mathbf{A}\boldsymbol{\Phi} \,. \tag{5.11}$$

We substitute (5.11) into (5.9) and we obtain

$$(\mathbf{I} - h\alpha\Lambda)\,\mathbf{z}_{i+1} = (\mathbf{I} + h(1-\alpha)\Lambda)\,\mathbf{z}_i \,. \tag{5.12}$$

The system of equations (5.12) is decoupled, and can be written as a set of n single equations

$$(1 - h\alpha\lambda_k)z_{i+1} = (1 + h(h(1-\alpha)\lambda_k)\,z_i, \;\; k = 1, 2, \ldots, n \,. \tag{5.13}$$

Getting rid of the index k, we write a single equation with the ratio of the transition

$$z_{i+1} = \frac{1 + h(1-\alpha)\lambda}{1 - h\alpha\lambda} z_i \,. \tag{5.14}$$

The passage rate modulus should be no greater than one

$$\left|\frac{1+h(1-\alpha)\lambda}{1-h\alpha\lambda}\right| \leq 1\,. \tag{5.15}$$

When $h \to \infty$, the condition (5.15) is fulfilled at $\alpha \geq 1/2$. The case $\alpha = 0$ diverges then.

Single-step methods, well described in the literature for first-order equations, are also applicable to equations of the second order. The second-order equation is converted into two equations of the first order. In the case of matrix equations with many degrees of freedom, the size of the task dramatically increases. Single step methods allow the use of a variable integration step. Nor does this require taking off procedures to generate an adequate set of solutions in the initial moments. The calculations consist of several operations performed as intermediates and one then calculates the final result of a single time step. The accuracy of the method can be changed by changing the number of intermediate steps. In many cases, the use of the fourth order Runge–Kutta method avoids the need for a non-linear iterative procedure for solving nonlinear equations. Runge–Kutta can be used to solve stiff equations. In static problems, this corresponds to ill-conditioned matrices of systems of equations.

The Euler method is inaccurate because we go to the next time t_{i+1} through a tangent drawn at the point $(\mathbf{y}_i, t_i)$. The idea is that the solution does not use the tangent $\mathbf{f}(\mathbf{y}_i, t_i)$, but the secant defined by the points $(\mathbf{y}_i, t_i)$ and $(\mathbf{y}_{i+1}, t_{i+1})$. The second point is set in an approximate manner using the Euler method and determined by y, and is called the predictor. Appropriate calculations are carried out using the secant $1/2\left[(\mathbf{f}(\mathbf{y}_i, t_i) + \mathbf{f}(\mathbf{y}^0_{i+1}, t_{i+1})\right]$. Ultimately, therefore, the corrector has the following form (Figure 5.3):

$$\mathbf{y}_{i+1} = \mathbf{y}_i + \frac{1}{2}\left[(\mathbf{f}(\mathbf{y}_i, t_i) + \mathbf{f}(\mathbf{y}^0_{i+1}, t_{i+1}))\right]\,. \tag{5.16}$$

Note that the method of Runge–Kutta is a method of order 2 type predictor–corrector. The prediction is made in the mid-step h, and then this value is used in the correction stage (Figure 5.4).

Now we present a simple algorithm for solving the problem using the fourth order Runge–Kutta method (Algorithm 2). The integration step is limited by the criteria of stability and accuracy. In the case of rigid equations, the integration step must be particularly short. The cost of the supporting steps and a short integration step makes this method attractive enough compared to multistep methods. Single step methods are normally applied to the initial stages of non-selfstarting multistep procedures.

Table 5.1 presents the results of sample calculations $\dot{y} + y^2 = 0$ of the equation, with initial condition $y(0) = 1$. The solution is obtained for a step $h = 0.1$. The strict solution is $y = 1/(1+t)$.

Below we will present selected methods in Butcher's tableau form. The simplest Euler method is represented by the tableau

$$\begin{array}{c|c} 0 & \\ \hline & 1 \end{array}$$

The second order Runge–Kutta method has the following form:

$$\begin{array}{c|cc} 0 & & \\ \frac{1}{2} & \frac{1}{2} & \\ \hline & 0 & 1 \end{array}$$

Heun's variant of the method is

$$\begin{array}{c|cc} 0 & & \\ 1 & 1 & \\ \hline & \frac{1}{2} & \frac{1}{2} \end{array}$$

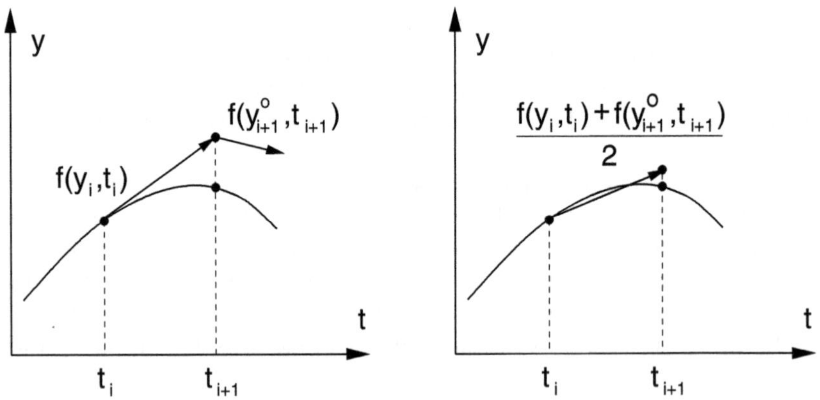

Fig. 5.3 Predictor and corrector in the two-stage method.

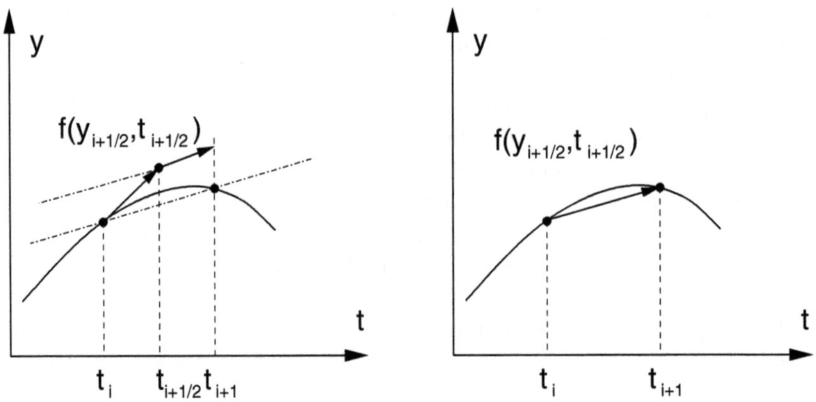

Fig. 5.4 Stage of prediction and correction in the second order Runge–Kutta method.

Algorithm 2. The fourth-order Runge–Kutta method.

- we calculate the auxiliary:

$$\bar{\ddot{y}}_0 = \ddot{y}(t_i, y_i),$$

$$\bar{\ddot{y}}_1 = \ddot{y}(t_{i+1/2}, y_i + \tfrac{h}{2}\dot{y}_i),$$

$$\bar{\ddot{y}}_2 = \ddot{y}(t_{i+1/2}, y_i + \tfrac{h}{2}\dot{y}_i + \tfrac{h^2}{4}\bar{\ddot{y}}_0),$$

$$\bar{\ddot{y}}_3 = \ddot{y}(t_{i+1}, y_i + h\dot{y}_i + \tfrac{h^2}{2}\bar{\ddot{y}}_1)$$

- we calculate:

$$y_{i+1} = y_i + h\dot{y}_i + \tfrac{h^2}{3}(\bar{\ddot{y}}_0 + \bar{\ddot{y}}_1 + \bar{\ddot{y}}_2) + \mathcal{O}(h^5)$$

- $\dot{y}_{i+1} = \dot{y}_i + \frac{h}{4}(\bar{\ddot{y}}_0 + \bar{\ddot{y}}_1 + \bar{\ddot{y}}_2 + \bar{\ddot{y}}_3) + \mathcal{O}(h^5)$

Table 5.1 Results of solutions of the equation $\dot{y} + y^2 = 0$ with chosen methods.

step	Euler method	RK 2nd order	RK 4th order	exactly
0	1.000000	1.000000	1.000000	1.000000
1	0.900000	0.909750	0.909091	0.909091
2	0.819000	0.834344	0.833334	0.833333
3	0.751924	0.770418	0.769231	0.769231
4	0.695385	0.715548	0.714286	0.714286
5	0.647029	0.667945	0.666667	0.666667
6	0.605164	0.626261	0.625000	0.625000
7	0.568542	0.589458	0.588236	0.588235
8	0.536218	0.556730	0.555556	0.555556
9	0.507465	0.527437	0.526316	0.526316
10	0.481713	0.501066	0.500000	0.500000

Another two-stage Runge–Kutta method has the following form:

$$\begin{array}{c|cc} 0 & & \\ \frac{2}{3} & \frac{2}{3} & \\ \hline & \frac{1}{4} & \frac{3}{4} \end{array}$$

The example of the integration of $\dot{y} = f(x,t)$ is depicted in Figure 5.5.

In addition to the previously listed methods, we can meet in the literature a lot of expressions from which we have reproduced a few:

Trapezoidal method. It is also known as the method of average acceleration. In parabolic problems it is known as the Cranck–Nicolson method. It is

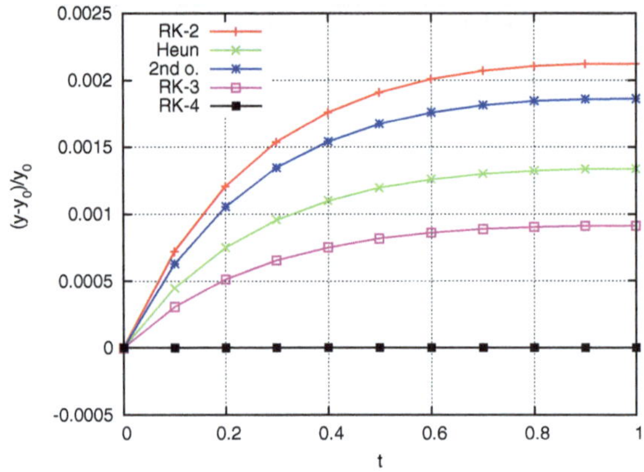

Fig. 5.5 Accuracy of the Runge–Kutta methods.

unconditionally stable. The following compounds are used: $\mathbf{u}_{n+1} = \mathbf{u}_n + h/2(\mathbf{v}_n + \mathbf{v}_{n+1})$ and $\mathbf{v}_{n+1} = \mathbf{v}_n + h/2(\mathbf{a}_n + \mathbf{a}_{n+1})$.

θ-Wilson method. We assume here a linear variation of the acceleration during the time interval $[t_n; t_{\theta\Delta t}]$: $\mathbf{a}_{n+\tau} = \mathbf{a}_n + \tau/(\theta h)(\mathbf{a}_{t+\theta\Delta t} - \mathbf{a}_t)$. It is an implicit method, unconditionally stable at $\theta \geq 1.37$. Usually we apply the value of $\theta = 1.40$.

Houbolt method. The accelerations $\mathbf{a}_{n+1}$ and velocities $\mathbf{v}_{n+1}$ are replaced with the expressions $\mathbf{a}_{n+1} = (2\mathbf{u}_{n+1} - 5\mathbf{u}_n + 4\mathbf{u}_{n-1} - \mathbf{u}_{n-2})/h^2$ and $\mathbf{v}_{n+1} = (11\mathbf{u}_{n+1} - 18\mathbf{u}_n + 9\mathbf{u}_{n-1} - 2\mathbf{u}_{n-2})/6h^2$. They are in fact backward differences with the error order $(\Delta t)^2$. The idea of the method is similar to the central difference method. The application, in turn, configures it to the methods of Adams. The balance of power in the equation of motion is fixed at t_{i+1} (in the central difference method, at time t_i) and hence in the final scheme a calculation system of algebraic equations with coefficients determined by the stiffness matrix $\mathbf{K}$ is obtained. The Houbolt method is classified as an implicit one. Also here, there is not a critical step of integration Δt. An interesting fact is that the in the case of zero inertia and damping matrices the scheme of calculation reduces to the solution of a problem of statics. A similar feature is shown in the space-element method in Chapter 6.4.3.

Additionally, one may consult the reviews [44, 45, 108, 119].

5.2 Single-Step Method SSpj

This method is described in [142, 154]. In special cases, you can turn it into a well-known classic methods of integration of the differential equation of motion.

We will expand the displacements at the end of the time step in the form of Taylor series

$$\mathbf{y}_{i+1} = \mathbf{y}_i + \Delta t\,\dot{\mathbf{y}}_i + \frac{1}{2}\Delta t^2 \ddot{\mathbf{y}}_i + \ldots + \frac{1}{p!}\alpha_i^{(p)}\Delta t^p \,. \tag{5.17}$$

Known values are the values $\mathbf{y}_i$ and the subsequent derivatives $\dot{\mathbf{y}}_i$, $\ddot{\mathbf{y}}_i$ etc. $\alpha_i^{(p)}$ contains unknown coefficients in terms of the remainder of the development. The development of (5.17) and the derivative are written in short as follows:

$$\mathbf{y}_{i+1} = \sum_{q=0}^{p-1} \frac{\Delta t^q}{q!}\mathbf{y}_i^{(q)} + \frac{\Delta t^p}{p!}\alpha_i^{(p)} \,, \tag{5.18}$$

$$\dot{\mathbf{y}}_{i+1} = \sum_{q=1}^{p-1} \frac{\Delta t^{q-1}}{(q-1)!}\mathbf{y}_i^{(q)} + \frac{\Delta t^(p-1)}{(p-1)!}\alpha_i^{(p)} \,.$$

The SSpj method (single-step) is characterized by two parameters: p – the number of words in a Taylor series, and j – the order of the differential equation. The unknown parameters $\alpha_i^{(p)}$ are determined based on the postulate of fulfillment of the global equations of motion in a weighted form:

$$\int_0^{\Delta t} \mathbf{W}(\mathbf{M}\ddot{\mathbf{y}} + \mathbf{C}\dot{\mathbf{y}} + \mathbf{K}\mathbf{y})\mathrm{d}t = 0 \,. \tag{5.19}$$

We define the quotient

$$\frac{\int_0^{\Delta t}\mathbf{W}t^q\mathrm{d}t}{\int_0^{\Delta t}\mathbf{W}\mathrm{d}t} = \theta_q \Delta t^q \,, \quad q = 1,2,\ldots,p, \quad 0 \le \theta_q \le 1 \,. \tag{5.20}$$

Using the definition (5.20), we can define the following integrals:

$$\frac{\int_0^{\Delta t}\mathbf{W}\mathbf{y}\mathrm{d}t}{\int_0^{\Delta t}\mathbf{W}\mathrm{d}t} = \sum_{q=0}^{p-1}\frac{\Delta t^q}{q!}\mathbf{y}_i^{(q)}\theta_q + \frac{\Delta t^p}{p!}\alpha_i^{(p)}\theta_p \,,$$

$$\frac{\int_0^{\Delta t}\mathbf{W}\dot{\mathbf{y}}\mathrm{d}t}{\int_0^{\Delta t}\mathbf{W}\mathrm{d}t} = \sum_{q=1}^{p-1}\frac{\Delta t^{q-1}}{(q-1)!}\mathbf{y}_i^{(q)}\theta_{q-1} + \frac{\Delta t^{p-1}}{(p-1)!}\alpha_i^{(p)}\theta_{p-1} \,, \tag{5.21}$$

$$\frac{\int_0^{\Delta t}\mathbf{W}\ddot{\mathbf{y}}\mathrm{d}t}{\int_0^{\Delta t}\mathbf{W}\mathrm{d}t} = \sum_{q=2}^{p-1}\frac{\Delta t^{q-2}}{(q-2)!}\mathbf{y}_i^{(q)}\theta_{q-2} + \frac{\Delta t^{p-2}}{(p-2)!}\alpha_i^{(p)}\theta_{p-2} \,,$$

$$\frac{\int_0^{\Delta t}\mathbf{W}\mathbf{F}\mathrm{d}t}{\int_0^{\Delta t}\mathbf{W}\mathrm{d}t} = \bar{\mathbf{F}} \,.$$

We divide the equation (5.19) by $\int_0^{\Delta t}$ and we use (5.21). We obtain

$$\mathbf{M}\left(\sum_{q=2}^{p-1}\frac{\Delta t^{q-2}}{(q-2)!}\mathbf{y}_i^{(q)}\theta_{q-2}+\frac{\Delta t^{(p-2)}}{(p-2)!}\alpha_i^{(p)}\theta_{p-2}\right)+ \tag{5.22}$$

$$+\mathbf{C}\left(\sum_{q=1}^{p-1}\frac{\Delta t^{q-1}}{(q-1)!}\mathbf{y}_i^{(q)}\theta_{q-1}+\frac{\Delta t^{(p-1)}}{(p-1)!}\alpha_i^{(p)}\theta_{p-1}\right)+ \tag{5.23}$$

$$+\mathbf{K}\left(\sum_{q=0}^{p-1}\frac{\Delta t^{q}}{q!}\mathbf{y}_i^{(q)}\theta_{q}+\frac{\Delta t^{p}}{p!}\alpha_i^{(p)}\theta_{p}\right)-\bar{\mathbf{F}}=\mathbf{0}\,. \tag{5.24}$$

Hence, we determine the coefficients $\alpha_n^{(p)}$:

$$\alpha_n^{(p)}=\left(\frac{\Delta t^{p-2}}{(p-2)!}\theta_{p-2}\mathbf{M}+\frac{\Delta t^{p-1}}{(p-1)!}\theta_{p-1}\mathbf{C}+\frac{\Delta t^{p}}{p!}\theta_p\mathbf{K}\right)\cdot$$
$$\cdot\left(\bar{\mathbf{F}}-\mathbf{M}\ddot{\tilde{\mathbf{y}}}_{i+1}-\mathbf{C}\dot{\tilde{\mathbf{y}}}_{i+1}-\mathbf{K}\tilde{\mathbf{y}}_{i+1}\right)\,. \tag{5.25}$$

Quantities $\tilde{\mathbf{y}}$, $\dot{\tilde{\mathbf{y}}}$ and $\ddot{\tilde{\mathbf{y}}}$ mean the average value of predictors in time interval and are defined as follows:

$$\begin{aligned}
\tilde{\mathbf{y}}_{i+1}&=\sum_{q=0}^{p-1}\frac{\Delta t^{q}}{q!}\mathbf{y}_i^{(q)}\theta_q\,,\\
\dot{\tilde{\mathbf{y}}}_{i+1}&=\sum_{q=1}^{p-1}\frac{\Delta t^{q-1}}{(q-1)!}\mathbf{y}_i^{(q)}\theta_{q-1}\,,\\
\ddot{\tilde{\mathbf{y}}}_{i+1}&=\sum_{q=2}^{p-1}\frac{\Delta t^{q-2}}{(q-2)!}\mathbf{y}_i^{(q)}\theta_{q-2}\,.
\end{aligned} \tag{5.26}$$

For starting the computations, we require the values $\mathbf{y}_0$ and the derivatives $\mathbf{y}_0^{(q)}$ up to order $p-1$. The simplified scheme of computations is depicted in Algorithm 3. It is important to select the coefficients θ_q and adopt an appropriate step Δt. You can construct more or less complex integration patterns, choosing the appropriate development (5.17). The study [154] shows in which cases the SSpj method becomes one of the methods of Newmark, Newmark–Bossak, Wilson, Houbolt, et al.

Algorithm 3. Scheme of computations with the SSpj method.

- Determine $\tilde{\mathbf{y}}_{i+1}$, $\dot{\tilde{\mathbf{y}}}_{i+1}$ and $\ddot{\tilde{\mathbf{y}}}_{i+1}$ from (5.26).
- Compute $\alpha_i^{(p)}$ from (5.25).
- Compute $\mathbf{y}_{i+1}$, $\dot{\mathbf{y}}_{i+1}$, ... from equation (5.18).
- Repeat the computations in the successive step.

5.3 Central Difference Method

This is the most commonly practiced method of integrating the differential equations of motion. What's more, it has been applied to almost all problems of the numerical resolution of initial value problems. The simplicity of its formulation is its main feature.

Let us consider the differential equation describing the motion of a simple oscillator:

$$m\ddot{q}(t) + c\dot{q}(t) + kq(t) = f(t)\,. \tag{5.27}$$

The derivatives are replaced with difference quotients:

$$\dot{q}_i = \frac{q_{i+1} - q_{i-1}}{2\Delta t}\,, \tag{5.28}$$

$$\ddot{q}_i = \frac{\dot{q}_{i+1/2} - \dot{q}_{i-1/2}}{\Delta t} = \frac{q_{i+1} - 2q_i + q_{i-1}}{\Delta t^2}\,. \tag{5.29}$$

We define the equation (5.27) at time $t = t_i$, then apply the difference quotients defined above (5.28) and (5.29). After re-arranging the terms, we obtain the following stepping formula

$$q_{i+1} = \left(m + \frac{h}{2}c\right)^{-1}\left[\left(2m - k\Delta t^2\right)q_i + \left(\frac{h}{2}c - m\right)q_{i-1} + \Delta t^2 f_i\right]\,. \tag{5.30}$$

With solutions at two successive moments, $i-1$ and i, we can determine the solution for the next time-step $i+1$. A difficulty may occur in the first step of calculations, during the taking into account of the initial conditions. Assume that the equation (5.27) can be solved with the following initial conditions: $q(0) = q_0$, $\dot{q}(0) = \dot{q}_0$. In this case, we introduce the first condition directly into the equation (5.30). In the second condition, the derivative is replaced with the backward difference quotient $\dot{q}_0 = (q_o - q_{-1})/\Delta t$. So we have a missing value of $q_{-1} = q_0 - \dot{q}_0 h$. This auxiliary quantity is formally required to run the calculation procedure. The Algorithm 4 shows the computational process in the case with initial displacement value q_0 and the velocity $\dot{q}_0$.

Let us now turn to the matrix form of 5.30. In the case of the full matrix of inertia, we obtain the following equation

---

Algorithm 4. Central difference method.

---

1. Using the initial condition $\dot{q}(0) = \dot{q}_0$ we determine the missing value $q_{-1} = q_0 - \dot{q}_0 h$.
2. Knowing the values of q_{-1} and q_0 we determine q_1 (i.e. q_{i+1} at $i = 0$): $q_{i+1} = \left(2 - \frac{k}{m}h^2\right)q_i - q_{i-1} + \frac{h^2}{m}f_i$.
3. We go to the next step and we repeat the calculation on p. 2, with $i = 1,2,3....,n$.

$$\mathbf{M}\mathbf{q}_{i+1} = (2\mathbf{M} - h^2\mathbf{K})\,\mathbf{q}_i - \mathbf{M}\mathbf{q}_{i-1} + h^2\mathbf{F}_i\,. \tag{5.31}$$

If the inertia of the structure can be represented as a diagonal matrix, then the inversion of $\mathbf{M}$ is immediate and simplifies the calculation scheme to the following form:

$$\mathbf{q}_{i+1} = \left(2\mathbf{I} - \mathbf{M}^{-1}\mathbf{K}\,h^2\right)\mathbf{q}_i - \mathbf{q}_{i-1}\,. \tag{5.32}$$

We obtain a system of algebraic equations with the separated equations. We solve it equation after equation, at minimal numerical cost. We see that for a full inertia matrix $\mathbf{M}$, we have to solve the system of equations with a full or band coefficient matrix. Therefore, the central difference method is usually used with a diagonal inertia matrix. The method of constructing the diagonal matrix of inertia is relatively simple. The easiest way to obtain a diagonal matrix is by placing the material particles of respective masses m_i, so that their sum is equal to the total mass of the object m. Material particles have no rotational inertia. The rotational inertia can be included in the matrix of inertia of beams or plates, adding quite freely diagonal coefficients, corresponding to the rotational inertia of the part of the discretized structure. In the case of a beam, it will be a moment of rotational inertia relative to the mid-element node with mass $m/2$ and length $l/2$, so $I = (m/2)\cdot(l/2)^2/3$.

This method of mass granulation is sufficient for most tasks. In a situation where we deal with both translational and rotational degrees of freedom, these first determine the displacements and stresses of structural elements. An imprecise description of the rotational inertia only affects the results minimally. Natural frequencies can be determined with an error not exceeding a few percent. Higher frequencies, with a large share of bending in the corresponding modes, have a double-digit error. For this reason, more accurate ways of substituting consistent matrices by diagonal inertia ones have been sought. The method proposed in [66] (the HRZ method) is equally simple. The principle is to take into account only the diagonal values of the full matrix. They are appropriately scaled. The procedure can be summarized in the following steps:

1. We calculate only the diagonal terms of the consistent inertia matrix.
2. We calculate the total mass m of the discrete element.
3. We calculate the value s equal to the sum of the diagonal terms m_{ii}, corresponding only to translational degrees of freedom in the direction of motion. We omit here rotational degrees of freedom.
4. We scale all diagonal coefficients by multiplying by the number m/s. Thus we keep the total weight of the element.

Example 5.1. The consistent inertia matrix of the Bernoulli–Euler beam element has the following form:

$$\mathbf{M} = \frac{m}{420}\begin{bmatrix} 156 & 22l & 54 & -13l \\ 22l & 4l^2 & 13l & -3l^2 \\ 54 & 13l & 156 & -22l \\ -13l & -3l^2 & -22l & 4l^2 \end{bmatrix}.$$

Here, m is the total mass element $m = \rho A l$. The matrix diagonalized directly has the form

$$\mathbf{M} = \frac{m}{2}\,\mathrm{diag}[1,\ l^2/12,\ 1,\ l^2/12]\ ,$$

and diagonalized by the HRZ method

$$\mathbf{M} = \frac{m}{2}\,\mathrm{diag}[1,\ l^2/78,\ 1,\ l^2/78]\ .$$

The HRZ method allows obtaining better results for the higher natural frequencies and related phenomena. We note that the coefficients corresponding to the rotary degrees of freedom in two ways of mass granulation vary considerably.

5.3.1 Stability of the Method

The problem of stability was already discussed in Chap. 5 for the method of 2nd order Runge–Kutta and generally – trapezoidal methods. Now consider the problem of stability of the central difference method. Let us write (5.30) as a single-layer scheme based on the double number of degrees of freedom:

$$\begin{Bmatrix} q_{i+1} \\ q_i \end{Bmatrix} = \begin{bmatrix} 2 - \frac{k}{m}h^2 & -1 \\ 1 & 0 \end{bmatrix} \begin{Bmatrix} q_i \\ q_{i-1} \end{Bmatrix} \tag{5.33}$$

or in short

$$\mathbf{q}_{i+1} = \mathbf{T}\,\mathbf{q}_i\ . \tag{5.34}$$

Using the above record we create a solution after n computing steps

$$\begin{aligned} \mathbf{q}_1 &= \mathbf{T}\,\mathbf{q}_0\ , \\ \mathbf{q}_2 &= \mathbf{T}\,\mathbf{q}_1 = \mathbf{T}\cdot\mathbf{T}\,\mathbf{q}_0 = \mathbf{T}^2\,\mathbf{q}_0\ , \\ &\ldots\ldots \\ \mathbf{q}_n &= \mathbf{T}^n\,\mathbf{q}_0\ . \end{aligned} \tag{5.35}$$

The stability of the system (5.34) requires that the total energy of the system does not grow from step to step, and therefore the amplitude of vibration as $n \to \infty$ remains limited. This in turn leads to the necessary condition of stability,

$$|\lambda_i| \le 1, \quad i = 1,2, \tag{5.36}$$

where, with the notation $\omega^2 = k/m$, we have the eigenvalues of the matrix $\mathbf{T}$

$$\lambda_{1/2} = \frac{\omega h}{2}\sqrt{\omega^2 h^2 - 4} \mp \frac{\omega^2 h^2}{2} \pm 1\ . \tag{5.37}$$

The condition (5.36) allows of determining the maximum integration step

$$h_{max} = \frac{2}{\omega} \,. \tag{5.38}$$

This criterion applies to a single degree of freedom. In a system with many degrees of freedom the determining frequency is the highest frequency ω_{max} of the system, as this frequency increases the spectral radius of $\mathbf{T}$. The eigenvalue appears outside the unit circle on the complex plane. We see that with an increase of the eigenfrequency ω, associated with the selected degree of freedom, the allowable time step h decreases. In practice, this happens with an increase in the stiffness or a reduction of the inertia of the finite elements. An important tip is provided by a simple analysis of the stiffness and inertia of the finite element rod of length l oscillating axially. The frequency of vibrations ω is proportional to the expression $\sqrt{E/\rho}/l$. We see that with a refining of the grid of the spatial partition, the element length decreases and the frequency ω increases. In the same way, the critical time step value h_{max} decreases.

Let us apply this stability criterion to the finite element of an axially vibrating rod. The nonzero eigenvalue is the number $\omega^2 = 4E/\rho/l^2$, and $c^2 = E/\rho$ is the velocity of wave propagation in the rod. Then $\omega = 2c/l$. Let us now consider the condition (5.38) in the form of the inequality $h \le 2/\omega$. We obtain the stability criterion, determining the ratio of the spatial size of the mesh grid l to the length of the time step h:

$$l/h \geq c \,. \tag{5.39}$$

This criterion is known as the Courant condition [40].

5.3.2 *Accuracy of the Method*

Consider two basic problems about the vibrations of an oscillator with parameter values $k = 1$ and $m = 1$, with different initial conditions: $q_0 = 0$, $\dot{q}_0 = 1$ and $q_0 = 1$, $\dot{q}_0 = 0$. The solution of the first of these problems is the function $\sin(t)$, and the second, the function $\cos(t)$. In every other case, the solution is a combination of these two.

The conditions $q_0 = 0$, $\dot{q}_0 = 1$ can be reduced, at step $\Delta t = h$, to the displacement conditions $q_{-1} = -h$, $q_0 = 0$. The solution is $y(h) = q_1 = h$. We can develop an accurate solution in Taylor series

$$\sin(h) = h - \frac{1}{3!}h^3 + \dots \,. \tag{5.40}$$

The maximum error in this case is

$$\varepsilon_1(h) = |\sin(x) - y(h)| = \frac{1}{3!}h^3 - \dots \,. \tag{5.41}$$

In the second case, $q_0 = 1$, $\dot{q}_0 = 0$, the initial conditions are reduced to the following displacement conditions: $q_{-1} = q_1$, $q_0 = 1$. Their inclusion in (5.30) gives the solution $y(h) = 1 - h^2/2$.

Knowing that $\cos(h) = 1 - 1/2!\,h^2 + 1/4!\,h^4 - \ldots$, we can determine in this case the maximum error:

$$\varepsilon_2(h) = |\cos(x) - y(h)| = \frac{1}{4!}h^4 - \ldots \;. \tag{5.42}$$

The conclusion is: the maximum error of the method of the central difference method, depending on the integration step h, is

$$\varepsilon(h) = \frac{1}{6}h^3 \;. \tag{5.43}$$

The error of one step is $1/6\,\omega^3 h^3$. The number of steps is proportional to $1/h$ (i.e. $n = T/h$). Hence the error of the result after time T is $\varepsilon = 1/6\,\omega^3 h^3 \cdot T/h$. It is therefore proportional to h^2. The method is second order.

5.4 The Adams Methods

The group of Adams methods, well known and used in recent years, is becoming less common. Although characterized by attractive properties, is slowly disappearing from academic textbooks. We will briefly present the basic algorithms of the methods of Adams and their advantages. An essential advantage is the possibility of applying the method of high order (k) without an increase in the cost of computing. The calculations require storage in the memory of a sufficient number of solutions from previous steps of the calculations, specifically, the functions $\mathbf{f}(\mathbf{y}_i, t_i)$. The current calculation step requires only a single computation of such a function and its summation, with the relevant factors, with the $k-1$ values of the functions determined in the previous steps and stored in memory. With an adequate number of stored values of $\mathbf{f}$, counting backwards, we can completely freely choose the order of the method. What's more, we can use a different accuracy with selected degrees of freedom or for the entire sub-area of a structure. These advantages were used to study the dynamics of engineering problems.

Consider the differential equation

$$\begin{cases} \dfrac{\mathrm{d}y}{\mathrm{d}t} = f(y,t) \;, \\ y(0) = y_0 \;. \end{cases} \tag{5.44}$$

After expansion of $y(t)$ in a Taylor series, we obtain

$$y(t+\Delta t) = y(t) + \Delta t\, y'(t) + \frac{\Delta t^2}{2!}y''(t) + \frac{\Delta t^3}{3!}y'''(t) + \ldots \;. \tag{5.45}$$

Then (5.44) results in

$$y'(t) = f(y,t), \;\; y''(t) = f'(y,t), \;\ldots \;. \tag{5.46}$$

The development of (5.45) at t_i can be written

$$y_{i+1} = y_i + \Delta t\, f_i + \frac{\Delta t^2}{2!} f_i' + \frac{\Delta t^3}{3!} f_i'' + \dots \; . \tag{5.47}$$

Depending on the number of terms of the expansion, we obtain the next Taylor's formula of higher order.

5.4.1 Explicit Adams Formulas (Open)

Explicit formulas for the Adams methods are also called Adams–Bashord formulas. Below we will show the idea of obtaining the formulas up to the third order, together with a general formula and coefficients given in a table.

First order formula

Leaving the two terms in the series (5.47), we obtain the Adams formula of first order, identical to the Euler formula:

$$y_{i+1} = y_i + \Delta t\, f_i \; . \tag{5.48}$$

Second order formula

We include three terms of the development (5.47)

$$y_{i+1} = y_i + \Delta t\, f_i + \frac{\Delta t^2}{2!} f_i' + O(\Delta t)^3 \; . \tag{5.49}$$

We replace the derivative f_i' by the backward difference quotient

$$f_i' = \frac{f_i - f_{i-1}}{\Delta t} \; . \tag{5.50}$$

Finally we obtain the second order formula

$$y_{i+1} = y_i + \frac{\Delta t}{2}(3f_i - f_{i-1}) + O(\Delta t)^3 \; , \tag{5.51}$$

with the error of a single step being of the order of Δt^3.

Third order formula

We proceed as in the case of the lower-order formulas. We include four terms of the expansion (5.47)

$$y_{i+1} = y_i + \Delta t\, f_i + \frac{\Delta t^2}{2!} f_i' + \frac{\Delta t^3}{3!} f_i'' + O(\Delta t)^4 \; . \tag{5.52}$$

After replacing derivatives by differences and then rearranging the relevant factors, we obtain the third-order Adams formula

$$y_{i+1} = y_i + \frac{\Delta t}{12}(23f_i - 16f_{i-1} + 5f_{i-2}) + O(\Delta t)^4 . \tag{5.53}$$

Explicit Adams method of arbitrary order

The explicit Adams method of order $n+1$ is described by the scheme

$$y_{i+1} = y_i + \Delta t \sum_{k=0}^{n} f_{i-k}\,\alpha_{nk} . \tag{5.54}$$

The coefficients of the formulas are placed in Table 5.2.

Table 5.2 Coefficients of explicit Adams methods.

n	k						method's order
	0	1	2	3	4	5	
0	1						1
1	$\frac{3}{2}$	$-\frac{1}{2}$					2
2	$\frac{23}{12}$	$-\frac{16}{12}$	$\frac{5}{12}$				3
3	$\frac{55}{24}$	$-\frac{59}{24}$	$\frac{37}{24}$	$-\frac{9}{24}$			4
4	$\frac{1901}{720}$	$-\frac{2774}{720}$	$\frac{2616}{720}$	$-\frac{1274}{720}$	$\frac{251}{720}$		5
5	$\frac{4277}{1440}$	$-\frac{7923}{1440}$	$\frac{9982}{1440}$	$-\frac{7298}{1440}$	$\frac{2877}{1440}$	$-\frac{475}{1440}$	6

The beginning of the calculation of Adams formulas of higher order than the first requires the use of other computational methods for its first step. The first order Adams method can be used, or else the first order, the second order and so on successively, until all values f_i, f_{i-1}, ... required to continue the calculation are obtained. Table 5.3 presents the results of the solution of the elementary differential equation $\dot{y} + y^2 = 0$, with initial condition $y(0) = 1$, by the Adams methods. The first steps were performed by lower order Adams methods. Hence, the final accuracy does not fully depend on the accuracy of the formulas of higher order, and is burdened with a higher error in its initial steps.

Figure 5.6 gives an example of the results for the displacements of an oscillator as a function of time obtained with the Adams method of the first and the second order. We see that the first order method is divergent, although the degree of accumulation

Table 5.3 Sample results of calculation with Adams methods.

step	order of the method					
	1	2	3	4	5	6
1	0.900000					
2	0.819000	0.835124				
3	0.751924	0.771831	0.768758			
4	0.695385	0.717345	0.713642	0.714436		
5	0.647029	0.669943	0.665892	0.666860	0.666612	
6	0.605164	0.628349	0.624185	0.625246	0.624933	0.625022
7	0.568542	0.591567	0.587412	0.588480	0.588143	0.588261
8	0.536218	0.558815	0.554749	0.555804	0.555474	0.555595
9	0.507465	0.529472	0.525538	0.526552	0.526227	0.526344
10	0.481713	0.503034	0.499257	0.500227	0.499923	0.500038

of the error is moderate. A shorter time step improves the results. We recall at this point that the popular Runge–Kutta method of the second order is also not stable. The second-order Adams method with an appropriately selected time step allows us to obtain a stable solution.

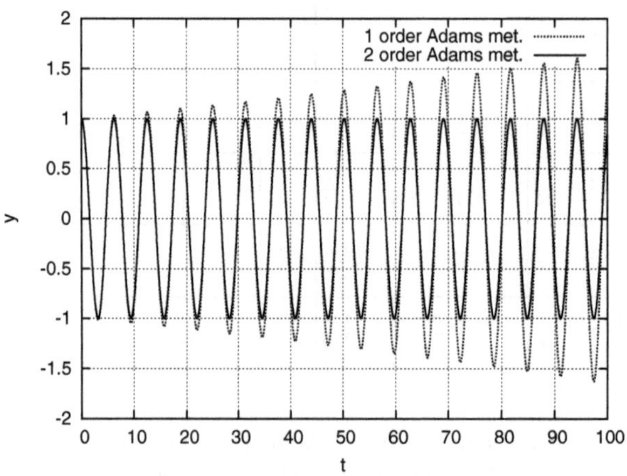

Fig. 5.6 Oscillator displacements obtained by explicit Adams first and second order methods.

5.4.2 *Implicit Adams Formulas (Closed)*

The function $y(t)$ can be expanded in a backward Taylor series at the point $t+\Delta t$

$$
\begin{aligned}
y(t) = y(t+\Delta t-\Delta t) = & y(t+\Delta t) + (-\Delta y)\, y'(t+\Delta t) + \frac{\Delta t^2}{2!} y''(t+\Delta t) + \\
& + \frac{-\Delta t^3}{3!} y'''(t+\Delta t) + \dots \; .
\end{aligned}
\tag{5.55}
$$

Assuming the formulas $y' = f$, $y'' = f'$, etc. at t_i, we can write

$$y_i = y_{i+1} - \Delta t\, f_{i+1} + \frac{\Delta t^2}{2!} f' - \frac{\Delta t^3}{3!} f''_{i+1} + \dots \ . \tag{5.56}$$

After re-arranging terms, we obtain the final form of the implicit Adams formulas:

$$y_{i+1} = y_i + \Delta t\, f_{i+1} - \frac{\Delta t^2}{2!} f'_{i+1} + \frac{\Delta t^3}{3!} f''_{i+1} + \dots \ . \tag{5.57}$$

First order formula

Already in its simplest form, the Adams implicit formula differs from the explicit formula (5.48)

$$y_{i+1} = y_y + \Delta t\, f_{i+1} \ . \tag{5.58}$$

The formula is called closed, because to obtain the unknown value of y_{i+1} we must determine $f_{i+1} = f(y_{i+1}, t_{i+1})$, depending on y_{i+1}. We use this iterative process:

- we define the first approximate value $y^{(0)}_{i+1}$ of the quantity y_{i+1},
- we consider it in the equation and we compute the new value $y^{(1)}_{i+1}$

$$y^{(1)}_{i+1} = y_i + \Delta t\, f(y^{(0)}_{i+1}, t_{i+1}) \ ,$$

- iterate.

The implicit Adams method of arbitrary order

The implicit Adams method of order $n+1$, as in the case of the explicit methods, is described by the scheme

$$y_{i+1} = y_i + \Delta t \sum_{k=0}^{n} f_{i+1-k}\, \beta_{nk} + O(\Delta t)^{n+2} \ . \tag{5.59}$$

The coefficients of the formulas are placed in Table 5.4.

The Adams methods are characteristized by their simplicity of use. A place in the memory is created for each of the vectors $\mathbf{f}_i$ of solutions at k successive times. Then, depending on the order k of the method, the solution at time t_i is calculated, for which a vector $\mathbf{y}_i$ was determined. Each time step is a summation of vectors $\mathbf{f}_i$ with the relevant factors and then a calculation of the vector $\mathbf{f}_{i+1}$. At each stage of the calculation, we can change the order of the method. The method of a higher order, applied at any stage, does not result in an increase of the cost of calculation, if a sufficient number of consecutive vectors $\mathbf{f}_i$ is kept in memory. This is important especially at the early stage, when the calculations start using the first order method and continue increasing the method's order from step to step, until the required accuracy is reached.

Table 5.4 Factors of implicit Adams methods.

n	k						method's order
	0	1	2	3	4	5	
0	1						1
1	$\frac{1}{2}$	$\frac{1}{2}$					2
2	$\frac{5}{12}$	$-\frac{8}{12}$	$-\frac{1}{12}$				3
3	$\frac{9}{24}$	$\frac{19}{24}$	$-\frac{5}{24}$	$\frac{1}{24}$			4
4	$\frac{251}{720}$	$\frac{646}{720}$	$-\frac{264}{720}$	$\frac{106}{720}$	$-\frac{19}{720}$		5
5	$\frac{475}{1440}$	$\frac{1427}{1440}$	$-\frac{798}{1440}$	$\frac{482}{1440}$	$-\frac{173}{1440}$	$\frac{27}{1440}$	6

5.5 The Newmark Method

This method was published in 1959 [103]. Its description can now be found in any book devoted to discrete structure vibrations. It is also the most commonly used method in commercial computing packages. Its advantage is clearly its unconditional stability with properly selected parameters. The Newmark method was modified [64, 143] to improve some properties of the original. Despite this, the original version of the method is usually chosen for integrating differential equations of motion. This choice is without a doubt determined, in addition to its good numerical property, by its simplicity and the availability of its procedures.

The derivation of the equations of the method is simple. We develop the functions of displacements $u(t)$ and velocities $v(t)$ in Taylor series. The remainder, as the last component of the expansion of displacement u, is determined at an intermediate point $t_\beta = (1-2\beta)t_n + 2\beta t_{n+1}$, $0 \le \beta \le 1/2$. In turn, the remainder of the development of the velocity v is determined for time $t_\gamma = (1-\gamma)t_n + \gamma t_{n+1}$. Algorithm 5 formulates the Newmark method. Suitable simple transformations lead to a computational step scheme. The parameter γ affects the numerical damping properties (Figure 5.7) and the parameter β affects the stability of the method. Both parameters are coupled together and must be selected with caution if we decide to select values other than those commonly used and described in the literature. $\gamma = 1/2$ provides no numerical damping. In this case, with $\beta \ge 1/4$, we obtain an unconditionally stable scheme. The maximum damping of high frequencies is obtained with $\beta = 1/4(\gamma + 1/2)^2$ and $\gamma > 1/2$. The eigenvalues of the transfer matrix are as follows:

$$\lambda_{1/2} = \frac{1}{4}\left|\frac{4\beta - 2\gamma - 1 \pm \sqrt{-16\beta + 4\gamma^2 + 4\gamma + 1}}{\beta}\right| . \tag{5.60}$$

Algorithm 5. The Newmark method.

Formulation

- $\mathbf{u}_{n+1} = \mathbf{u}_n + h\mathbf{v}_n + h^2(1/2-\beta)\mathbf{a}_n + h^2\beta\mathbf{a}_{n+1}\,,$
- $\mathbf{v}_{n+1} = \mathbf{v}_n + h(1-\gamma)\mathbf{a}_n + h\gamma\mathbf{a}_{n+1}\,,$
- $\mathbf{M}\mathbf{a}_{n+1} + \mathbf{C}\mathbf{v}_{n+1} + \mathbf{K}\mathbf{u}_{n+1} = \mathbf{F}_{n+1}\,.$

Solution

- Build a system of equations

$$\left(\mathbf{K} + \frac{1}{\beta h^2}\mathbf{M} + \frac{\gamma}{\beta h}\mathbf{C}\right)\mathbf{u}_{i+1} = \mathbf{F}_{i+1} + \mathbf{M}\left(\frac{1}{\beta h^2}\mathbf{u}_i + \frac{1}{\beta h}\mathbf{v}_i + (\frac{1}{2\beta}-1)\mathbf{a}_i\right) +$$

$$+\mathbf{C}\left(\frac{\gamma}{\beta h}\mathbf{u}_i + (\frac{\gamma}{\beta}-1)\mathbf{v}_i + \frac{h}{2}(\frac{\gamma}{\beta}-2)\mathbf{a}_i\right),$$

- Solve the system of equations in calculating $\mathbf{u}_{i+1}$,
- Calculate the missing accelerations and velocities

$$\mathbf{a}_{i+1} = \frac{1}{\beta h^2}(\mathbf{u}_{i+1}-\mathbf{u}_i) - \frac{1}{\beta h}\mathbf{v}_i - (\frac{1}{2\beta}-1)\mathbf{a}_i\,,$$

$$\mathbf{v}_{i+1} = \mathbf{v}_i + h(1-\gamma)\mathbf{a}_i + \gamma h\mathbf{a}_{i+1}\,.$$

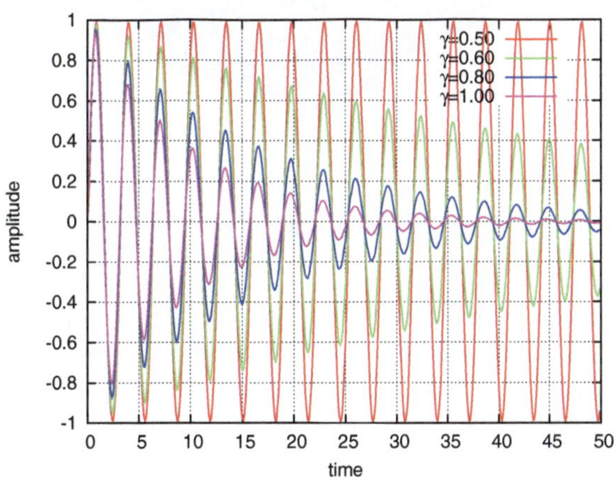

Fig. 5.7 Numerical damping of the Newmark method with different parameters γ.

The influence of both parameters on the value of the spectral radius (and thus the stability or divergence of the method) is shown in Figure 5.8.

In the group of Newmark methods we can include several derived methods, brought out for special values of parameters β.

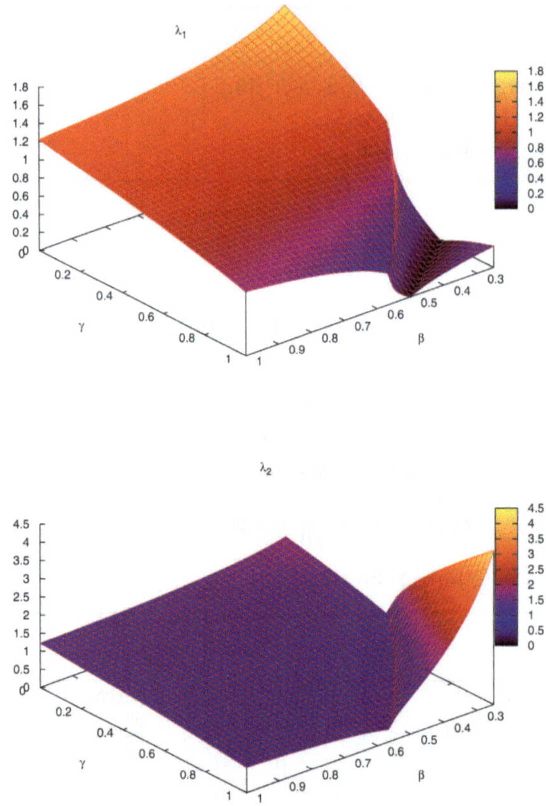

Fig. 5.8 Eigenvalues depending on the parameters β and γ.

The constant acceleration method. Is also called the average acceleration method. This is the case of $\beta = 1/4$. The method is implicit, unconditionally stable, and second order in accuracy. This is the most popular variant of the Newmark method, much used in practice. It is also known by the name of the trapezoid method.

Linear acceleration method. This is the case of $\beta = 1/6$. The method is implicit, with an accuracy of the second order, and conditionally stable.

The Fox–Goldwin method. Here, $\beta = 1/12$, and an implicit method is obtained, with second order accuracy, conditionally stable. In the absence of damping, i.e. with $\mathbf{C} = 0$, we obtain a fourth order method.

The central difference method. This is the simplest case of the Newmark method, with $\beta = 0$. The method is explicit.

5.6 The Bossak Method

Most of the schemes of integration of differential equation of motion are or are derived from the expansion of the functions of displacements and speed in Taylor series, or can be expressed in terms of such an expansion. Several methods were established by modifying the equations of motion. An example of such a way is the modification introduced by Bossak [143]. The forces of inertia are here modified. The Bossak–Newmark method is an extension of the Newmark method. In the equation of motion we take into account the acceleration **a** before time t_{i+1}. The inertial forces are interpolated in the interval $[t_i, t_{i+1}]$. The formulation of the methods is shown below (Algorithm 6). In the case of $\alpha_B = 0$, we obtain the Newmark method.

Algorithm 6. The Bossak method.

Formulation

- $\mathbf{u}_{n+1} = \mathbf{u}_n + h\mathbf{v}_n + h^2(1/2 - \beta)\mathbf{a}_n + h^2\beta\mathbf{a}_{n+1}$,
- $\mathbf{v}_{n+1} = \mathbf{v}_n + h(1-\gamma)\mathbf{a}_n + h\gamma\mathbf{a}_{n+1}$,
- $\mathbf{M}(1-\alpha_B)\mathbf{a}_{n+1} + \mathbf{M}\alpha_B\mathbf{a}_n + \mathbf{C}\mathbf{v}_{n+1} + \mathbf{K}\mathbf{u}_{n+1} = \mathbf{F}_{n+1}$.

Solution

- Build a system of equations

$$\left(\mathbf{K} + \tfrac{1-\alpha_B}{\beta h^2}\mathbf{M} + \tfrac{\gamma}{\beta h}\mathbf{C}\right)\mathbf{u}_{i+1} = \mathbf{F}_{i+1} +$$
$$+\mathbf{M}\left(\tfrac{1-\alpha_B}{\beta h^2}\mathbf{u}_i + \tfrac{1-\alpha_B}{\beta h}\mathbf{v}_i + (\tfrac{1}{2\beta} - \tfrac{\alpha_B}{2\beta} - 1)\mathbf{a}_i\right) +$$
$$+\mathbf{C}\left(\tfrac{\gamma}{\beta h}\mathbf{u}_i + (\tfrac{\gamma}{\beta} - 1)\mathbf{v}_i + \tfrac{h}{2}(\tfrac{\gamma}{\beta} - 2)\mathbf{a}_i\right),$$

- Solve the equations by calculating $\mathbf{u}_{i+1}$,
- Calculate the missing acceleration and velocity

$$\mathbf{a}_{i+1} = \tfrac{1}{\beta h^2}(\mathbf{u}_{i+1} - \mathbf{u}_i) - \tfrac{1}{\beta h}\mathbf{v}_i - (\tfrac{1}{2\beta} - 1)\mathbf{a}_i,$$
$$\mathbf{v}_{i+1} = \mathbf{v}_i + h(1-\gamma)\mathbf{a}_i + \gamma h\mathbf{a}_{i+1}.$$

The stability conditions are satisfied at

$$\alpha_B \leq 1/2\,, \quad \beta_B \geq \gamma_B/2 \geq 1/4\,, \quad \alpha_B + \gamma_B \geq 1/4\,.$$

The Bossak method is characterized by the very important property of absorbing the higher frequency vibrations.

Another modification, based on a similarly simple concept, was introduced by the authors of [64]. In this case, a change of potential forces was introduced, shifting the period of time at which they are defined in the equation of motion. The equation of motion takes the form listed in the Algorithm 7. Algorithms based on this modification are very sensitive to the parameters chosen. Despite the theoretical possibility

Algorithm 7. Hilber, Hughes, and Taylor method, according to [64].

Formulation

- $\mathbf{u}_{n+1} = \mathbf{u}_n + h\mathbf{v}_n + h^2(1/2-\beta)\mathbf{a}_n + h^2\beta\mathbf{a}_{n+1}$,
- $\mathbf{v}_{n+1} = \mathbf{v}_n + h(1-\gamma)\mathbf{a}_n + h\gamma\mathbf{a}_{n+1}$,
- $\mathbf{M}\mathbf{a}_{n+1} + \mathbf{C}\mathbf{v}_{i+1} + (1+\alpha_H)\mathbf{K}\mathbf{u}_{n+1} - \alpha_H\mathbf{K}\mathbf{u}_n = \mathbf{F}_{n+1}$.

Solution

- Build a system of equations

$$\left((1+\alpha_H)\mathbf{K} + \tfrac{1}{\beta h^2}\mathbf{M} + \tfrac{\gamma}{\beta h}\mathbf{C}\right)\mathbf{u}_{i+1} = \mathbf{F}_{i+1} +$$
$$+\mathbf{M}\left(\tfrac{1}{\beta h^2}\mathbf{u}_i + \tfrac{1}{\beta h}\mathbf{v}_i + (\tfrac{1}{2\beta}-1)\mathbf{a}_i\right) +$$
$$+\mathbf{C}\left(\tfrac{\gamma}{\beta h}\mathbf{u}_i + (\tfrac{\gamma}{\beta}-1)\mathbf{v}_i + \tfrac{h}{2}(\tfrac{\gamma}{\beta}-2)\mathbf{a}_i\right) + \alpha_H\mathbf{K}_i\mathbf{u}_i,$$

- Solve the equations by calculating $\mathbf{u}_{i+1}$,
- Calculate the missing acceleration and velocity

$$\mathbf{a}_{i+1} = \tfrac{1}{\beta h^2}(\mathbf{u}_{i+1} - \mathbf{u}_i) - \tfrac{1}{\beta h}\mathbf{v}_i - (\tfrac{1}{2\beta}-1)\mathbf{a}_i,$$
$$\mathbf{v}_{i+1} = \mathbf{v}_i + h(1-\gamma)\mathbf{a}_i + \gamma h\mathbf{a}_{i+1}.$$

of influencing the properties of the method by selecting these parameters, a stable solution is obtained only within a very narrow range of values, given by the authors. For this reason, its application has not spread very widely.

5.7 The Park Method

The method described by Park, [107], is designed for stiff equations. Differential equations are called stiff, when, despite the very small step in integrating computational numerical methods, they present unstable solutions. The primary task of the method of calculation in this case is to adequately accurately integrate the variables responsible for the behaviour of a rigid body, i.e. one characterized by high frequencies of vibration, without loss of computational efficiency of the process. The construction of methods for resolving these issues began with Gear [59].

5.8 The Park–Housner Method

The Park–Housner method [110] is an example of a semi-implicit method. It combines the advantages of both explicit methods (low numerical cost, low memory consumption) and implicit methods (unconditional stability). In this method, the diagonal form of the matrix **M** is assumed. We underline this assumption here, because it can considerably simplify the procedure and increase its efficiency. The method in the general case can also be applied with a full matrix **M**. In this case, however, an expansive inversion of it is required. The matrix **K** is decomposed into the sum of triangular matrices. As a result we obtain two systems of algebraic equations to solve. Finally we compute the displacement and velocity vectors (Algorithm 8).

Algorithm 8. Park–Housner method

- Construction of the matrix **K** and the diagonal matrix **M**,
- Decomposition of **K** into $\mathbf{K}_L$ and $\mathbf{K}_U$ ($\mathbf{K} = \mathbf{K}_L + \mathbf{K}_U$, $\mathbf{K}_L = \mathbf{K}_U$),
- Construction of the matrix of coefficients of the system of equations:

  $\mathbf{L} = \mathbf{M}(\mathbf{I} + \alpha\beta h^2\mathbf{M}^{-1}\mathbf{K}_L)$,

  $\mathbf{U} = (\mathbf{I} + \alpha\beta h^2\mathbf{M}^{-1}\mathbf{K}_U)$,

  $\mathbf{g}_{n+1} = \alpha\beta h^2[\beta\mathbf{f}_{n+1} + (1-\beta)\mathbf{f}_n] + \mathbf{M}(\mathbf{u}_n + \beta h\mathbf{v}_n)$.
- The solution of the system of equations (with triangular matrices):

  $\mathbf{L}\mathbf{y}_{n+1} = \mathbf{g}_{n+1}$, $\mathbf{U}\bar{u}_{n+1} = \mathbf{y}_{n+1}$.
- Final solution:

  $\mathbf{u}_{n+1} = 1/\beta\,[\bar{\mathbf{u}}_{n+1} - (1-\beta)\mathbf{u}_n]$,

  $\mathbf{v}_{n+1} = 1/(\alpha h)\,(\mathbf{u}_{n+1} - \mathbf{u}_n) - (1-\alpha)/\alpha\,\mathbf{v}_n$.

5.8.1 Stability of the Park–Housner Method

The analysis of the eigenvalues of the transition matrix gives a graph shown in Figure 5.9. The eigenvalues of the transfer matrix are as follows:

$$\lambda_{1/2} = \frac{1}{2}\left|\frac{\alpha\beta^2 - 2\alpha^2\beta^2 \mp \sqrt{\alpha^2\beta^2(\alpha-\beta)^2 + \alpha^2\beta}}{\alpha^2\beta^2}\right| . \tag{5.61}$$

Values shown in both graphs have a value equal to one only at the point $\alpha = \beta = 1/2$. Moreover, when α, $\beta > 1/2$, the method exhibits damping. In the remaining range of parameters, it is divergent (Figure 5.10). Figure 5.11 shows the time function of the axial displacements of the free end of the rod, subjected to an initial impulse. The initial conditions were imposed as a zero displacement and zero velocity, with the exception of the free end, at which the initial velocity v_0 was imposed. The graph of displacements suffers from a large share of parasitic oscillations. Their amplitudes reach values of 100% of the amplitude of the theoretical result. Although the averaged, smoothed graph coincides quite well with the theoretical line, this example of the numerical analysis of undamped vibration shows the trap in a situation when some local destruction of the material may occur.

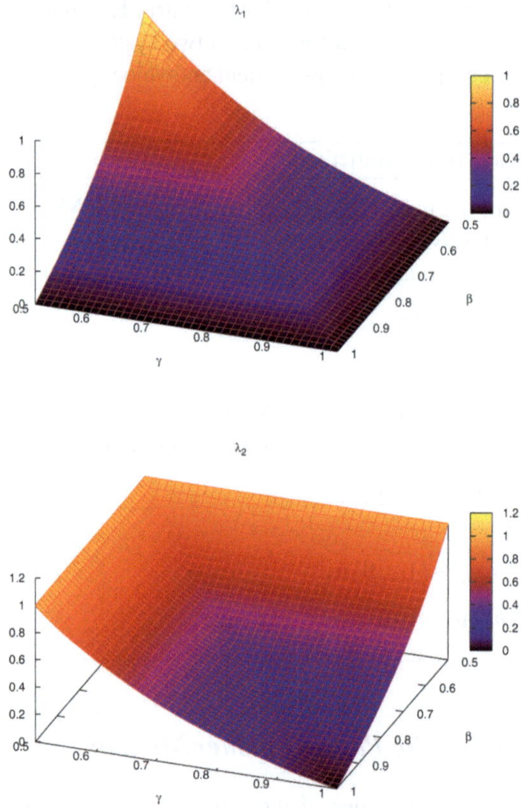

Fig. 5.9 Eigenvalues of the Park–Housner method, depending on the parameters α and β.

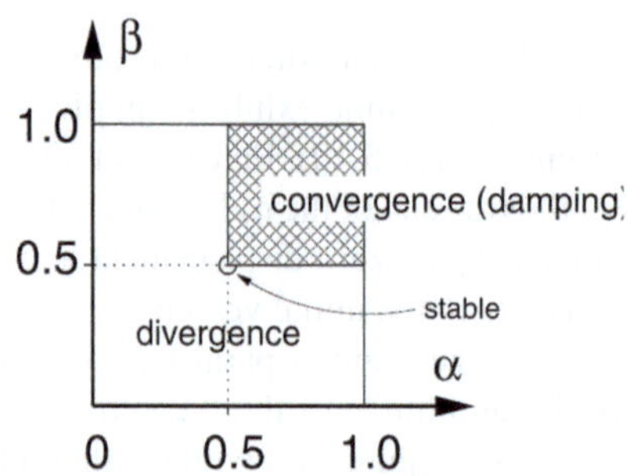

Fig. 5.10 Area of stability of the Park–Housner method.

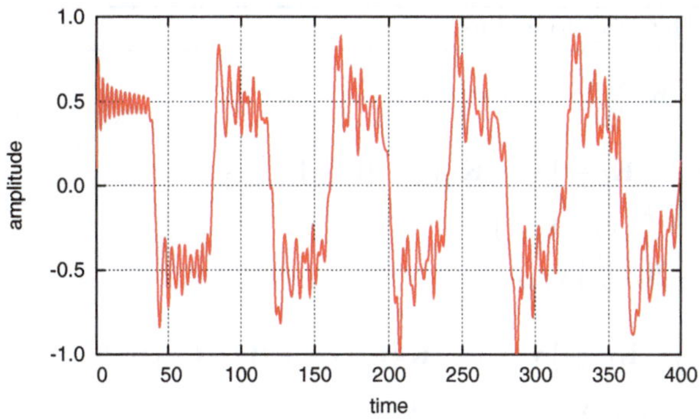

Fig. 5.11 Displacements of the free end of the rod, obtained by the Park–Housner method.

5.9 The Trujillo Method

The method developed by Trujillo [139] is also based on the concept of decomposing the damping **C** and stiffness **K** matrices to sums of upper and lower triangular matrices. The inertia matrix **M** may be full, although in the case of a diagonal form the resulting algebraic equations simplify, and their matrices of coefficients take triangular form. Then the alternating system of equations is solved with a lower triangular matrix (Algorithm 9 p. 1) and an upper triangular matrix (see p. 3). Each of the solutions allows of computing the speed, and then the displacements at a time after Δt from the previously considered time. As a result, the implementation of both phases of the calculations shifts the time of analysis by $h = 2\Delta t$.

The method usually operates with triangular matrices of coefficients of equations. It significantly accelerates the calculations. The Trujillo method is an explicit method. Unconditional stability is obtained if the decomposition of the matrix is symmetric. In the case of an asymmetrical decomposition, the algorithm is conditionally stable, but has higher accuracy.

An additional important advantage is the possibility of using only finite element matrices. There is no need for the formation of the global matrix. This significantly imporves the effectiveness of the Trujillo method. Another advantage is its lack of control parameters needing to have their values set. The disadvantage is a two-stage algorithm and the need to use, for reasons of efficiency of calculations, the diagonal mass matrix.

Figure 5.12 shows the displacements in time of the free end of a rod, under the initial impulse. The Trujillo method better reflects the theoretical rectangular plot of vibration than the Park–Housner method. The data in the task were chosen so that the period of oscillation is equal to 80. In the respective moments the theoretical graph should be vertical, because the solution is discontinuous there. In subsequent cycles, parasitic oscillations have slightly smaller amplitudes and preserve the rectangular nature of the plot. Unfortunately, we observe a worsening performance with

Algorithm 9. The Trujillo method.

1. $\left(\mathbf{M}+\mathbf{C}_L\frac{h}{2}+\mathbf{K}_L\frac{h^2}{8}\right)\mathbf{v}_{j+1/2}=$

$=\left(\mathbf{M}-\mathbf{C}_U\frac{h}{2}-\mathbf{K}_L\frac{h^2}{8}\right)\mathbf{v}_j-\mathbf{K}\frac{h}{2}\mathbf{u}_j+\left(\mathbf{f}_j+\mathbf{f}_{j+1}\right)\frac{h}{4},$

2. $\mathbf{u}_{j+1/2}=\mathbf{u}_j+\frac{h}{4}\left(\mathbf{v}_j+\mathbf{v}_{j+1}\right),$
3. $\left(\mathbf{M}+\mathbf{C}_U\frac{h}{2}+\mathbf{K}_U\frac{h^2}{8}\right)\mathbf{v}_{j+1}=$

$=\left(\mathbf{M}-\mathbf{C}_L\frac{h}{2}-\mathbf{K}_U\frac{h^2}{8}\right)\mathbf{v}_{j+1/2}-\mathbf{K}\frac{h}{2}\mathbf{u}_{j+1/2}+\left(\mathbf{f}_j+\mathbf{f}_{j+1/2}\right)\frac{h}{4},$

4. $\mathbf{u}_{j+1}=\mathbf{u}_{j+1/2}+\frac{h}{4}\left(\mathbf{v}_{j+1/2}+\mathbf{v}_{j+1}\right),$

$\mathbf{K}=\mathbf{K}_L+\mathbf{K}_U,\ \mathbf{C}=\mathbf{C}_L+\mathbf{C}_U,\ t=jh=2j\Delta t\,.$

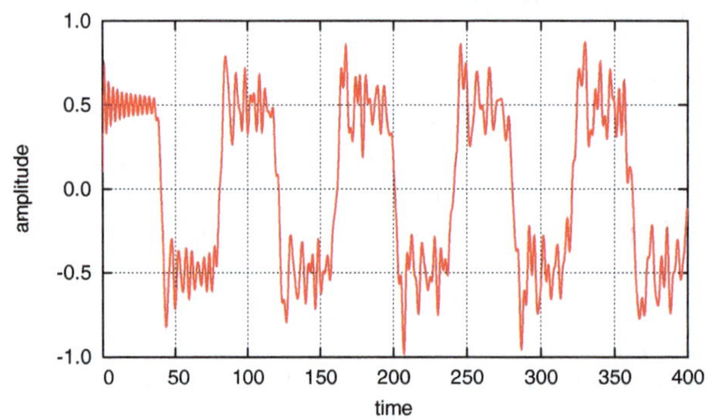

Fig. 5.12 Displacements of the free end of a rod obtained by the Trujillo method.

increasing time steps. This is due to the limited speed of information flow in a single step of calculation, in comparison with the fully explicit methods (see Chapter 6.2).

Chapter 6
Space–Time Finite Element Method

In the previous section we discussed some classical methods for the time integration of the differential equations of motion. They have interesting properties, not appreciated by researchers and software developers. In this section we will present the space–time element method. We will give its basic concepts and how to derivate the stepwise equations for this method. We will present the displacement formulation, used in the early stages of the development of the method, and the velocity formulation, which is currently being successfully used for difficult or atypical tasks.

An essential feature which differentiates the space–time element method from traditional approaches to solving initial-boundary problems is its discretization of the differential equation. Classically, two-step interpolation is used, the spatial variables separately from the time. Therefore, in the first stage, the system of partial differential equations is transformed into a system of ordinary differential equations in time of the basic state variables, and further purely numerical procedures are applied for the integration of the differential equations in the time. This approach has its advantages, mainly due to the possibility of selecting the most effective tools separately for each stage, including strict methods. In addition, the passage from the static to the dynamic solution is simpler. Most numerical procedures for static analysis are easy to use in this situation. In fact, the solution of the dynamics reduces to the solution of a set of static tasks (Figure 6.1). The control of the estimation error and the unconditional stability of the solution due to the time step involved in the integration of the differential equation are also significant. All this makes the most popular methods, i.e., the finite element method combined with the method of integration over time, e.g. the Newmark method, a permanent part of the practice of simulation calculations.

One of the drawbacks of the classical approach is the need for partition (discretization) of the considered spatial area which is constant in time (stationary). In this case, the local adaptation of the mesh to the processes involved (e.g., the development of plastic zones, zones of contact, propagation of cracks, the movement of the load) is very difficult. The existing methods of adaptation, including multigrid methods, or moving meshes, are an attempt to adapt a numerical process to a phenomenon, and to remove this defect. However, the first of these techniques is

C.I. Bajer and B. Dyniewicz: Numerical Analysis of Vibrations of Structures, LNACM 65, pp. 123–180.
springerlink.com

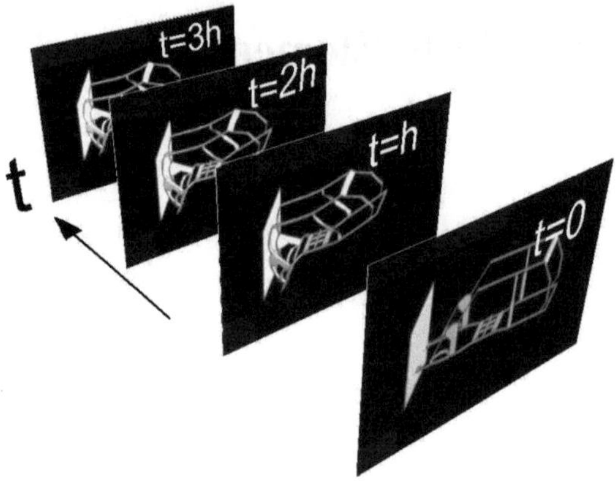

Fig. 6.1 Dynamic problem as a series of static tasks.

generally used for static tasks, while the other, at the present stage of development, can effectively model the task of transport of heat, mass, etc. Certain couplings of the concept of the moving mesh method presented here by space elements should be noted.

The second major drawback is usually the use of the same procedure in time integration for all mesh nodes of the structure. Although it is possible to use locally procedures for the time integration of differential equations with higher accuracy, in order to describe all variables of the physical phenomena within the same time interval (time step), we do not improve the numerical model of the problem. An unconditionally stable scheme of integration of the equations of motion allows of increasing the efficiency of the calculations by lengthening the time step, but at the same time introduces significant amplitude and phase errors, which vary in different zones of the area of the structure. Mesh refining brings, on the one hand, an improvement of the spatial accuracy, but on the other hand can worsen the results of the integration over time. Local interference in the ways of integration in time at the level of a finite element, depending on its size and the characteristics of the material, usually produces difficulties.

Of course all the complexities of the calculations can not categorically determine the advantages and disadvantages of the different groups of methods. Artificial damping of higher frequency vibrations by one method of calculation is a defect in wave problems, but an advantage in the analysis of structural vibrations. The selection of a computational tool is carried out based on the type of task, the phenomena under examination, the required accuracy of the arithmetical operations which can be performed on the computer, and non-substantive factors (e.g., the availability of numerical procedures, one's own experiences, etc.).

The space–time approximation of solutions of dynamic problems lacks some of these drawbacks of the classical numerical methods. However, the method is not

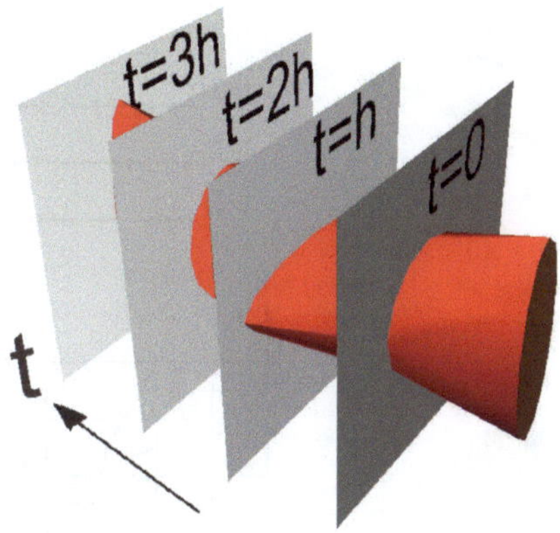

Fig. 6.2 Continuous representation time of a dynamic problem.

perfect. In essence, it reduces to the fact that functions $\mathbf{w}$, characterizing a solution, are described in space–time sub-areas through nodal parameters $\mathbf{q}_e$ (Figure 6.2)

$$\mathbf{w}(\mathbf{x},t) = \mathbf{N}(\mathbf{x},t)\,\mathbf{q}_e. \tag{6.1}$$

The matrix $\mathbf{N}(\mathbf{x},t)$ is a matrix of interpolation functions, depending on the spatial variables and the time[1]. This approach assumes a continuous distribution of characteristic functions, such as displacement or velocity, in the whole space–time domain $\Omega = \{\mathbf{x},t : \mathbf{x} \in V(t), 0 \le t < \infty\}$ where the structure is considered. In discrete time t_i, i =0, 1, 2, ..., you can use different bases of nodes (with certain restrictions), and therefore adapt the mesh to current requirements. This has the following possibilities, also shown in Figure 6.3:

- the possibility of redistributing the mesh, depending on the changing distribution of the approximation error,
- moving the zone of a mesh refinement together with a travelling load,
- the ability to adjust the sides of the elements to characteristic lines determined in the space–time domain: the front of the plastic area, the front of the material phase change, in particular—the possibility of modelling a moving edge of a body,
- the use of mesh shapes other than the space-mesh multiplexed meshes: multiplexed networks are the result of an evolution of the spatial grid in the layer of time and the corresponding elements have the same number of nodes in the initial and final time; other meshes, such as simplices, have new, important properties; simplex elements with dimension n have $n+1$ nodes at the initial time and $i+1$ nodes at the final time ($i = 0, 1, \ldots, n$),

[1] For the purpose of comparison, the spatial discretization uses the interpolation formula $\mathbf{u}(\mathbf{x},t) = \mathbf{N}(\mathbf{x})\,\mathbf{q}_e(t)$.

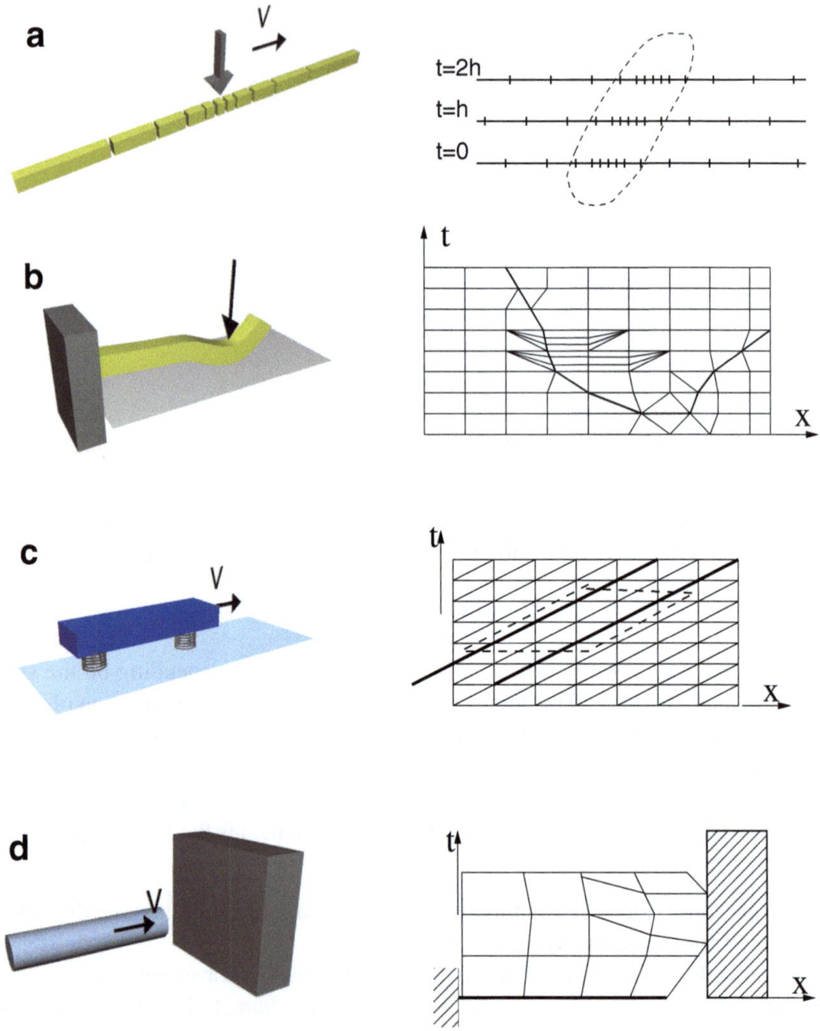

Fig. 6.3 Examples of problems solved in a time-space.

- the possibility of individual formulation of the time integration in an active manner for each spatial element,
- the particular case of space–time approximation can give a classical method of solution, based on fixed grid nodes (evolving only with the material).

This last point can be expanded to a statement that space–time approximation and the resulting space–time element method are a generalization of the finite element method, classically referred to a real space. By real space we mean the space of spatial variables x, y, z, in contrast to space–time, which is described by the spatial variables x, y, z, and the time t.

The first attempts at space–time modelling of physical problems were published in 1964 by Gurtin [60, 61] and Herrera [63]. Defining a minimized functional, resulting from the theory of convolution, enabled the derivation of relationships between the time variable and spatial variables in space–time domains. These areas can be interpreted as space–time finite elements. Later, in 1969, Oden [105] proposed a generalization of the finite element method. He extended the interpretation of the image of the structure over the time domain. Unfortunately, the interesting idea of a non-stationary partition of a structure into sub-domains proposed in that paper was not continued later on. Fried [55] and Argyris, Scharpf, and Chan [5, 6, 7] began to treat the temporal and spatial variables equally in the formulation of problems. Nevertheless, in the papers of Kuang and Atluri [76] the final digitization in space and time were conducted separately. Dynamic problems were solved with the separation of the temporal and the spatial variables. The physical area of a structure was discretized by one method (finite element method, finite difference method), while the temporal derivatives were integrated with other methods (central difference method, Newmark method, etc.). Numerous papers appeared on the direct integration of the differential equations of motion, assuming a stationary discretization.

Regardless of this direction of research, some elaborations by Kączkowski appeared [77, 79], in which for the first time physical interpretations of certain meanings previously considered in real space were introduced to structural mechanics. This is for example the time-work equation, a mass as a vector, or a space–time stiffness. Simple cases of the vibration of a rod axis and strings were considered.

Another issue raised was the problem of stability [18] and an attempt at synthesis of the space–time formulation [81]. A major contribution was the indication of possibilities for building an unconditionally stable solution by modifying virtual shape functions [73]. Unfortunately, the use of that technology is confined to rectangular space–time elements, whose shape functions can be expressed as a product of terms defining the interpolation functions in space and in time. Further studies were focused on elements other than space–time rectangular shapes. The next important step was to move away from the stationary partition of the spatial structure, and thus to introduce non-rectangular elements in time [17, 18]. This step allowed the application of the method to an entirely new group of issues: contact problems [10] and processes with an adaptation of the mesh [19].

In addition to work on developing the same method, there were many attempts to assess its accuracy and efficiency and to use the space-time element method in various technical issues [12, 13, 14, 114, 136]. In addition to problems of mechanics, issues of heat propagation have been treated [15].

The next step was to incorporate nonlinear effects: geometrical [33, 115, 141] and material nonlinearities [21, 116]. It is also worth mentioning [62], in which the authors use harmonic functions for interpolation in time.

In the paper by Podhorecki [117], rectangular space–time elements known from the paper of Kączkowski (e.g. [79]) were applied to viscoelastic problems. Their considerations were limited to one-dimensional structures. Models of Hooke,

Kelvin–Voigt, Maxwell, and Zener material were used. However, no concrete proposals supported by real examples were given. A group of reviews on the space–time element method are [16, 23].

In this large number of papers, not one has tried to use the opportunities that exist in the method of space–time elements. The considerations were limited to the use of STEM as a method of integration of differential equations. The following sections will demonstrate the equivalence of such procedures using other methods, such as for example the Newmark method. Furthermore, further development of the use of the STEM for modelling dynamic problems is reasonable.

In this chapter, we will present the new opportunities created by changing the spatial approximation continuously in time. In Chapters 6.1 and 6.3, the formulation of the space–time element method will be presented. The resulting equation can be expressed both through displacements and velocities. However, the formulation differs between the two cases. The displacement description leads ultimately to the Galerkin method, applied to space and time, and the velocity description does not lead directly to the Galerkin method. This needs emphasizing. The two formulations have different properties and have different utilities. The derivation of the characteristic matrices of space–time finite elements in the velocity formulation was given. They allow users to easily change the number of nodes in the spatial mesh, and thus to use adaptive techniques. The topological properties of this group of elements allow us to obtain significant computational benefits, thoroughly discussed in the next chapter.

Another, separate, characteristic group of space–time elements are the elements with simplicial shape. The properties of derived with their use in the solutions will be discussed in Chapters 6.7 and 6.8. The most important feature here is to obtain the final triangular matrix of the system of algebraic equations. This results in some interesting applications, such as the possibility of limiting the speed of propagation of information in a discrete system. This allows us in some cases to restrict an infinite domain to a small number of finite elements. An example of this approach is presented in [11].

The space–time formulation was used to model contact zones variable in time [20, 34]. Dynamic contact conditions were written, which were applied to lightweight engineering tasks. A dense digitization of time was used. This allowed an accurate determination of the contact forces. A way of eliminating discontinuities of the velocity function at the ends of the time domain of a contact was presented. The experience gained was used to model the rail–wheel contact area and the generation of corrugations. With material nonlinearities, iterative solutions of a nonlinear equation of motion were linked with the integration of this equation in time. Relevant numerical schemes were presented.

Examples of calculations of both test and real engineering problems confirm the effectiveness of the techniques described. However, many issues will certainly arise. This is not the final state of the development of the space–time description of the dynamics of structure.

In numerical examples, as the primary system of units we use the SI system. In order to improve the conditioning the of resulting matrices of algebraic equations in computer calculations, the respective multipliers were scaled by multipliers: the length by 10^{-2}, the mass by 10^{-3}, and time by 10^{-6}. This allowed us to avoid severe range differences in solid material properties and thus get rid of certain numerical difficulties. Due to the fact that many numerical examples are only illustrative,we used non-dimensional values. In the comparative examples and more complex tasks, complete, actual values of all parameters are given.

6.1 Formulation of the Method—Displacement Approach

Although the first mention of the possibility of nonstationary discretization of a structure appeared in the paper of Oden[105], the truly systematic introduction of the space–time element method was by Kączkowski [77, 78]. He introduced terms which look unusual in engineering. A material point appearing in the discussion at time t_0 and at place x_0, moves to the position x_1 at time t_1. In the system of coordinates (x,t), these two positions of the material point designate the section. To describe the motion of the mass as part of the oscillator, at both ends of the so called vector imaging the life of the mass, we define some physical quantities such as momentum or force from an elastic element, and between the two ends we take into account the impact of external effects. Writing the corresponding equation of virtual work, we can make the state at t_1 depend on the state at t_0. This way we get the stepping scheme of calculating the differential equation of motion.

Described in a series of new papers, one look at the dynamics detects a fault—it was conceptually complicated. The dynamics was not seen as a process of binding phenomenon in both space and time. While spatial discretization could be performed relatively easily, even in complex problems, time was treated with great care. During the rapid development of the process automation of the calculation, when the resulting classic finite element models took into account increasingly complex physical phenomena, it was difficult to produce adequate interest in a new and difficult technique.

The failure by the scientific world to model difficult tasks of structural dynamics and wave phenomena in the full sense of the time-space, and the underdevelopment of efficient algorithms in some types of problems, had for many years left undone tasks such as the description of continuous problems involving a point mass moving over an assumed trajectory or alterations in the local rigidity.

Further on we will derive the basic equations of the displacement version of the space–time finite element method as presented in papers published slightly later. We will show the typical course of treatment of an axially vibrating rod. We will obtain sample solutions and show the simplicity of the numerical calculations. Anticipating the content, we will show the identity of some solutions with the solutions obtained using the finite difference method applied to the spatial derivatives when the central difference method is applied to the velocity and acceleration. We will describe the finite space–time elements of some selected types of structures. Taking these descriptions as a point of departure one could, repeating these procedures, determine the characteristic matrices according to one's own needs.

We will consider a continuous solid, closed in the domain $\overline{V}$, and being a sub-area of the Euclidean space E^3. V denotes the interior of the area, and ∂V its boundary, which, in turn, is the sum of ∂V_t and ∂V_u, where, respectively, the displacement and stress boundary conditions will be imposed. We will consider the motion of the body during the interval $[0,T]$. The variables included in the description, such as the displacement vector $\mathbf{u}$, the velocity $\mathbf{v}$, the vector of mass forces $\rho\mathbf{f}$, the symmetric tensor of the stress field σ, and the deformation field ε, are defined on the Cartesian product set $V \times [0,T]$. The vector of surface forces $\widehat{\mathbf{t}}$ is defined on the product $\partial V \times [0,T]$. We assume that all functions are sufficiently continuous. Geometrically and physically linear problems are described by the following system of equations:

- geometric equations

$$\varepsilon(\mathbf{x},t) = \frac{1}{2}\left(\operatorname{grad}\mathbf{w} + \operatorname{grad}^T\mathbf{w}\right), \quad (\mathbf{x},t) \in V \times [0,T], \tag{6.2}$$

- physical equations

$$\sigma(\mathbf{x},t) = \mathbf{E}\varepsilon, \quad (\mathbf{x},t) \in V \times [0,T], \tag{6.3}$$

- equations of dynamical equilibrium

$$\operatorname{div}\sigma^T + \rho\mathbf{f} = \rho\frac{\partial \mathbf{v}}{\partial t}, \quad (\mathbf{x},t) \in V \times [0,T], \tag{6.4}$$

- boundary conditions

$$\sigma\mathbf{n} = \widehat{\mathbf{t}}(\mathbf{x},t), \quad (\mathbf{x},t) \in \partial V_t \times [0,T], \tag{6.5}$$

$$\mathbf{w}(\mathbf{x},t) = \widehat{\mathbf{u}}(\mathbf{x},t), \quad (\mathbf{x},t) \in \partial V_u \times [0,T], \tag{6.6}$$

- initial conditions

$$\mathbf{w}(\mathbf{x},t) = \mathbf{w}^0, \quad (\mathbf{x},t) \in \overline{V} \times \{0\}, \tag{6.7}$$

$$\dot{\mathbf{w}}(\mathbf{x},t) = \mathbf{v}^0, \quad (\mathbf{x},t) \in \overline{V} \times \{0\}. \tag{6.8}$$

In the above relationship, $\mathbf{w}^0$ and $\mathbf{v}^0$ define the initial displacement and the velocity, respectively, and $\widehat{\mathbf{w}}^0$, the displacements on the boundary ∂V_u. This system is a local formulation. The existence and uniqueness of the solution can be proved as in [104]. A global formulation can be achieved by multiplying (6.4) by the variation of the virtual function of displacements $\delta\mathbf{w}(\mathbf{x},t)$

$$\delta\mathbf{w}(\mathbf{x},t) = \begin{cases} 0, & (\mathbf{x},t) \in \partial V_u \times [0,T]\,, \\ \text{any}, & (\mathbf{x},t) \in (\overline{V} - \partial V_u) \times [0,T]\,. \end{cases} \tag{6.9}$$

After integration we obtain

$$\int_{t_0}^{t_1}\int_V \left(\operatorname{div}\sigma^T + \rho\mathbf{f} - \rho\dot{\mathbf{v}}\right)\delta\mathbf{w}\,\mathrm{d}V\,\mathrm{d}t + \int_{t_0}^{t_1}\int_{\partial V_t} \widehat{\mathbf{t}}\,\delta\mathbf{w}\,\mathrm{d}(\partial V)\,\mathrm{d}t = 0. \tag{6.10}$$

Integrating by parts, we obtain

$$\int_{t_0}^{t_1}\int_V \rho\,(\mathbf{f}\delta\mathbf{w}+\dot{\mathbf{w}}\,\delta\dot{\mathbf{w}})\,\mathrm{d}V\,\mathrm{d}t+\int_{t_0}^{t_1}\int_{\partial V_t}\widehat{\mathbf{t}}\,\delta\mathbf{w}\,\mathrm{d}(\partial V)\,\mathrm{d}t=\int_{t_0}^{t_1}\int_V \sigma\delta\varepsilon\,\mathrm{d}V\,\mathrm{d}t. \quad (6.11)$$

The area in which (6.11) is described should now be discretized. The area $\{\overline{V}, 0 \leq t \leq T\}$ is divided into subareas, which are in fact space–time finite elements. The simplest possible divisions are shown in Figure 6.4*a*. The space–time layers $[t_i \leq t \leq t_{i+1}]$, $i=0, 1, \ldots, n-1$, are cut in a semi infinite band, where n is the number of layers in the interval [0, T], and inside them there are isolated elements of the geometry, fixed in time. A more complex partitioning of the space–time layer is also possible (Figure 6.4*b*). We will return to this issue in later chapters.

The simplest element is separated from the time-space by planes $t = t_i$ and $t = t_{i+1}$. They are hyper-prisms (hyper-prisms are prisms in space of dimension higher than 3), which are finite elements, stretched in time (Figure 6.5). The values of the unknowns, in our case the displacements $\mathbf{w}$ and their derivatives ($\dot{\mathbf{w}}$, ε, σ), real and

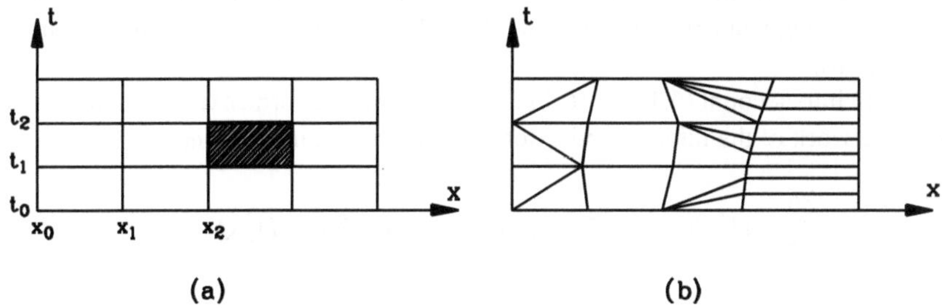

Fig. 6.4 Examples of space–time element mesh: (a) stationary, (b) nonstationary.

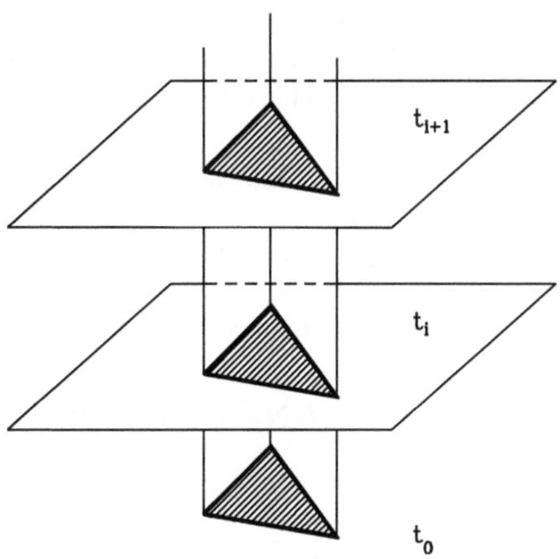

Fig. 6.5 Space–time element, separated from the time-space.

virtual, in the area of the space–time element are interpolates on the basis of the values of the nodal displacements $\mathbf{q}$:

$$\begin{aligned} \mathbf{w}(\mathbf{x},t) &= \mathbf{N}(\mathbf{x},t)\,\mathbf{q}\,, \quad \delta\mathbf{w}(\mathbf{x},t) = \mathbf{N}^*(\mathbf{x},t)\,\delta\mathbf{q}\,, \\ \dot{\mathbf{w}}(\mathbf{x},t) &= \dot{\mathbf{N}}(\mathbf{x},t)\,\mathbf{q}\,, \quad \delta\dot{\mathbf{w}}(\mathbf{x},t) = \dot{\mathbf{N}}^*(\mathbf{x},t)\,\delta\mathbf{q}\,, \\ \varepsilon(\mathbf{x},t) &= \mathbf{B}(\mathbf{x},t)\,\mathbf{q}\,, \quad \delta\varepsilon(\mathbf{x},t) = \mathbf{B}^*(\mathbf{x},t)\,\delta\mathbf{q}\,, \\ \sigma(\mathbf{x},t) &= \mathbf{E}\mathbf{B}(\mathbf{x},t)\,\mathbf{q}\,. \end{aligned} \tag{6.12}$$

These relationships apply separately to each element of time-space. Quantities marked by $(.)^*$ refer to a virtual state. The matrix $\mathbf{B}$ can easily be created by letting a suitable differential operator $\mathscr{D}$ act on the matrix of shape functions $\mathbf{N}$: $\mathbf{B} = \mathscr{D}\mathbf{N}$, where $\mathscr{D} = \frac{1}{2}\left(\text{grad} + \text{grad}^T\right)$. It should be emphasized that the vector $\mathbf{q}$ consists of the displacements in the space–time element nodes. These nodes have different time coordinates.

Linear (affine) shape functions $\mathbf{N}(\mathbf{x},t)$ express the boundary conditions on the free end $\partial\mathbf{w}/\partial\mathbf{n} = 0$ in a natural way. In the case of the element of length b, with time step Δt, they are satisfied approximately. Refining the digitization ($b \to 0$) in the extreme element $\partial\mathbf{w}/\partial\mathbf{n} \to 0$ ($\partial\mathbf{w}/\partial\mathbf{n} \sim 1/N$, where N is the number of nodes in the mesh).

Taking into account (6.12) in (6.11) gives a quadratic form, expressing the equality of the work of the internal and external forces in the interval $[t_0, t_1]$:

$$\sum_{e=1}^{NE} \left((\Pi_e^T \delta\mathbf{q}_e)^T \Pi_e^T \mathbf{K}_e^* \Pi_e \cdot \Pi_e^T \mathbf{q}_e - (\Pi_e^T \delta\mathbf{q}_e)^T \Pi_e^T \mathbf{Q}_e\right) = 0. \tag{6.13}$$

Here, NE is the number of space–time elements in the structure. The matrices Π_e are zero–one matrices assigning the places of the elemental matrices and vectors to places in the global matrix and vector of the structure. Their form depends on the topology of the mesh discretization. The number of rows equals the number of unknowns associated with a single element, and the number of columns equals the number of unknowns in the structure. These matrices define an aggregation of the matrix elements.

The matrix $\mathbf{K}_e^*$ is the space–time stiffness element matrix

$$\mathbf{K}_e^* = \mathbf{K}_e + \mathbf{M}_e. \tag{6.14}$$

We give here the form of the stiffness matrix $\mathbf{K}_e$ and the inertia matrix $\mathbf{M}_e$ of the element

$$\mathbf{K}_e = \int_{t_0}^{t_1} \int_V (\mathscr{D}\mathbf{N})^T\, \mathbf{E}\, \mathscr{D}\mathbf{N}\, \mathrm{d}V\, \mathrm{d}t, \tag{6.15}$$

$$\mathbf{M}_e = -\int_{t_0}^{t_1} \int_V \left(\frac{\partial N}{\partial t}\right)^T \mathbf{R} \frac{\partial N}{\partial t}\, \mathrm{d}V\, \mathrm{d}t, \tag{6.16}$$

where $\mathbf{E}$ is a matrix of elasticity, $\mathscr{D}$ is a matrix of differential operators, and $\mathbf{R}$ is the unitary matrix of inertia.

If we assume a viscoelastic body described by the Kelvin–Voigt model, where the relationship between the stress and the strain is written

$$\sigma = \left(\mathbf{E} + \eta_w \frac{\partial}{\partial t}\right) \varepsilon \tag{6.17}$$

(η_w—viscous damping coefficient) and if into the equation (6.10) we introduce a dissipative term, depending on the speed of displacement [17], then (6.14) takes the form

$$\mathbf{K}_e^* = \mathbf{K}_e + \mathbf{M}_e + \mathbf{W}_e + \mathbf{Z}_e, \tag{6.18}$$

where $\mathbf{W}_e$ and $\mathbf{Z}_e$ are the terms of internal damping and external damping, respectively:

$$\mathbf{W}_e = \int_{t_0}^{t_1} \int_V (\mathscr{D}\mathbf{N})^T \eta_w \frac{\partial}{\partial t} \mathscr{D}\mathbf{N} \, \mathrm{d}V \, \mathrm{d}t, \tag{6.19}$$

$$\mathbf{Z}_e = \int_{t_0}^{t_1} \int_V \mathbf{N}^T \eta_z \frac{\partial}{\partial t} \mathbf{N} \, \mathrm{d}V \, \mathrm{d}t. \tag{6.20}$$

$\mathbf{Q}_e$ is the vector of the external load acting on the space–time element e:

$$\mathbf{Q}_e = \int_{t_0}^{t_1} \int_V \mathbf{N}_e(\mathbf{x},t) \, \widehat{\mathbf{t}}(\mathbf{x},t) \, \mathrm{d}V \, \mathrm{d}t. \tag{6.21}$$

It can be obtained from (6.5). Since (6.13) must be satisfied with any variation of displacements and for the entire space–time area, we can write

$$\sum_{e=1}^{E} \left(\Pi_e^T \mathbf{K_e} \Pi_e \cdot \Pi_e^T \mathbf{q}_e - \Pi_e^T \mathbf{Q}_e\right) = \mathbf{0}. \tag{6.22}$$

The above system of algebraic equations includes the entire space–time domain $[0,T]$. The solution is obtained using the initial conditions (6.7) and (6.8). The displacement condition (6.7) can easily be discretized:

$$\mathbf{q}_0 = \sum_{e=1}^{E} \Pi_e^T \int_{V_e} \mathbf{N}_e(\mathbf{x},0) \, \mathbf{w}(\mathbf{x},0) \, \mathrm{d}V_e. \tag{6.23}$$

In turn, the velocity condition (6.8 requires the use of an additional difference formula to express velocities at the initial moment by means of two displacement vectors $\mathbf{q}_0$ given by the expression (6.23) and $\mathbf{q}_{-1}$ (when $t = -h$), for example, as follows:

$$\mathbf{q}_{-1} = \mathbf{q}_0 - \dot{\mathbf{q}}_0 h, \;\; \dot{\mathbf{q}}_0 = \sum_{e=1}^{E} \Pi_e^T \int_{V_e} \mathbf{N}_e(\mathbf{x},0) \, \dot{\mathbf{w}}(\mathbf{x},0) \, \mathrm{d}V_e. \tag{6.24}$$

Consider now one time layer, $t_i \le t \le t_{i+1}$. Divide the unknown displacements by the displacements with respect to time t_i and the displacements with respect to time t_{i+1}. Then the coefficients matrix of equation (6.22)) can be divided into four submatrices, separating a submatrix $\mathbf{A}_i$ with rows and columns relating to the time t_i, the submatrix $\mathbf{B}_i$ with rows relating to time t_i and columns relating to time t_{i+1} and similarly $\mathbf{C}_i$ and $\mathbf{D}_i$, but with rows relating to time t_{i+1}. We can therefore write

$$\begin{bmatrix} \mathbf{A}_i & \mathbf{B}_i \\ \mathbf{C}_i & \mathbf{D}_i \end{bmatrix} \begin{Bmatrix} \mathbf{q}_i \\ \mathbf{q}_{i+1} \end{Bmatrix} = \begin{Bmatrix} \mathbf{Q}_i \\ \mathbf{Q}_{i+1} \end{Bmatrix}. \tag{6.25}$$

Now the process of summation in (6.22) refers to the addition of the matrix (6.25). A tridiagonal block matrix of a system of equations is formed, with size corresponding to the number of moments into which the space–time region was partitioned. The following is the block form corresponding to (6.22)

$$\begin{bmatrix} \mathbf{A}_1 & \mathbf{B}_1 & & & & \\ \mathbf{C}_1 & \mathbf{D}_1+\mathbf{A}_2 & \mathbf{B}_2 & & & \mathbf{0} \\ & \mathbf{C}_2 & \mathbf{D}_2+\mathbf{A}_3 & \mathbf{B}_3 & & \\ & & \cdots & \cdots & \cdots & \\ & \mathbf{0} & & \mathbf{C}_{i-1} & \mathbf{D}_{i-1}+\mathbf{A}_i & \mathbf{B}_i \\ & & & & \cdots & \cdots\cdots \end{bmatrix} \begin{Bmatrix} \mathbf{q}_1 \\ \mathbf{q}_2 \\ \mathbf{q}_3 \\ \vdots \\ \mathbf{q}_{i-1} \\ \mathbf{q}_i \\ \mathbf{q}_{i+1} \\ \vdots \end{Bmatrix} = \begin{Bmatrix} \mathbf{Q}_1 \\ \mathbf{Q}_2 \\ \mathbf{Q}_3 \\ \vdots \\ \mathbf{Q}_{i-1} \\ \mathbf{Q}_i \\ \mathbf{Q}_{i+1} \\ \vdots \end{Bmatrix}. \tag{6.26}$$

The system (6.26) can be solved in stages, with one line of this system for each step. So we have to solve

$$\mathbf{C}_{i-1}\,\mathbf{q}_{i-1} + (\mathbf{D}_{i-1} + \mathbf{A}_i)\,\mathbf{q}_i + \mathbf{B}_i\,\mathbf{q}_{i+1} = \mathbf{Q}_i, \quad i = 1,2,\dots\,. \tag{6.27}$$

We can now return to the conditions (6.23) and (6.24) and to the solutions of the system (6.26) in the first steps. At $\mathbf{q}_0 \neq \mathbf{0}$ and $\dot{\mathbf{q}}_0 \neq \mathbf{0}$ we have:

$$\mathbf{C}_{-1}\mathbf{q}_{-1} + (\mathbf{D}_{-1} + \mathbf{A}_0)\,\mathbf{q}_0 + \mathbf{B}_0\,\mathbf{q}_1 = \mathbf{Q}_0, \;\; \mathbf{C}_{-1} = \mathbf{C}_0, \;\; \mathbf{D}_{-1} = \mathbf{D}_0. \tag{6.28}$$

The simplest case occurs when $\mathbf{q}_0 = \mathbf{0}$ and $\dot{\mathbf{q}}_0 = \mathbf{0}$. Then we have:

$$\mathbf{B}_1\,\mathbf{q}_2 = \mathbf{0}. \tag{6.29}$$

The only unknown displacement vector in (6.27) is $\mathbf{q}_{i+1}$. The vectors $\mathbf{q}_{i-1}$ and $\mathbf{q}_i$ were obtained in the previous steps, i.e., $i-2$ and $i-1$. An efficient construction of the system of equations and its solution can be carried out following the scheme below, using the auxiliary vector $\mathbf{r}$:

$$\begin{aligned} \mathbf{r} &= \mathbf{C}_{i-1}\,\mathbf{q}_{i-1} + \mathbf{D}_{i-1}\,\mathbf{q}_i\,, \\ \mathbf{A}_i\mathbf{q}_i + \mathbf{B}_i\,\mathbf{q}_{i+1} &= \mathbf{Q}_i - \mathbf{r}\,. \end{aligned} \tag{6.30}$$

Here we must mention [72], in which the author gives a way of constructing the so-called space–time macroelement. It consists of eliminating every second (or, more generally, every 2^n) layer of nodes, by removing every second row in the system (6.26). In this way, the time step increases to $2^n h$. This technique requires the inversion of a submatrix and is effective only in the case of constant matrices $\mathbf{A}_i$, ..., $\mathbf{D}_i$. par The stability of numerical schemes with a stationary mesh, based on the displacements, is described in [73, 88]. Depending on the selection of the virtual shape functions we can obtain unconditionally or conditionally stable variants of the method. These considerations can be summarized by saying that the formulations in the papers cited above are equivalent to Newmark methods, and have all their properties, including those concerning stability.

6.1.1 Space–Time Finite Elements in the Displacement Description

6.1.1.1 Bar Element

We will deal with the derivation of the fundamental equations of the equilibrium of forces in the space–time element method with a stationary spatial mesh. This issue will be treated for the example of the simplest equation describing the axial vibration of a bar. The differential equation of an axially vibrating bar is identical to the equation of a vibrating string

$$\rho A \frac{\partial^2 w(x,t)}{\partial t^2} = EA \frac{\partial^2 w(x,t)}{\partial x^2} + p(x,t)\,. \tag{6.31}$$

Here, E is the elastic modulus, A—the cross sectional area, ρ—mass density, and p—external load. Multiplying this equation by a virtual displacement function $w^*(x,t)$ and integrating over the present area $x \in [0,b]$ and $0 \le t \le h$ we get the equation of virtual work

$$\int_0^h \int_0^b w^*(x,t)\rho A \frac{\partial^2 w(x,t)}{\partial t^2} \mathrm{d}x\mathrm{d}t = \int_0^h \int_0^b w^*(x,t) EA \frac{\partial^2 w(x,t)}{\partial x^2} \mathrm{d}x\mathrm{d}t + \\ + \int_0^h \int_0^b w^*(x,t) p(x,t) \mathrm{d}x\mathrm{d}t\,. \tag{6.32}$$

We lower the degree of the derivative by integrating the first term by parts with respect to t and the second term with respect to x. We recall that w^* has equals zero at the points $x = 0$ and $x = b$ and also $t = 0$ and $t = h$. We do not need to do so, if at a later stage we can provide a non-zero second derivative of the solutions, and in practice a sufficiently high degree polynomial for approximation of the solutions. Then we can write

$$\int_0^h \int_0^b \frac{\partial w^*(x,t)}{\partial x} EA \frac{\partial w(x,t)}{\partial x} \mathrm{d}x\mathrm{d}t - \int_0^h \int_0^b \frac{\partial w^*(x,t)}{\partial t} \rho A \frac{\partial w(x,t)}{\partial t} \mathrm{d}x\mathrm{d}t = \\ = \int_0^h \int_0^b w^*(x,t) p(x,t) \mathrm{d}x\mathrm{d}t\,. \tag{6.33}$$

Now we suggest a way of interpolating the values of $w(x,t)$ and its virtual equivalent, $w^*(x,t)$, based on the nodal values of $\mathbf{q}_e$ in the element. Take a linear interpolation in the form

$$w(x,t) = [N_1(x,t), N_2(x,t), N_3(x,t), N_4(x,t)] \begin{Bmatrix} q_1 \\ \vdots \\ q_4 \end{Bmatrix} = \mathbf{N}\mathbf{q}_e \,. \tag{6.34}$$

The real shape functions can be written in local coordinates (ξ,τ). Then $N_i = 1/4\,(1+\xi\,\xi_i)(1+\tau\,\tau_i)$, where ξ_i, τ_i are the coordinates of the vertices of the rectangle $\{\xi,\tau : -1 \le \xi \le 1, -1 \le \tau \le 1\}$. The transition between a local and global coordinate system is defined by $x = \sum_{i=1}^{4} N_i x_i$ and $t = \sum_{i=1}^{4} N_i t_i$. Then we assume virtual shape functions in the same way. After the substitution of w and w^* into (6.33), we get the total energy of the system, comprising terms describing the internal energy, the kinetic energy, and the work of the external forces on the virtual displacements. The energy minimization condition reduces to zero the derivatives with respect to all unknowns q_i, $i = 1,\ldots,4$. In the end, we obtain a system of four equations

$$\left(\frac{EAh}{6b} \begin{bmatrix} 2 & -2 & 1 & -1 \\ -2 & 2 & -1 & 1 \\ 1 & -1 & 2 & -2 \\ -1 & 1 & -2 & 2 \end{bmatrix} - \frac{\rho ab}{6h} \begin{bmatrix} 2 & 1 & -2 & -1 \\ 1 & 2 & -1 & 2 \\ -2 & -1 & 2 & 1 \\ -1 & -2 & 1 & 2 \end{bmatrix} \right) \begin{Bmatrix} q_1 \\ q_2 \\ q_3 \\ q_4 \end{Bmatrix} = \\ = \begin{Bmatrix} F_1 \\ F_2 \\ F_3 \\ F_4 \end{Bmatrix} . \tag{6.35}$$

The first matrix is the stiffness matrix, with coefficients

$$K_{ij} = \frac{EAh}{12b} \xi_i \xi_j (3 + \tau_i \tau_j) \,, \tag{6.36}$$

and the second matrix is the inertia matrix, with coefficients

$$M_{ij} = -\frac{\rho Ab}{12h} \tau_i \tau_j (3 + \xi_i \xi_j) \,. \tag{6.37}$$

By introducing the notation $k = c^2h^2/b^2$ and $c^2 = E/\rho$, we can write the symmetric space–time stiffness matrix $\mathbf{K}_e$ contained in the equation (6.35) in parentheses

$$\mathbf{K}_e = \frac{\rho ab}{6h} \begin{bmatrix} 2k-2 & -2k-1 & k+2 & -k+1 \\ & 2k-2 & -k+1 & k+2 \\ & & 2k-2 & -2k-1 \\ & & & 2k-2 \end{bmatrix} . \tag{6.38}$$

6.1.1.2 Beam Element

Consider a rectangular beam element of medium thickness. Nodes may be in the general case shifted in space within time interval. One case is when the element is a rectangle in space.

Assume a linear displacement

$$\begin{Bmatrix} w \\ \theta \end{Bmatrix} = \begin{Bmatrix} a_1xt + a_2x + a_3t + a_4 \\ b_1xt + b_2x + b_3t + b_4 \end{Bmatrix} = \begin{Bmatrix} \mathbf{ag} \\ \mathbf{bg} \end{Bmatrix}, \tag{6.39}$$

where $\mathbf{g}(x,t) = [xt, x, t, 1]$ is a vector of monomials. If we denote by $\mathbf{G}$ the matrix composed of the vectors $\mathbf{g}(x_i,t_i)$ defined at element nodes, then we can extract from the inverse matrix $\mathbf{G}^{-1}$ the columns marked by $\mathbf{r}_i$:

$$\mathbf{G}^{-1} = \begin{bmatrix} \mathbf{g}(x_1,t_1) \\ \mathbf{g}(x_2,t_2) \\ \mathbf{g}(x_3,t_3) \\ \mathbf{g}(x_4,t_4) \end{bmatrix}^{-1} = \begin{bmatrix} x_1t_1 & x_1 & t_1 & 1 \\ x_2t_2 & x_2 & t_2 & 1 \\ x_3t_3 & x_3 & t_3 & 1 \\ x_4t_4 & x_4 & t_4 & 1 \end{bmatrix}^{-1} = [\mathbf{r}_1, \mathbf{r}_2, \mathbf{r}_3, \mathbf{r}_4] . \tag{6.40}$$

We obtain the shape functions $\mathbf{N}$ as

$$\mathbf{N} = [\mathbf{N}_1, \mathbf{N}_2, \mathbf{N}_3, \mathbf{N}_4] \ , \ \ \mathbf{N}_i = \mathbf{gr}_i \begin{bmatrix} 1 & 0 \\ 0 & 1 \end{bmatrix} . \tag{6.41}$$

The differential operator $\mathscr{D}$ describes deformations depending on displacements

$$\mathscr{D} = \begin{bmatrix} \frac{\partial}{\partial x} & 1 \\ 0 & \frac{\partial}{\partial x} \end{bmatrix} , \ \ \varepsilon = \mathscr{D}\mathbf{Nq} . \tag{6.42}$$

The corresponding matrix of elasticity, in turn, is

$$\mathbf{E} = \begin{bmatrix} \frac{GA}{K} & 0 \\ 0 & EI \end{bmatrix} . \tag{6.43}$$

In the particular case where the space–time element is a rectangle, the shape functions are simpler:

$$\mathbf{N} = \left[\left(1 - \frac{x}{b}\right)\left(1 - \frac{t}{h}\right), \frac{x}{b}\left(1 - \frac{t}{h}\right), \left(1 - \frac{x}{b}\right)\frac{t}{h}, \frac{x}{b}\frac{t}{h} \right] . \tag{6.44}$$

The stiffness matrix $\mathbf{K}$ and the inertia matrix $\mathbf{M}$ are determined as integrals over the element Ω (see (6.15), (6.16), (6.19), and (6.20))

$$\mathbf{K} = \int_{\Omega} (\mathscr{D}\mathbf{N})^T \mathbf{E} \mathscr{D}\mathbf{N} \mathrm{d}\Omega , \tag{6.45}$$

$$\mathbf{M} = \int_{\Omega} \left(\frac{\partial}{\partial t}\mathbf{N}\right)^T \mathbf{E} \frac{\partial}{\partial t}\mathbf{N} \mathrm{d}\Omega . \tag{6.46}$$

6.2 Properties of the Integration Schemes

The classic methods of the analysis of the vibrations of structures usually use one and the same method of integration for the whole structure and all its degrees of freedom. Often, some sub-areas, or some kinds of degrees of freedom do not require a high accuracy in the calculations and a small step-size for the integration. An example might be the axial movement of frames of tall buildings. From the engineering point of view, it is important to analyse the flexural motion of the rod components. In this direction, the stiffness of the structural elements is much lower than the axial stiffness. The axial vibrations of bars have minimal amplitudes and less practical importance. Therefore, we can analyse them with less accuracy. The same occurs for structures placed on the ground, for the interaction between the structure and the ground or with a surrounding fluid. The inertia and stiffness of the two subdomains differ significantly. A short time step in the explicit procedures, chosen accouding to the criterion of stability or accuracy, greatly lengthens the calculations. A reasonable procedure would be to use explicit methods for modelling the ground and implicit methods for modelling the structure. The application of mixed operators to integration opens new opportunities for improving the accuracy efficiency of the solving of complex tasks. A description of the issues concerning both the statics and dynamics are presented, for example, in [109]. Techniques based on the algebraic partition of the operator are called operator splitting techniques and based on the structure partition – domain splitting.

Operator splitting methods and semi-implicit methods can be classified as being part of the same group of methods. A characteristic feature here is a willingness to preserve the unconditional stability of the integration, or a desire to achieve a significant lengthening of the critical step of integration. Such an increase in the integration step can be achieved by, for example, macro-elements, where in the description of consecutive time layers, intermediate algebraic degrees of freedom are eliminated. In this way, the time span of the integration step increases. The same technique applies to spatial variables.

6.2.0.3 Information Flow

Calculation methods may differ in another important feature: the information flow diagram in a mesh in space and time. Consider, for example, the system of algebraic equations arising (at every step of the calculation) from the method of central differences with a diagonal inertia matrix. An external impulse acting as a load on the last degree of freedom in the mesh produces a non-zero displacement of this degree of freedom (Figure 6.6). In the next step, the product of the matrix $2\mathbf{M} - h^2\mathbf{K}$ with the vector of displacements $\mathbf{q}_i$ results in a vector with nonzero last two entries. This in turn gives the nonzero last two entries of the solution vector $\mathbf{q}_{i+1}$. In this way, in each successive step in the calculation, the influence of the initial single pulse of the external force $\mathbf{F}_0$ propagates to the neighbouring nodes in the finite element mesh at a rate of one node in one step. The speed of the information flow is thus the average of $\Delta x/\Delta t$, where Δx is the spatial distance between neighbouring nodes.

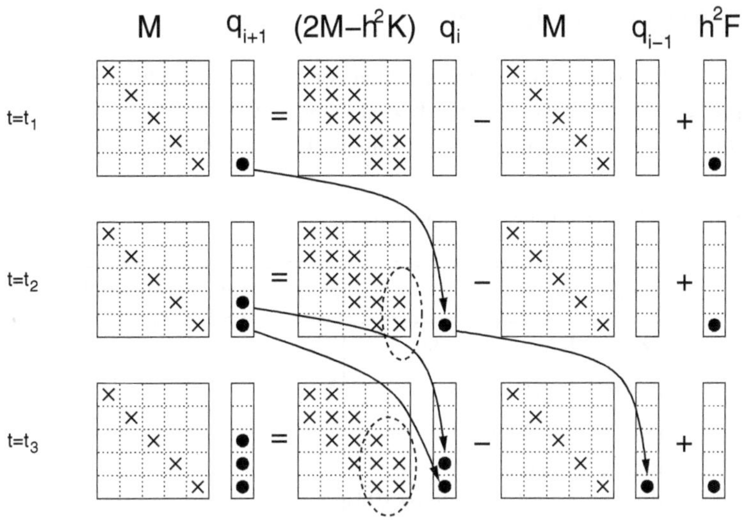

Fig. 6.6 Mechanism for information flow between successive steps in the central difference method.

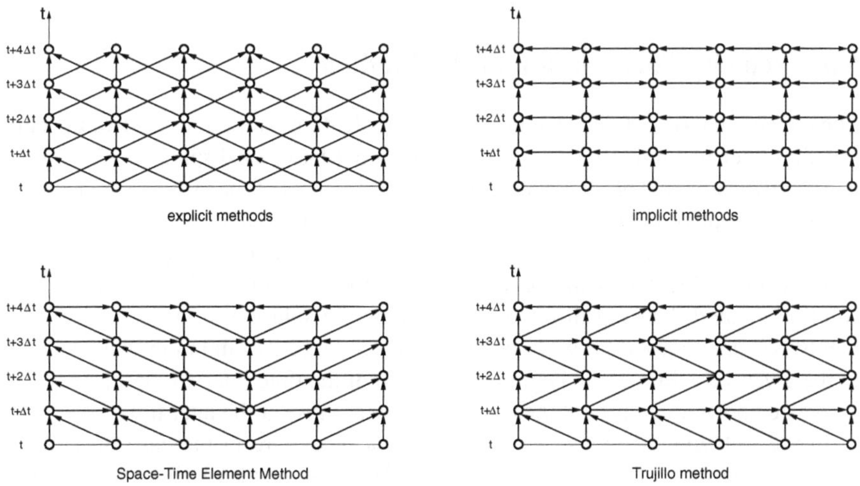

Fig. 6.7 Flow of information in various calculation methods.

In the case of implicit methods we deal with the solution of algebraic equations with a full matrix of coefficients. A single external impulse transfers immediately to all unknowns of the system. In turn, as will be discussed in subsequent chapters, the space–time finite element method with simplicially shaped elements has a reduced speed of information flow to the sloping sides of the space–time mesh. The Trujillo method has a limited speed of information flow among the grid nodes. A schematic of the flow of information in selected calculation methods is given in Figure 6.7.

6.2.1 Accuracy of Methods

In Table 6.1, the global error of the selected computational methods is estimated. In principle, all methods have a global error of the second order. The estimates included in Chapters 6.3.4 and 6.4.1, especially Equation (6.103), are relevant here.

Table 6.1 Global error of the calculation methods.

method	error degree	coefficient
Runge–Kutta 2nd order	h^2	1/6
central difference m.	h^2	1/6
space–time elem. m.	h^2	1/12
space–time elem. m. higher order formulae (6.103)	h^3	1/12
Newmark m. β=1/4	h^2	$(\sqrt{2}-1)/2$
Runge–Kutta 4th order	h^4	1/120

6.3 Velocity Formulation of the Method

The method in terms of displacements has many features similar to those of the central difference method. The schematic calculation is based on displacements in three consecutive moments of time. Depending on the features of the virtual displacement functions adopted, the method in its displacement variant may be conditionally or unconditionally stable. The solution scheme for the equation of motion can be obtained using velocities as the basic quantities for description of the system. This solution method does not differ from obtaining the solution in terms of displacements. We create a virtual work equation, which takes its minimum value subject to the given velocities at the beginning and end of the time interval. In this way, we associate known quantities at the earlier time with those at the later time.

In the following sections, we will present a procedure in the case of a single degree of freedom. Then we will repeat it for the case of a string. We will use a variety of shapes for the velocity virtual functions. We will compare the schemes obtained from the Newmark time integration method with those from other methods of calculation.

6.3.1 One Degree of Freedom System

Let us elaborate a solution scheme for the vibrations of a material point described by

$$m\frac{\mathrm{d}v}{\mathrm{d}t}+kx=0\,. \tag{6.47}$$

We assume a linear distribution of the real velocity v in the time interval h $(0 \leq t \leq h)$

$$v = \left(1 - \frac{t}{h}\right) v_0 + \frac{t}{h} v_1 \ . \tag{6.48}$$

The displacement $x(t)$ is described by the integral

$$x(t) = \int_0^t v\,\mathrm{d}t = x_0 + \frac{h}{2}\left[1 - \left(1 - \frac{t}{h}\right)^2\right] v_0 + \frac{t^2}{2h} v_1 \ . \tag{6.49}$$

Displacement $x(t)$ is linearly dependent on the speed v_0 and v_1 set at the ends of the interval $[0,h]$. Now we take the form of the virtual function. We can choose it among many possible functions. The selected function at the ends of the time interval must have value equal to zero. Below we take a virtual velocity function as Dirac distribution, depending on parameters α $(0 \leq \alpha \leq 1)$ and speed v_1:

$$v^* = v_1 \cdot \delta\left(\frac{t}{h} - \alpha\right) \ . \tag{6.50}$$

If we multiply the differential equation of motion (6.47), which in fact is the equation of equilibrium of forces acting on a material point, by the virtual velocity (6.50), we get the equation of virtual power. After integration it in time interval $[0,h]$ we obtain the equation of virtual work

$$\int_0^h v^* \frac{1}{h}(v_1 - v_0)\,\mathrm{d}t + \int_0^h v^* \frac{k}{m} x(t)\mathrm{d}t = 0 \ . \tag{6.51}$$

Finally we get the equation which allows us to determine the speed at the next while, based on the velocity and displacement at the previous while

$$v_1 = \frac{1 - \frac{kh^2}{2m}[1 - (1-\alpha)^2]}{1 + \frac{k\alpha^2 h^2}{2m}} v_0 - \frac{k}{m} \frac{h}{\left(1 + \frac{k\alpha^2 h^2}{2m}\right)} x_0 \ . \tag{6.52}$$

The same can be written symbolically

$$v_1 = T\,v_0 + B\,x_0 \ . \tag{6.53}$$

It remains to determine from the velocity v_0 and v_1 the missing displacement x_1. We use the relationship

$$x_1 = x_0 + h[(1-\beta)v_0 + \beta v_1] \ . \tag{6.54}$$

A little further on we shall show that the stable solution is obtained in a certain range of parameter α at $\beta = 1 - \alpha$. Taking it into account we can write the final formula

$$x_1 = x_0 + h[\alpha v_0 + (1-\alpha)v_1] \ . \tag{6.55}$$

In the particular case $\alpha = 1/2$ the equation (6.55) is identical to the equation (6.49) adopted at $t = h$, i.e., $x_1 = x_0 + h(v_0 + v_1)/2$.

Introducing the notation $\kappa = h^2k/m$ we can describe the transition to the next moment as follows:

$$\left\{\begin{matrix} v_1 \\ x_1 \end{matrix}\right\} = \begin{bmatrix} 1-\frac{2\alpha\kappa}{2+\alpha^2\kappa} & -\frac{2\kappa}{h(\alpha^2\kappa+2)} \\ 3h-\frac{2h(\alpha\kappa+2)}{\alpha^2\kappa+2} & \frac{2\kappa(\alpha-1)}{\alpha^2\kappa+2}+1 \end{bmatrix} \left\{\begin{matrix} v_0 \\ x_0 \end{matrix}\right\} . \tag{6.56}$$

The transition matrix $\mathbf{T}$ is a 2×2 matrix. We can use it to determine the stability criterion for any value of the time step, and so as $h \to \infty$. Let us determine the eigenvalues of this matrix:

$$\lim_{h\to\infty} \lambda_{1/2} = \frac{\alpha^2-1}{\alpha^2} \pm \frac{i\sqrt{2\alpha^2-1}}{\alpha^2} , \tag{6.57}$$

and their moduli:

$$\lim_{h\to\infty} |\lambda_{1/2}| = \begin{cases} 1, & \text{if } \sqrt{2}/2 \le \alpha \le 1 , \\ \frac{1}{\alpha^2}\sqrt{\alpha^4-4\alpha^2+2}, & \text{if } 0 \le \alpha < \sqrt{2}/2 . \end{cases} \tag{6.58}$$

Both eigenvalues have modulus unity when $\alpha \ge \sqrt{2}/2$. In this range of the parameter α, we obtain an unconditionally stable solution scheme. This important feature allows us to safely carry out the calculations for systems with a large number of degrees of freedom, or for systems with evolving material properties and geometry. In such cases, the unconditional stability of the method is necessary.

The ability to damp higher frequency vibrations while leaving the basic frequencies undamped is also important. Many authors have devoted papers to this problem (for example [64, 67]). If we modify the formula (6.54) so that the speed will be determined at a slightly later time than $\alpha-1$, we get this damping effect. The parameter β should be modified to

$$\beta = 1 - \frac{\alpha}{1+\gamma}, \qquad 0 \le \gamma \le 1. \tag{6.59}$$

Figure 6.8 shows the value of the spectral radius in its dependence on γ with increasing values of relative the time step h/T. They are made with $\alpha = 0.8$ and $\alpha = 0.9$. The bold lines are the levels plotted for each increment of 0.02. With $\gamma = 0.0$, the spectral radius ρ is unity, by the previous considerations. The next Figure 6.9 shows the logarithmic damping decrement at $\alpha = 0.8$ and 0.9, depending on the size of the relative time step h/T. The levels of both figures are plotted for each increment of 0.05.

Sample calculations of the vibrations of a single point, with the initial conditions $x_0 = 0$ and $v_0 = 1$ and at $\alpha = 0.5$ are shown in Figure 6.10, and at $\alpha = 1.0$ in Figure 6.11.

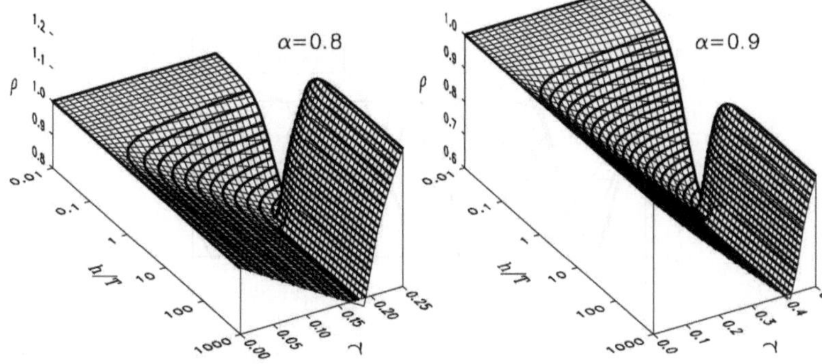

Fig. 6.8 Spectral radius ρ depending on the parameter γ for $\alpha = 0.8$ and $\alpha = 0.9$.

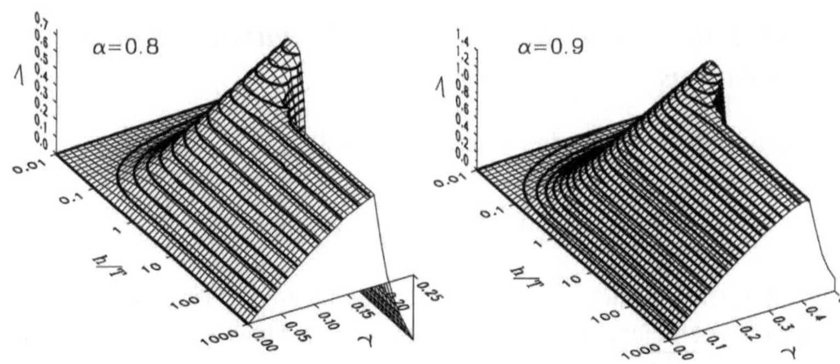

Fig. 6.9 Logarithmic damping decrement Λ depending on the parameter γ at $\alpha = 0.8$ and $\alpha = 0.9$.

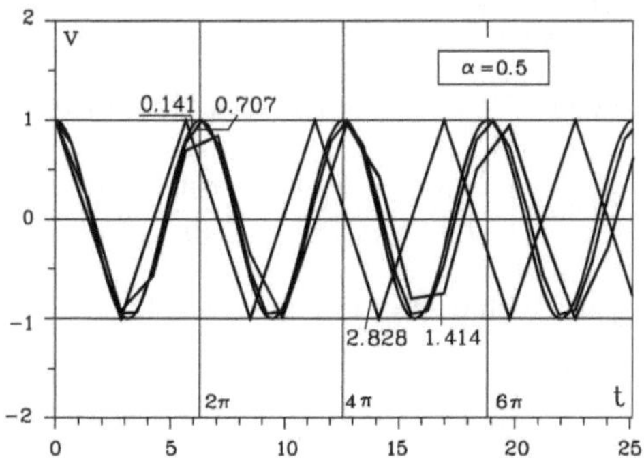

Fig. 6.10 Velocity v determined with various time step, at $\alpha = 0.5$.

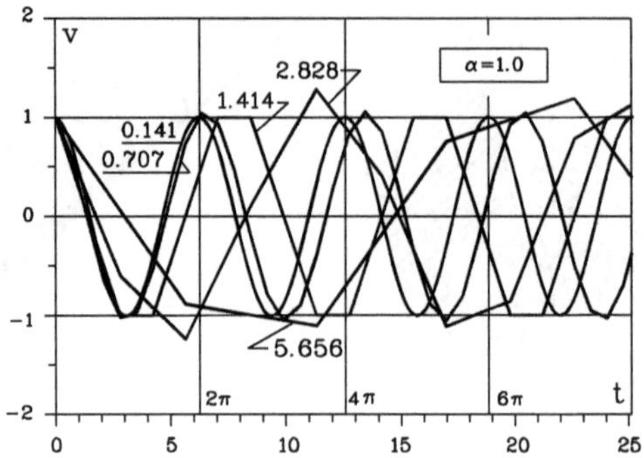

Fig. 6.11 Velocity v determined with various time step, at $\alpha = 1.0$.

6.3.2 Discretization of the Differential Equation of String Vibrations

The differential equation of string vibration is as follows:

$$N\frac{\partial^2 w}{\partial x^2} - \rho A\frac{\partial^2 w}{\partial t^2} - \eta\frac{\partial w}{\partial t} - q(x,t) = 0\,. \tag{6.60}$$

Here, N is the string tension, A is the cross sectional area, ρ is the mass density, η is the damping coefficient, and q is the external load. The wave speed in the string at $\eta = 0$ is $c = \sqrt{N/\rho}$.

Consider the equation in the region $\Omega = \{(x,t)\colon 0 \le x \le b,\ 0 \le t \le h\}$. The virtual power equation is created by multiplying (6.60) by the virtual speed $v^*(x,t)$

$$\int_0^b v^*(x,t)\left(\frac{\partial^2 w}{\partial x^2} - \frac{1}{c^2}\frac{\partial^2 w}{\partial t^2} - \eta\frac{\partial w}{\partial t} + \tilde{q}(x,t)\right)\mathrm{d}x = 0\,. \tag{6.61}$$

$\tilde{q}$ denotes the load scaled by N: $\tilde{q} = q/N$. The total virtual work in the area Ω is

$$\int_0^h\int_0^b v^*(x,t)\left(\frac{\partial^2 w}{\partial x^2} - \frac{1}{c^2}\frac{\partial^2 w}{\partial t^2} - \eta\frac{\partial w}{\partial t} + \tilde{q}\right)\mathrm{d}x\,\mathrm{d}t = 0\,. \tag{6.62}$$

Integrating (6.62) by parts with respect to x and t we obtain

$$\iint_\Omega v^*\frac{\partial v}{\partial t}\mathrm{d}\Omega + \iint_\Omega \frac{\partial v^*}{\partial x}\frac{\partial w}{\partial x}\mathrm{d}\Omega + \iint_\Omega \frac{\partial v^*}{\partial x}\varepsilon_0\,\mathrm{d}\Omega - \eta\iint_\Omega v^* v\,\mathrm{d}\Omega = 0\,. \tag{6.63}$$

Here, ε_0 is the initial strain.

We assume a linear variation in velocity $v = \partial w/\partial t$ in terms of x and t:

$$v(x,t) = \sum_{i=1}^{4} N_i(x,t)\, v_i \,. \tag{6.64}$$

In the area of Ω, the shape functions $\mathbf{N} = [N_1, \ldots, N_4]$ have the form

$$\begin{aligned} N_1 &= \frac{1}{bh}(x-b)(t-h) \,, \\ N_2 &= -\frac{1}{bh}x(t-h) \,, \\ N_3 &= -\frac{1}{bh}(x-b)t \,, \\ N_4 &= \frac{1}{bh}xt \,. \end{aligned} \tag{6.65}$$

The displacements are determined by integration:

$$w(x,t) = w(x,0) + \int_0^t (N_1 v_1 + \ldots + N_4 v_4)\,\mathrm{d}t \,. \tag{6.66}$$

The result is

$$w(x,t) = w(x,0) + \frac{xt^2}{2bh}(v_1 - v_2 - v_3 + v_4) + \frac{xt}{b}(-v_1 + v_2) + \frac{t^2}{2h}(-v_1 + v_3) + v_1 t \,. \tag{6.67}$$

The derivative $\partial w/\partial x$ is obtained from (6.67):

$$\frac{\partial w}{\partial x} = \frac{t^2}{2bh}(v_1 - v_2 - v_3 + v_4) + \frac{t}{b}(-v_1 + v_2) + \left.\frac{\mathrm{d}w}{\mathrm{d}x}\right|_{t=0} \,, \tag{6.68}$$

where $\varepsilon_0 = \mathrm{d}w/\mathrm{d}x|_{t=0}$. It is essential to select the appropriate virtual velocity function. Following the derivation for the oscillator, we assume

$$v^*(x,t) = \delta(t - \alpha h)\left(\left(1 - \frac{x}{b}\right)v_3 + \frac{x}{b}v_4\right) \,. \tag{6.69}$$

Here, δ is the Dirac delta function. The required derivatives of the virtual function v^* and real function v are obtained from (6.69) and (6.64)

$$\frac{\partial v^*}{\partial x} = \frac{1}{b}(-v_3 + v_4) \,, \tag{6.70}$$

$$\frac{\partial v}{\partial t} = \frac{x}{bh}(v_1 - v_2 - v_3 + v_4) + \frac{1}{h}(-v_1 + v_3) \,. \tag{6.71}$$

We note that due to the Dirac delta function part of the integrand, the integration over Ω is, in terms of x, reduced to an integration over the interval $[0,\, b]$. Taking into account the above relationships, the final form of the equation (6.63) can be written in matrix form:

$$\left\{\rho A \int_0^b \begin{bmatrix} 0 \\ 0 \\ -(\frac{x}{b}-1) \\ \frac{x}{b} \end{bmatrix} \left[\frac{x}{bh}-\frac{1}{h},\ -\frac{x}{bh},\ -\frac{x}{bh}+\frac{1}{h},\ \frac{x}{bh}\right] \mathrm{d}x + \right.$$

$$+N \int_0^b \begin{bmatrix} 0 \\ 0 \\ -\frac{1}{b} \\ \frac{1}{b} \end{bmatrix} \left[\frac{t^2}{2bh}-\frac{t}{b},\ -\frac{t^2}{2bh}+\frac{t}{b},\ -\frac{t^2}{2bh},\ \frac{t^2}{2bh}\right] \mathrm{d}x \bigg|_{t=\alpha h} -$$

$$-\eta \int_0^b \begin{bmatrix} 0 \\ 0 \\ -(\frac{x}{b}-1) \\ \frac{x}{b} \end{bmatrix} \left[\frac{(x-b)(t-h)}{bh},\ -\frac{x(t-h)}{bh},\ -\frac{(x-b)t}{bh},\ \frac{xt}{bh}\right] \mathrm{d}x \bigg|_{t=\alpha h} + \tag{6.72}$$

$$\left. + N\,\varepsilon_0 \int_0^b \begin{bmatrix} 0 \\ 0 \\ -\frac{1}{b} \\ \frac{1}{b} \end{bmatrix} \mathrm{d}x \right\} \cdot \begin{Bmatrix} v_1 \\ \vdots \\ v_4 \end{Bmatrix} = \mathbf{Q}\,.$$

We see that the first two rows of all the matrix products are zero. From now on, we will operate only with the lower half of the matrix obtained after integration. The resulting matrices take the following form:

$$\mathbf{M} = \frac{\rho A b}{h} \left[\begin{array}{cc|cc} -\frac{1}{3} & -\frac{1}{6} & \frac{1}{3} & \frac{1}{6} \\ -\frac{1}{6} & -\frac{1}{3} & \frac{1}{6} & \frac{1}{3} \end{array}\right] = \frac{1}{h}\left[-\mathbf{M}_{stat} \mid \mathbf{M}_{stat}\right], \tag{6.73}$$

$$\mathbf{K} = \frac{Nh}{b} \left[\begin{array}{cc|cc} \alpha(1-\frac{\alpha}{2}) & -\alpha(1-\frac{\alpha}{2}) & \frac{\alpha^2}{2} & -\frac{\alpha^2}{2} \\ -\alpha(1-\frac{\alpha}{2}) & \alpha(1-\frac{\alpha}{2}) & -\frac{\alpha^2}{2} & \frac{\alpha^2}{2} \end{array}\right] =$$

$$= h\left[\alpha\left(1-\frac{\alpha}{2}\right)\mathbf{K}_{stat} \mid \frac{\alpha^2}{2}\mathbf{K}_{stat}\right], \tag{6.74}$$

$$\mathbf{C} = \eta b \left[\begin{array}{cc|cc} \frac{1-\alpha}{3} & \frac{1-\alpha}{6} & \frac{\alpha}{3} & \frac{\alpha}{6} \\ \frac{1-\alpha}{6} & \frac{1-\alpha}{3} & \frac{\alpha}{6} & \frac{\alpha}{3} \end{array}\right] = \left[(1-\alpha)\mathbf{C}_{stat} \mid \alpha\mathbf{C}_{stat}\right], \tag{6.75}$$

$$\mathbf{e} = N\varepsilon_0 \begin{Bmatrix} -1 \\ 1 \end{Bmatrix}. \tag{6.76}$$

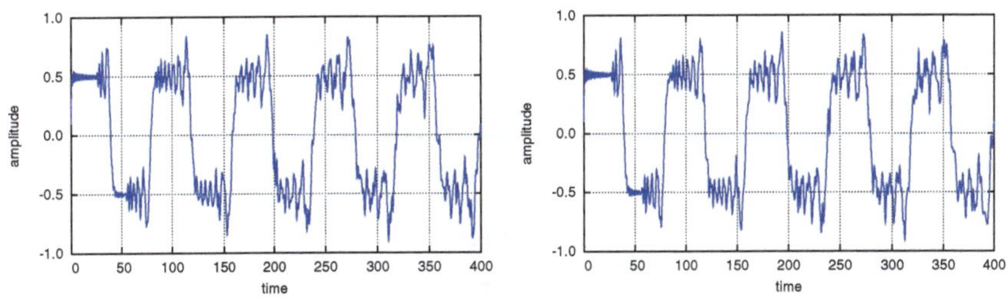

Fig. 6.12 Displacements of the free end of the bar, obtained by the space–time elements of the full matrix of mass coefficients at $\alpha = 0.5$ and $\alpha = 1.0$ at $\alpha = 0.5$ and $\alpha = 1.0$.

The final form of the equations of motion describes the balance of forces on the boundary of the domain Ω

$$(\mathbf{M}+\mathbf{C}+\mathbf{K})\left\{\begin{array}{l}\dot{\mathbf{q}}_a\\ \dot{\mathbf{q}}_p\end{array}\right\}+\mathbf{e}=\mathbf{Q}\,. \tag{6.77}$$

The nodal velocity vector $\dot{\mathbf{q}}$ contains the velocities $\dot{\mathbf{q}}_a$ at the initial time $t = 0$ and $\dot{\mathbf{q}}_p$, those at the final time $t = h$. We get a matrix equation in which $\dot{\mathbf{q}}_p$ is the unknown. Figure 6.12 shows a graph of the axial displacement plotted over time at the free end of the rod, forced with the initial pulse. Figure 6.13 shows the result of an identical task, but using a diagonal inertia matrix.

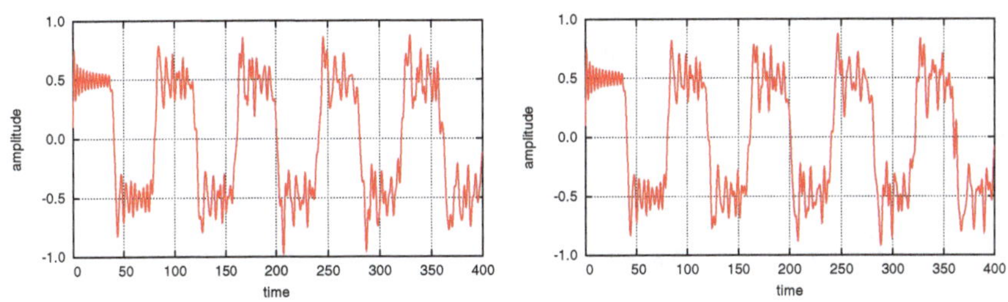

Fig. 6.13 Displacements of the free end of the bar obtained by the space–time element method with a diagonal mass matrix, for $\alpha = 0.5$ and $\alpha = 1.0$.

6.3.2.1 Example of a Vibrating String Loaded with a Moving Force

The diagram of the task (Figure 6.14) and the results for comparison are taken from [31]. The authors performed calculations using the methods of Fourier and of d'Alembert. Below, the same problem is solved by the with the velocity formulation of the space–time finite element method.

The following values were assumed: length of string $l = 100$ m, tension $T_0 = 10$ kN, mass density $\rho A = 0.89$ kg/m, point force $P_0 = 90$ N, concentrated mass $m = 10$ kg, damping $c_1 = c_2 = 0$. A conditionally stable variant of the method, with parameter $\alpha = 0.5$ and numerical damping $\gamma = 0.1$, was assumed.

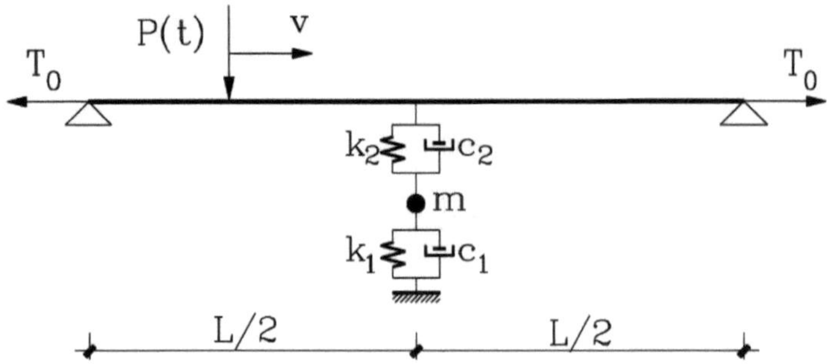

Fig. 6.14 String loaded by a moving force.

In the first case we assumed $m = c_1 = c_2 = 0$. The force $P(t) = P_0 \sin \omega t$ was harmonic with a frequency of $\omega = 0.4$ (this corresponds to 40 cycles over 200 m). The velocity v was 79.5 m/s (corresponding to $0.75c$). Figure 6.15 shows the lateral displacement at selected moments. The vertical lines on the first three graphs mark the position of the force. The thick line shows the results obtained by the space–time element method, and the thin lines those by the Fourier method. We can see a very good agreement of both solutions.

The second example is more complex. In the mid-span it includes a support consisting of two springs, k_1, k_2, and a mass m, which are non-zero. The force is constant and moves with the speed v = 79.5 m/s. Figure 6.16 presents the results, a thick vertical bar indicates a support. The solutions obtained using the space–time element method and the Fourier method coincide. In the cited paper [31], the results obtained by qualitatively different wave methods differ from both the results shown in Figure 6.16.

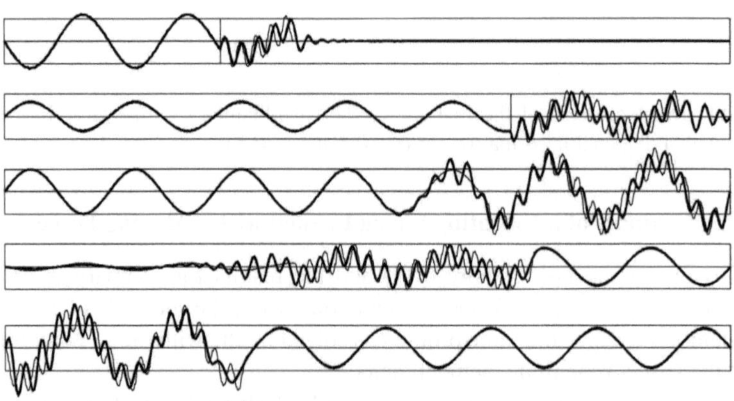

Fig. 6.15 Lateral displacements of the string at moments equal to 0.3, 0.7, 1.0, 1.2, and 1.5 of the total time of passage of the load (thin line: Fourier method, thick line: space–time element method).

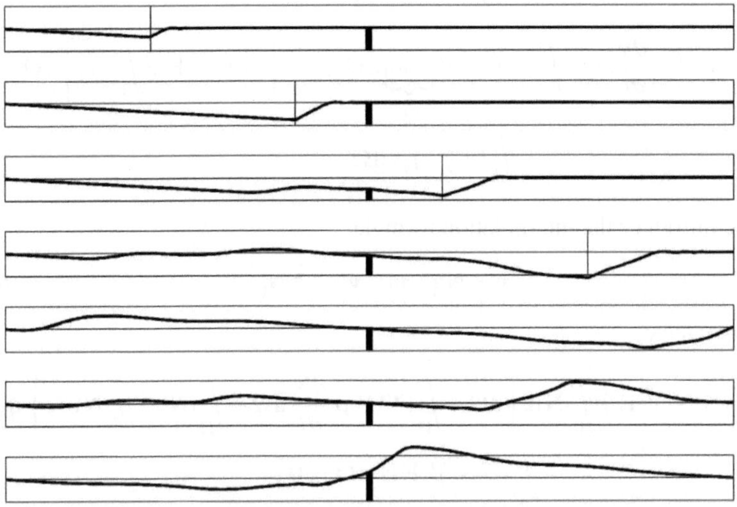

Fig. 6.16 Lateral displacements of the string at each 0.2 of the total time of passage of the load (thin line: Fourier method [32], thick line: space–time element method).

6.3.3 General Case of Elasticity

We now turn to a more general approach that allows the discretization of any dynamics problem of a continuous system.

If we define the deformations ε by

$$\varepsilon = \mathscr{D}\mathbf{w} , \tag{6.78}$$

where $\mathscr{D}$ is a differential operator, and the tension σ by

$$\sigma = \mathbf{E}\varepsilon , \tag{6.79}$$

and if we assume a certain distribution of virtual velocity $\mathbf{v}^*$, the equation of virtual work, expressed in terms of velocities, takes the following form

$$\int_\Omega (\mathbf{v}^*)^T \rho \frac{\partial \mathbf{v}}{\partial t} \mathrm{d}\Omega + \int_\Omega (\dot{\varepsilon}^*)^T \sigma \mathrm{d}\Omega + \int_\Omega (\mathbf{v}^*)^T \eta_z \mathbf{v} \mathrm{d}\Omega = 0 . \tag{6.80}$$

The inertia of the system is given by the matrix ρ and the damping is described by η. The displacements $\mathbf{w}(t)$ are given by the integral

$$\mathbf{w}(t) = \mathbf{w}_0 + \int_0^t \mathbf{v}\, \mathrm{d}t . \tag{6.81}$$

After taking into account (6.78), (6.79), and (6.81), we get

$$\int_{\Omega} (\mathbf{v}^*)^T \rho \frac{\partial \mathbf{v}}{\partial t} \mathrm{d}\Omega + \int_{\Omega} (\mathscr{D}\mathbf{v}^*)^T \mathbf{E} \underbrace{\mathscr{D}\mathbf{w}_0}_{\varepsilon_0} \mathrm{d}\Omega + \int_{\Omega} \left[(\mathscr{D}\mathbf{v}^*)^T \mathbf{E} \mathscr{D} \int_0^t \mathbf{v}\, \mathrm{d}t \right] \mathrm{d}\Omega + \\ + \int_{\Omega} (\mathbf{v}^*)^T \eta_z \mathbf{v}\, \mathrm{d}\Omega = 0 \,. \tag{6.82}$$

Now we introduce the interpolation formula

$$\mathbf{v} = \mathbf{N}\dot{\mathbf{q}} \quad \text{and} \quad \mathbf{v}^* = \mathbf{N}^*\dot{\mathbf{q}} \,. \tag{6.83}$$

Finally we have

$$\left\{ \int_{\Omega} \left[(\mathscr{D}\mathbf{N}^*)^T \mathbf{E} \mathscr{D} \int_0^t \mathbf{N}\, \mathrm{d}t \right] \mathrm{d}\Omega + \int_{\Omega} (\mathbf{N}^*)^T \rho \frac{\partial \mathbf{N}}{\partial t} \mathrm{d}\Omega + \int_{\Omega} (\mathbf{N}^*)^T \eta_z \mathbf{N}\, \mathrm{d}\Omega \right\} \dot{\mathbf{q}} + \\ + \int_{\Omega} (\mathscr{D}\mathbf{N}^*)^T \mathbf{E} \varepsilon_0\, \mathrm{d}\Omega = \mathbf{0} \,. \tag{6.84}$$

We assumed before the distribution of virtual displacements depending on nodal parameters in which $t = h$. In such a case, the expression (6.84) has zeroes in the upper halves of the matrices **M**, **K** and of the vector **e**. As before, we can control the properties of the procedure by choosing the parameter α.

Concerning the cost of computing the resulting formula, we need to go back to the given virtual distributions of the nodal parameters. A Dirac distribution with respect to time reduces the problem of integration in the space–time element volume Ω to an integral on the surface which is the section in time $t = \alpha h$, in terms of spatial variables x, y, z. This reduces the calculational cost in comparison with the cost of the procedures obtained by the classical, linear interpolation of virtual parameters with respect to time.

In the case of (6.84), the areas of integration are reduced, from the space–time volume Ω to the space–time surface (real spatial domain) $V(\alpha h)$. The first integral contains a component integrated within the limits $[0, \alpha h]$. Due to the above conclusion concerning the integrals of the distribution, it is sufficient to integrate within the limits $[0, \alpha h]$. Assuming linearity of the integrands **N** we can determine the average value of the integrand at $t = \alpha h/2$ and multiply by the length of the interval αh. Then the stiffness, inertia, and damping matrices, and the initial stress vector describing the space–time element have the following forms:

$$\mathbf{K} = \int_{V_{\alpha h}} (\mathscr{D}\mathbf{N}_{\alpha h}(\mathbf{x}))^T \mathbf{E} \mathscr{D}\mathbf{N}(\mathbf{x}, \alpha h/2)\, \mathrm{d}V \cdot \alpha h \,, \tag{6.85}$$

$$\mathbf{M} = \int_{V_{\alpha h}} \mathbf{N}_{\alpha h}^T(\mathbf{x}) \rho \frac{\partial \mathbf{N}(\mathbf{x}, \alpha h)}{\partial t} \mathrm{d}V \,, \tag{6.86}$$

$$\mathbf{Z} = \int_{V_{\alpha h}} \mathbf{N}_{\alpha h}^T(\mathbf{x}) \eta_z \mathbf{N}(\mathbf{x}, \alpha h)\, \mathrm{d}V \,, \tag{6.87}$$

$$\mathbf{e} = \int_{V_{\alpha h}} (\mathscr{D}\mathbf{N}_{\alpha h}(\mathbf{x}))^T \mathbf{E} \varepsilon_0\, \mathrm{d}V \,. \tag{6.88}$$

Here, $V_{\alpha h}$ is the element which is the cross-space at $t = \alpha h$, $\mathbf{N}_{\alpha h}$ is the matrix of the interpolated functions defined on the surface $V_{\alpha h}$, and $\mathbf{N}(\mathbf{x},\cdot)$ is the matrix interpolation function in $\boldsymbol{\Omega}$ established at a specific time. Changing the limits of integration makes the formula even simpler for numerical calculation of the characteristic matrices. The matrices (6.85), (6.86), and (6.87) have dimensions $N \times 2N$ (N is the total number of degrees of freedom). They bind together the time-moments t_i and t_{i+1}.

If the actual velocity distribution functions are linear, this formulation is approximately equivalent to the procedure previously discussed when $\alpha = 0.5$. Identical compounds are obtained for small deformations, when $\mathbf{x}_i = \mathbf{x}_{i+1}$. For large deformations the spatial geometry is taken at a time coinciding with the centre of gravity of the space–time element, and thus for positive deformations, i.e., extensions of the domain in time, above $\alpha = 0.5$, and for negative, below $\alpha = 0.5$.

6.3.4 Other Functions of the Virtual Velocity

The procedure here is the same as in the case of the virtual functions assumed according to (6.50). We review and examine the properties of different types of virtual functions. The accuracy and stability of the solution depend on the shape functions. In the following, we will give the stiffness and inertia matrices of the rod element obtained with different virtual functions. We will estimate the discretization error of the method [26].

6.3.4.1 Global Equilibrium (Hat Function)

We postulate a global equilibrium in the interval $[0,h]$. We assume a constant function (for example equal to 1) in the time interval (Figure 6.17a)

$$v^*(x,t) = \left(1 - \frac{x}{b}\right) v_3 + \frac{x}{b} v_4 \,. \tag{6.89}$$

The resulting stiffness and inertia matrices have the form

$$\mathbf{K} = \frac{EAh}{b} \left[\begin{array}{cc|cc} \frac{1}{3} & -\frac{1}{3} & \frac{1}{6} & -\frac{1}{6} \\ -\frac{1}{3} & \frac{1}{3} & -\frac{1}{6} & \frac{1}{6} \end{array} \right] , \tag{6.90}$$

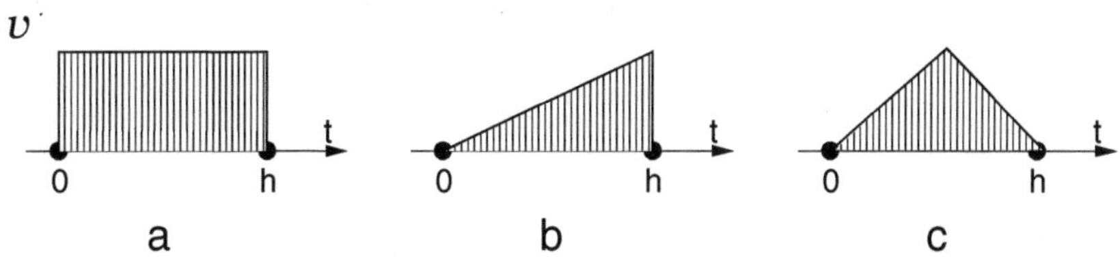

Fig. 6.17 Virtual functions: a – hat-type, b – triangular, c – roof-type.

$$\mathbf{M} = \frac{\rho A b}{h}\left[\begin{array}{cc|cc} -\frac{1}{3} & -\frac{1}{6} & \frac{1}{3} & \frac{1}{6} \\ -\frac{1}{6} & -\frac{1}{3} & \frac{1}{6} & \frac{1}{3} \end{array}\right]. \tag{6.91}$$

The displacements at the next point in time are determined based on the average speed rule $\mathbf{q}_{i+1} = \mathbf{q}_i + h(\mathbf{v}_i + \mathbf{v}_{i+1})/2$. In practice it is the expression (6.54) at $1/2 \leq \beta \leq 1$.

6.3.4.2 Triangular Function

In the interval $[0,h]$ we assume a virtual distribution time function in the shape of a triangle (Figure 6.17b)

$$v^*(x,t) = \left(1 - \frac{x}{b}\right)\frac{t}{h}v_3 + \frac{x}{b}\frac{t}{h}v_4 . \tag{6.92}$$

The stiffness and inertia matrices in this case have the form

$$\mathbf{K} = \frac{EAh}{b}\left[\begin{array}{cc|cc} \frac{5}{24} & -\frac{5}{24} & \frac{1}{8} & -\frac{1}{8} \\ -\frac{5}{24} & \frac{5}{24} & -\frac{1}{8} & \frac{1}{8} \end{array}\right], \tag{6.93}$$

$$\mathbf{M} = \frac{\rho A b}{h}\left[\begin{array}{cc|cc} -\frac{1}{6} & -\frac{1}{12} & \frac{1}{6} & \frac{1}{12} \\ -\frac{1}{12} & -\frac{1}{6} & \frac{1}{12} & \frac{1}{6} \end{array}\right]. \tag{6.94}$$

From the condition of stability we obtain the parameter $2/3 \leq \beta \leq 1$.

6.3.4.3 Roof Function

In this case, we assume a distribution of the virtual time function in the shape of a double triangle as shown in Figure 6.17c

$$v^*(x,t) = \begin{cases} (1-\frac{x}{b})\frac{2t}{h}v_3 + \frac{x}{b}\frac{2t}{h}v_4, & \text{at } 0 \leq t \leq t/2\,, \\ (1-\frac{x}{b})(-\frac{2t}{h}+2)\,v_3 + \frac{x}{b}(-\frac{2t}{h}+2)\,v_4, & \text{at } t/2 < t \leq h\,. \end{cases} \tag{6.95}$$

The stiffness and inertia matrices have in this case the form

$$\mathbf{K} = \frac{EAh}{b}\left[\begin{array}{cc|cc} \frac{17}{96} & -\frac{17}{96} & \frac{7}{96} & -\frac{7}{96} \\ -\frac{17}{96} & \frac{17}{96} & -\frac{7}{96} & \frac{7}{96} \end{array}\right], \tag{6.96}$$

$$\mathbf{M} = \frac{\rho A b}{h}\left[\begin{array}{cc|cc} -\frac{1}{6} & -\frac{1}{12} & \frac{1}{6} & \frac{1}{12} \\ -\frac{1}{12} & -\frac{1}{6} & \frac{1}{12} & \frac{1}{6} \end{array}\right]. \tag{6.97}$$

The stability condition requires the parameter $3/4 \leq \beta \leq 1$.

6.3.4.4 Point Equilibrium

Using the Dirac delta function as the virtual velocity distribution was described in Chapter 6.3.1. We will analyse the process described by equation (6.56). The velocity and displacement can be developed in Taylor series:

$$
\begin{aligned}
v_{i+1} &= \left(1-\alpha\omega^2h^2+\frac{1}{2}\alpha^3\omega^4h^4+\mathscr{O}(h^6)\right)v_i+\\
&+\left(-\omega^2h+\frac{1}{2}\alpha^2\omega^4h^3+\mathscr{O}(h^5)\right)w_i\,, \qquad (6.98)\\
w_{i+1} &= \left(h-\omega^2h^3\alpha(1-\alpha)+\mathscr{O}(h^5)\right)v_i+\\
&+\left(1-\omega^2h^2(1-\alpha)+\frac{1}{2}\omega^4h^4\alpha^2(1-\alpha)+\mathscr{O}(h^6)\right)w_i\,.
\end{aligned}
$$

The parameter $\omega^2 = k/m$ is the square of the eigenfrequency. Considering the expansions of the trigonometric functions, we can estimate the error of the method. To do this we must take into account two cases of initial conditions: $v(0) = 0$, $w(0) = 1$ and $v(0) = 1$, $w(0) = 0$, and compare the results (6.98) with the exact expansions of the solutions. We then obtain the velocity error ε^v

$$\varepsilon^v_{11} = \omega^2h^2\left(\alpha-\frac{1}{2}\right)+\omega^4h^4\left(\frac{1}{24}-\frac{\alpha^3}{2}\right)+\mathscr{O}(h^6)\,, \qquad (6.99)$$

$$\varepsilon^v_{12} = \omega^4h^3\left(\frac{1}{6}-\frac{\alpha^2}{2}\right)+\mathscr{O}(h^5)\,, \qquad (6.100)$$

and displacements error ε^u

$$\varepsilon^u_{21} = \omega^2h^3\left(\alpha(1-\alpha)-\frac{1}{6}\right)+\mathscr{O}(h^5)\,, \qquad (6.101)$$

$$\varepsilon^u_{22} = \omega^2h^2\left(\frac{1}{2}-\alpha\right)+\omega^4h^4\left(\frac{1}{24}-\frac{\alpha^2}{2}(1-\alpha)\right)+\mathscr{O}(h^6)\,. \qquad (6.102)$$

The lower indices indicate elements of the transition matrix (6.56). We see that when $\alpha = 1/2$, the second degree terms vanish. The error is in this case $1/12\,h^3+\mathscr{O}(h^4)$.

A similar error analysis can be carried out also in the case of the other virtual velocity functions, shown previously. Estimating this is left to the reader.

The results of the sample calculations are shown in Figure 6.18. The string is loaded by a force moving with constant speed. The graphs were plotted in the range of $0.1c$ to $1.0c$, for each tenth of the wave speed c. The vertical displacement at the point of the moving force and the vertical displacement in the centre of the string are depicted. We see that spurious vibrations occur in the case $\mathrm{v} = 1.0c$. This is the extreme case since the force moves with the speed of the wave. In the remaining cases, all the plots practically coincide with the theoretical results.

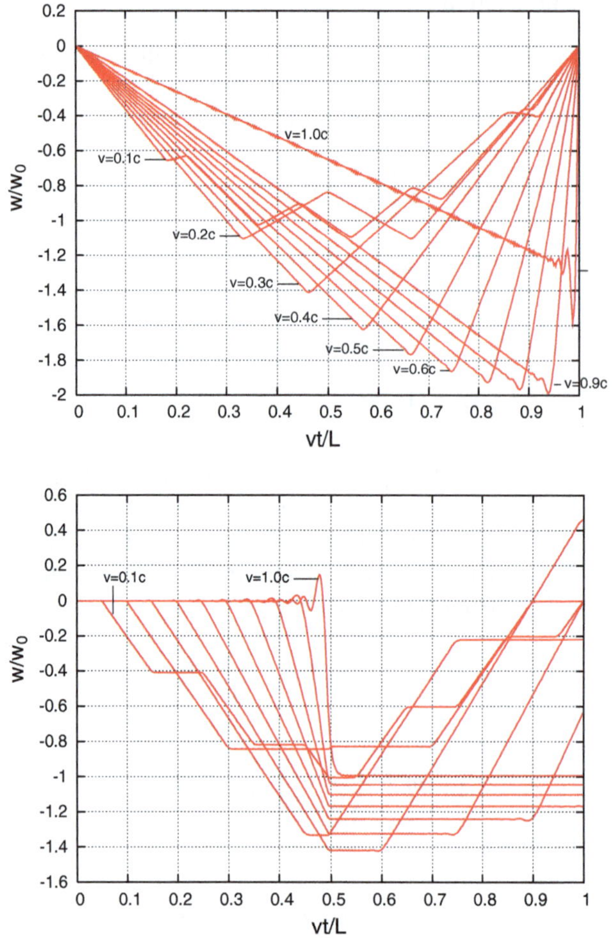

Fig. 6.18 Dimensionless displacements under the force (upper) and in the centre of the string (lower).

6.4 Space–Time Element Method and Other Time Integration Methods

6.4.1 Convergence

The error of the method is estimated as the difference between the exact and the approximate solution. The exact solution of the equation (6.47) is the sine wave $x = A\sin\omega t + B\cos\omega t$, with constants A and B derived from the initial conditions. Just subtract the development of our solution (6.56) from the exact solution developed in a Maclaurin series to determine the step error $[0,h]$. The case most preferred from the viewpoint of convergence is when $\alpha = 1/2$. The approximate solution converges to

Table 6.2 The convergence of the methods.

method	error degree	coefficient
s-t.e.m. $\alpha = 0$	Δt^2	$\frac{1}{2}$
s-t.e.m. $\alpha = \frac{1}{4}$	Δt^2	$\frac{1}{4}$
s-t.e.m. $\alpha = \frac{1}{2}$	Δt^3	$\frac{1}{12}$
s-t.e.m. $\alpha = \frac{\sqrt{2}}{2}$	Δt^2	$\frac{\sqrt{2}-1}{2}$
s-t.e.m. $\alpha = \frac{3}{4}$	Δt^2	$\frac{1}{4}$
s-t.e.m. $\alpha = 1$	Δt^2	$\frac{1}{2}$
s-t.e.m. $v^* = 1$ (6.104)	Δt^2	$\frac{1}{12}$
Runge–Kutta 2nd order	Δt^3	$\frac{1}{6}$
Runge–Kutta 4th order	Δt^5	$\frac{1}{120}$

the exact solution with an error proportional to the third power of the step Δt. Table 6.2 summarizes the different variants of the method. Using a linear combination of these variants defined with different α allows of improving the accuracy. Using virtual functions of the form $\sum_j \chi_j \delta(t/h - \alpha_j)$, where the α_j are the coordinates of the peaks, and χ_j the size (weight) of these peaks, the speed v_{i+1} is determined by

$$v_{i+1} = \frac{\sum_j \left(1 - \frac{kh^2}{2m}\left[1-(1-\alpha_j)^2\right]\right)\chi_j}{\sum_j \left(1 + \frac{kh^2\alpha_j^2}{2m}\right)\chi_j} v_i - \frac{\frac{k}{m}h}{\sum_j \left(1 + \frac{kh^2\alpha_j^2}{2m}\right)\chi_j} x_i \,. \tag{6.103}$$

If we take $\alpha_1 = 0$, $\alpha_2 = 1/2$, and $\alpha_3 = 1$, then we obtain the appropriate coefficients χ_j: $\chi_1 = 1/6$, $\chi_2 = 2/3$, $\chi_3 = 1/6$. The solution has an error $\mathcal{O}(h^4)$ (exactly $h^4/12 + \mathcal{O}(h^5)$). Choosing the location of successive peaks we can reset more terms describing the local solution error. In this way we can construct a system of algebraic equations in which the unknowns will be shares of Dirac peaks in the final function of the virtual velocity. Another problem is the selection of the location α_i in the function $\delta(t - \alpha_i h)$. This can be done optimally, reducing the error and yet enabling the meeting of a specific criterion for the resulting factors (e.g. that they have equal sign). This issue will not be further dealt with here.

In a similar manner as in Chapter 6.3.1, one can define a family of computational schemes, differing in the shape of the virtual function adopted on exit, and thus differing in the points where equilibrium is established. Yet noteworthy is the scheme expressing the global equilibrium, when the speed of a virtual function is equal to unity throughout the period of time h, except for the ends of the interval, where the values are zero. The resulting formula is the following:

$$v_{i+1} = \frac{1 - \frac{k}{m}\frac{h^2}{3}}{1 + \frac{k}{m}\frac{h^2}{6}} v_i - \frac{\frac{k}{m}h}{1 + \frac{k}{m}\frac{h^2}{6}} x_i, \quad x_{i+1} = x_i + \frac{1}{2}h(v_i + v_{i+1}) \, . \tag{6.104}$$

The scheme (6.104) has a local error of $\mathcal{O}(h^3)$ (exactly $h^3/12$) and is conditionally stable. The stability condition imposes in this case the restriction $\omega h \leq 2\sqrt{3}$.

The operators a_1 of the difference formula $y_{i+1} = a_1 y_i + a_2 y_{i-1}$ for the selected methods are presented in Table 6.3. a_2 is equal to minus one. Figure 6.19 illustrates the phase error of the chosen methods of integration of the differential equation of motion: the space-time element method (STEM), displacement formulation, with the parameter η, the space-time element method, velocity formulation, with the parameter α, the Newmark method (NM), and the trapezoidal method (TM). It can be shown that when $\alpha = \sqrt{2\beta}$ (where β is the parameter of the Newmark method) the scheme (6.56) corresponds to the Newmark method. It should be noted that the identity of the Newmark method and the space–time element method is limited to the case of a zero damping matrix. In general, a full analogy has so far failed to show.

Table 6.3 Operators of the difference schemes of the integration of the differential equation.

method		operator
central difference m.		$2-\kappa$
trapezoidal m.		$\frac{2(4-\kappa)}{4+\kappa}$
Newmark m.:	β=1/4, γ=1/2	$\frac{2(4-\kappa)}{4+\kappa}$
	β=1/6, γ=1/2	$\frac{4(3-\kappa)}{6+\kappa}$
	β=1/12, γ=1/2	$\frac{2(12-5\kappa)}{12+\kappa}$
	β=0, γ=1/2	$2-\kappa$
STEM—displacement formula:	classical (η=0)	$\frac{4(3-\kappa)}{6+\kappa}$
	η=5/4	$\frac{2(4-\kappa)}{4+\kappa}$
	η=5/2	$\frac{6-\kappa}{3+\kappa}$
STEM—velocity formula:	α=0	$2-\kappa$
	α=1/2	$\frac{16-6\kappa}{8+\kappa}$
	$\alpha=\sqrt{2}/2$	$\frac{2(4-\kappa)}{4+\kappa}$
	α=1	$\frac{4}{2+\kappa}$
	formula with v^*=1 (6.104)	$\frac{4(3-\kappa)}{6+\kappa}$

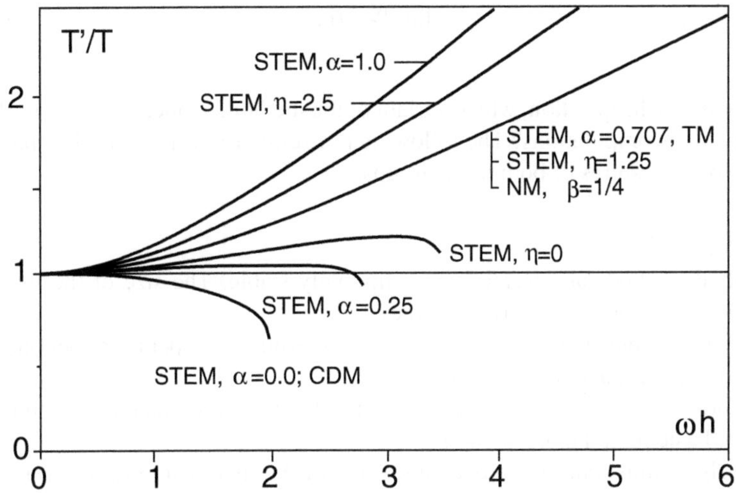

Fig. 6.19 The phase error of the time integration methods: CDM – central difference method, NM – Newmark method, TM – trapezoidal method, STEM – space–time element method.

6.4.2 *Phase Error*

The phase error in the integration scheme expressed in velocities is determined by the ratio

$$P = \frac{T'}{T} = \frac{\omega}{\omega'} \,. \tag{6.105}$$

The characteristic equation of the scheme described by (6.56) has the form

$$\lambda^2 - (T_{11} + T_{22})\lambda + (T_{12}T_{21} - T_{11}T_{22}) = 0 \,, \tag{6.106}$$

where T_{ij} are the elements of the transition matrix (6.56). The solution of (6.106) gives

$$\lambda_{1/2} = \frac{\alpha^2\kappa + 2 - \kappa \pm i\sqrt{\kappa(2\alpha^2\kappa + 4 - \kappa)}}{\alpha^2\kappa + 2}, \quad \kappa = \frac{k}{m}h^2 \,. \tag{6.107}$$

Since the frequency in the numerical scheme is determined by

$$\operatorname{tg}\omega' h = \frac{\Im(\lambda)}{\Re(\lambda)}, \quad \omega' = \frac{1}{h}\operatorname{arctg}\frac{\Im(\lambda)}{\Re(\lambda)} \,, \tag{6.108}$$

the phase error is

$$P = \frac{\sqrt{\kappa}}{\operatorname{arctg}\left(\frac{\sqrt{\kappa(2\alpha^2\kappa+4-\kappa)}}{\alpha^2\kappa+2-\kappa}\right)} \,, \tag{6.109}$$

and in the limit

$$\lim_{\kappa \to 0} P = 0 \,. \tag{6.110}$$

■

There remains the question, which variant of the method, explicit or implicit, should be adopted in computations? The following is a summary which partially characterizes the basic features of both these groups.

Implicit methods

- Implicit methods are usually unconditionally stable. The size of the time step depends only on the required accuracy.
- The calculations require a greater number of arithmetic operations per one computational step, and more memory.
- The precision criteria require that the time step be approximately one hundredth of the basic period of oscillation.
- The algorithms require the decomposition of the matrix into factors (factorization).
- Implicit methods are convenient in the analysis of inertial issues.
- Implicit methods are recommended for bent structures due to the high value of the highest vibration frequency compared to the basic period.
- The efficiency of this group of methods is lower in the analysis of multidimensional systems.

Explicit methods

- Stability conditions limit the time step of explicit methods.
- Calculations are effective with diagonal mass matrices. Consistent mass matrix improves the accuracy of bent structures but overestimates the frequency, while a diagonal matrix lowers the frequency. A diagonal modified matrix [66] (when the total mass of the finite element is distributed on the diagonal elements with the ratio of the consistent matrix) has the most advantageous properties.

6.4.3 Non-inertial Problems

Now let us look at the solution of massless tasks. We will apply the space–time element method. We will consider in fact static problems, although the process of material deformation is kinematically induced. This way we can obtain illustrative results for quasistatic deformation processes. We can apply this approach to the material shaping processes in forms, car stringers crushing, etc.

Let us examine the stability of the solution scheme based on the velocity space–time element method with zero inertia coefficients. In this scheme we use the Dirac delta function for the virtual speed. Assuming $m = 0$, we calculate the moduli of the eigenvalues of the transition matrix $\mathbf{T}$ for the problem of a single degree of freedom. Figure 6.20 shows the dependence of the two eigenvalues on the value of the parameter α for the massive and the massless tasks. In the case of an inertial system, the modulus of an eigenvalue does not exceed 1 when $\alpha \geq \sqrt{2}/2$. In non-inertial problems, the moduli of both eigenvalues are equal to 1 only when $\alpha = 1$.

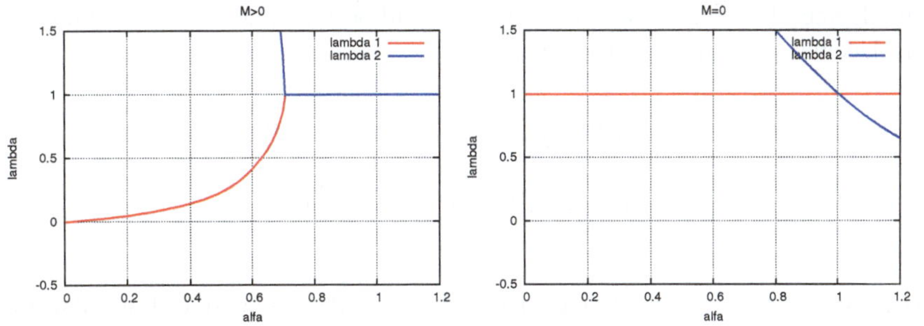

Fig. 6.20 Eigenvalues of the transition matrix for the massive ($m > 0$) and massless ($m = 0$) tasks.

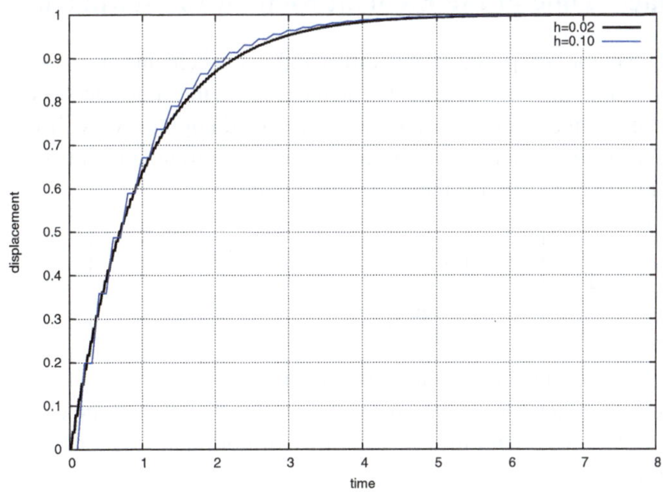

Fig. 6.21 Solution process for the massless spring subjected to a constant force, with time steps $h = 0.02$ and 0.1.

This is the only parameter of the procedure for which the solution is unconditionally stable when $m = 0$.

The space–time scheme applied to massless systems is advantageous since the same code can be applied both to inertial and non-inertial structures. Without needing reprogramming, the quasi-static problem can be transformed into a dynamic one. Below, the massless algorithm will be applied with the aim of comparing the numerical results for simple statical or quasi-statical problems. Let us look at the example.

Example 6.1. Let us consider an oscillator with zero mass. In practice, we consider a simple massless spring with stiffness $k = 1$, subjected to a force $P = -1$. All initial parameters, i.e., the velocity and displacement, are equal to zero. The system has one degree of freedom. We can write the equilibrium equation with formulas for displacement x_{i+1} and nodal force e_{i+1}

$$\begin{aligned}
&\left[\alpha\left(1-\frac{\alpha}{2}\right)v_i+\frac{\alpha^2}{2}v_{i+1}\right]+e_i=P_i\,,\\
&x_{i+1}=x_i+[\alpha v_i+(1-\alpha)v_{i+1}]h\,,\\
&e_{i+1}=x_{i+1}\;.
\end{aligned} \tag{6.111}$$

The results for $\alpha=1$ are presented in Figure 6.21. Other values of α, lower than one, involve divergence of the time stepping process. We can notice that the computed displacement converges to the exact value, equal to one.

6.5 Space–Time Finite Element Method vs. Newmark Method

We link the vector composed of velocities and displacements at the next time with the same vector at the present time. We will deal with the simplified case. Assuming a zero value for the load function and a lack of damping, we will consider free undamped vibrations. Using Algorithm 5 after appropriate transformations we can write the transition matrix for the Newmark method

$$\mathbf{T}=\begin{bmatrix}\frac{1}{2\beta(h^2\omega^2\beta+1)}-\frac{1}{2\beta}+1 & -\frac{h\omega^2(h^2\omega^2(4\beta-1)+4)}{4(h^2\omega^2\beta+1)}\\[2mm] \frac{h}{1+h^2\omega^2\beta} & \frac{1-h^2\omega^2(\frac{1}{2}-\beta)}{1+h^2\omega^2\beta}\end{bmatrix}. \tag{6.112}$$

If we expand the terms of the matrix (6.112) into Taylor series and act similarly to the matrix (6.56), then for the parameter $\alpha=1/2$ in the space–time element method and for $\beta=1/2$ we obtain the compatibility of both matrices with the error level $\Delta\mathbf{T}$ of the range $\mathcal{O}(h^3)$

$$\Delta\mathbf{T}=\begin{bmatrix}0 & \frac{h^3\omega^4}{8}\\[2mm] -\frac{h^3\omega^2}{4} & 0\end{bmatrix}. \tag{6.113}$$

Only in the case of the application of the virtual hat function, and $\mathbf{C}=\mathbf{0}$, can one easily compare the Newmark method with the space–time element method and see their identity. We obtain the transition matrix of the space–time element method in the form (6.104). We then easily see that both arrays (6.104) and (6.112) are equal if $\beta=1/6$ and $\gamma=1/2$ in the Newmark method. Therefore, unless we take into account a damping described by the damping matrix $\mathbf{C}$, we obtain the identity of both methods.

Identical considerations hold for comparing the central difference method with the method of trapezoids. They are a particular case of the group of Newmark methods.

6.6 Simplex Elements

The diagrams of finite element solutions reduce to systems of algebraic equations. Their solution has a large share in the cost of the whole process of calculation. The number of arithmetic operations increases significantly in non-linear tasks that require multiple solving of the system of equations at each time step. There are two groups of methods for solving systems of equations: direct and iterative. The first one leads to a solution with a single engagement of the matrix coefficients. The matrix can be split into a product of two triangular matrices, or directly triangularized. The calculation of the unknowns in the system with the triangular matrix is immediate. We can use the classical decomposition LU on the matrix product of lower and upper triangular matrices, the Cholesky–Banachiewicz decomposition $U^T U$ on the product of a triangular matrix and its transposed form, the decomposition of Doolittle or LU Croute with ones on the diagonal of the matrix L or U and decomposition LDL^T, with the diagonal matrix D and the lower triangular matrix L. These methods are characterized by relatively low computational cost. The decomposition LDL^T and the Croute and Doolittle decomposition require $n^3/6$ multiplications, and the Gaussian decomposition requires $n^3/3$ operations. The computational cost in a real practical case depends on the bandwidth of the coefficient matrix of the system of equations. The disadvantage of the group of matrix decomposition methods is the need to first use special algorithms to reduce the bandwidth of the matrix. In the case of two-dimensional problems discretized with a mesh with the same number of nodes in both directions, the cost of the solution is proportional to the fourth power of the number of nodes in one direction of the grid. And in the case of a three-dimensional cube-shaped domain, covered with a mesh evenly in three directions, this cost increases with the seventh power of the number of nodes on one edge. Nonlinear effects and the need for iteration, slowed by an increase in the number of nodes, achieving the balance of forces, prolongs even more the computational process.

The second group, a group of iterative methods (e.g., Jacobi, Gauss–Jordan), rarely used in practice, allows the incorporation of nonlinear effects of the task into an iterative solving process. In this way, the approach to the solution is based on simultaneously updated matrix elements. This process can be carried out without forming a global matrix (or even part of it). This is the procedure performed 'element by element'. Although these are undoubtedly advantageous features, the whole computation process is very expensive and vulnerable to problems with its convergence and stability.

Another group of iterative methods of the 'element by element' type are presented, e.g., in the papers [68]. Their computational cost, however, is large and methods of this type start to be attractive only in sufficiently large tasks. Their convergence is relatively slow and depends on the conditioning of the matrix (and thus the bandwidth of the matrix, the shape of the test object, and the material parameters of the elements). It was concluded that they had a high sensitivity to nonlinear effects. The increase of the time step in nonlinear tasks worsens the convergence dramatically.

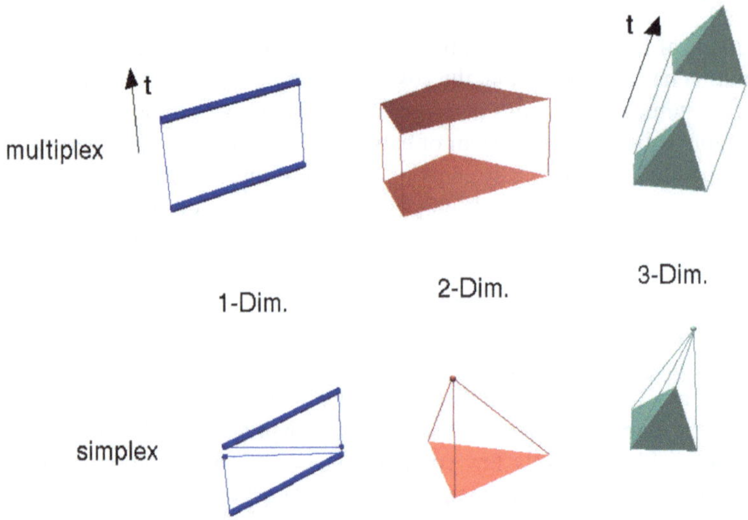

Fig. 6.22 Examples of multiplex and simplex elements of one-, two-, and three-dimensional objects.

The above drawbacks and particularly the difficulty in predicting the rate of convergence of iterative methods make direct methods more likely to be applied in practice.

The space–time approximation opens up new opportunities by applying simplicial space–time subdomains (Figure 6.22). These elements in one-dimensional real space and in time have the shapes of triangles. The elements of this shape can model axially vibrating rods, bending rods, or strings. Two-dimensional objects such as discs, plates, and shells in space are discretized with triangular spatial elements, and in time–space tetrahedra, created from these triangles. Three-dimensional blocks in space translate into hyper-tetrahedral simplices in time-space. Hyper-tetrahedra are creations in the $n+1$ dimensional space, and have $5-i$ ($i=1,\ldots,4$) nodes in time t_i and i nodes in time t_{i+1}.

6.6.1 *Property of Space Division*

A space–time layer bounded by planes t_i and t_{i+1} can be filled with simplicial elements in many ways. The special filling of the space–time layer with simplices allows of gaining a triangular coefficient matrix for the system of equations directly in forming the global matrix (Figure 6.23). The creation of a space partition by triangles should be guided by the principle that skew edges run from the point $(x_i,\Delta t)$ to the point $(x_j,0)$, if $i<j$. In this way, we obtain skew edges in time-space. Otherwise, i.e., $i \geq j$, we obtain the edges parallel to the time axis. We proceed similarly in a task of higher spatial dimensionality. The procedure is described in

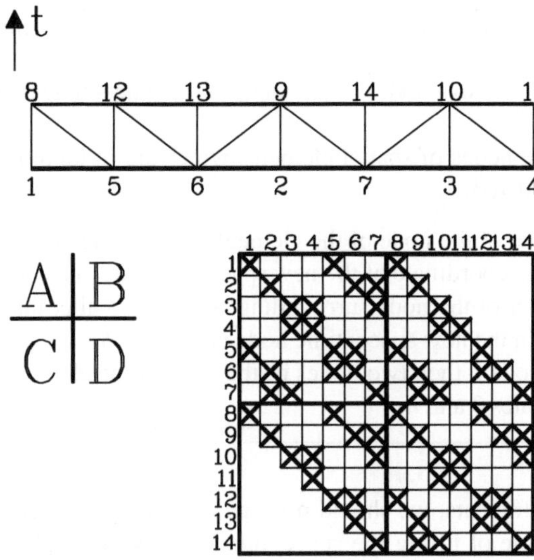

Fig. 6.23 Filling up the global matrix in the one-dimensional domain with a triangular grid of space–time elements.

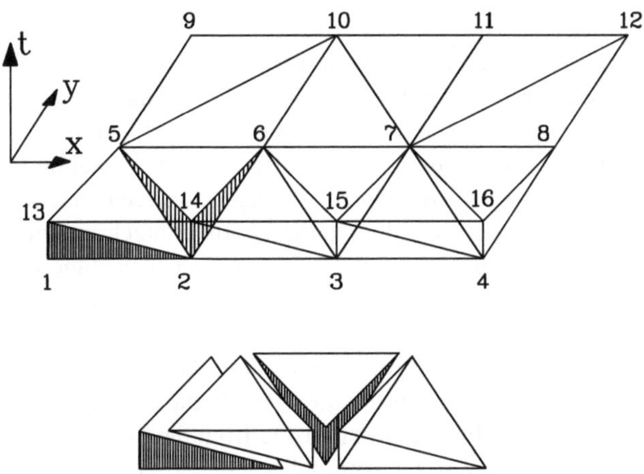

Fig. 6.24 Filling in the two-dimensional time-space.

Algorithm 10. The example of filling the volume of the space–time layer in a one-dimensional task is shown in Figure 6.24. In this way, a lower triangular matrix of coefficients is obtained.

In the same way, we create a layer of tetrahedral elements in two-dimensional spatial areas. This is illustrated in Figure 6.24. We can also obtain an upper triangular matrix of coefficients. When creating the space–time grid and constructing

Algorithm 10. The method of partitioning the time-space into simplex-shaped elements.

1. We consider successive nodes of the mesh of the space–time partition (for example node i).
2. We number the nodes of spatial elements which are in topological contact with the node considered:
   - if the number of the node of the element surrounding node i is greater than i, then the time coordinate of the new space–time node is equal to 0,
   - if the number of the node surrounding node i is smaller than i, then the time coordinate in the new space–time node is equal to h,
   - if it is the node i, then two nodes of the space–time element are created with the coordinates 0 and h.

the matrix of coefficients, we should proceed oppositely to Algorithm 10. We can extend the principle of filling the space–time layer with simplex elements by the following statement: any partition of the space–time layer with simplicial elements enables solving the resulting system of equations node by node. This should be done by choosing the appropriate sequence of nodes resulting from the method of filling the space–time layer with elements. This is obvious because the reordering of equations and variables in the vector of unknowns always brings such a system of equations to triangular form (upper or lower).

Solving the system of equations with a triangular matrix of coefficients is effective because we can proceed directly to solving it equation after equation. The characteristic property of such a procedure can be underlined. Is the flow of information in time in one spatial direction only: from the first equation to the last, i.e., from the spatial point associated with the first degree of freedom to the spatial point governed by the last degree of freedom in the system of algebraic equations (or vice versa, if we are dealing with an upper triangular matrix). In the time stepping process it is expressed in this way: the flow of information between the nodes of the spatial grid in one direction has a limited speed, but is unlimited in the opposite direction (Figure 6.25a). Numerical tests have shown that in the initial phase of the calculation, some differences in the results are observed. In the long term simulation, these differences are dominated by the errors arising from the discrete model, properties of the method of calculation of parasitic vibrations. Space–time anisotropy affects the performance of wave problems (shock, reflections, etc.). There is no significant effect of anisotropy on the results of the structural dynamics. An anisotropy can be removed by introducing a special division of the layer, without preference for any direction (Figure 6.25b).

Another practical property of having simplices for elements is the possibility of separating the parts of the system of equations. It allows sub-dividing the construction into substructures and solving the system of equations in separate batches. The whole structure, with the exception of nodes common to neighbouring substructures, can be modelled with elements of 'multiplex' type, and so in the classical

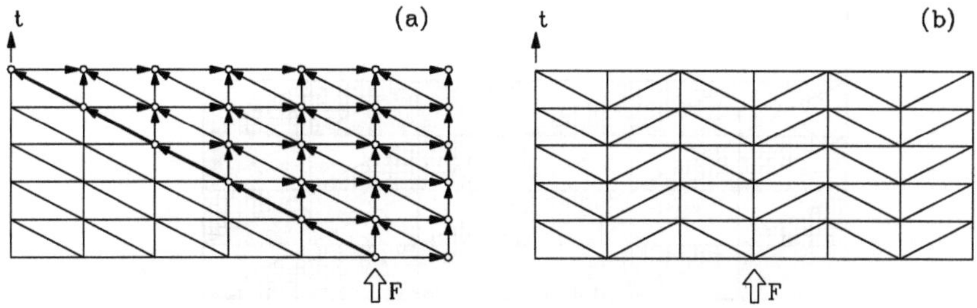

Fig. 6.25 Limit of the speed of information flow *(a)* and isotropic properties of time-space *(b)*.

way. Only elements connecting substructures should be simplicial (Figure 6.26). The solution is obtained in two stages: first the subsystem No. I is solved and then using the known solution in the node at the junction of two subdivisions is subsystem No. II solved. Thanks to this, the calculation can be carried out on parallel processors. Figure 6.27 shows for the example of the simplest element mesh, how the computational process can be divided between four processors. Their areas of involvement are outlined. Starting from the node nearest the origin of the coordinate system, we can move with time, carrying out calculations for the selected subsystems of elements. A larger number of processors allows even greater acceleration of the computational process.

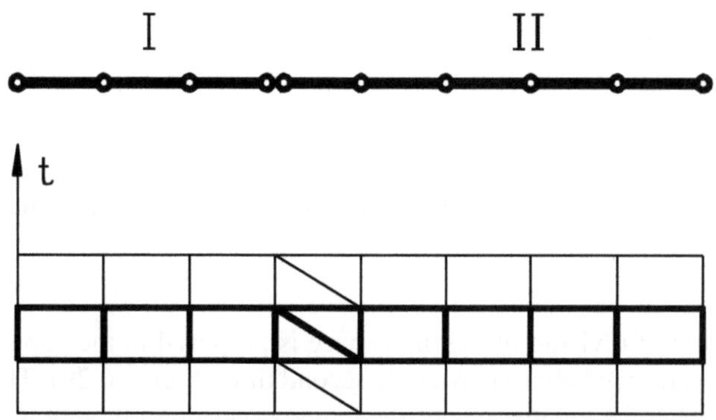

Fig. 6.26 The division of the domain into sub-areas using simplices.

The elements of higher order, with intermediate nodes both along the x-axis and the t-axis, also allow us to construct a system of equations which can be solved 'node after node'. This means that we can separate groups of nodes (Figure 6.28) that enter into the calculations sequentially. In this way, properties similar to the basic, low order elements are achieved, where the nodes are taken into the computations individually.

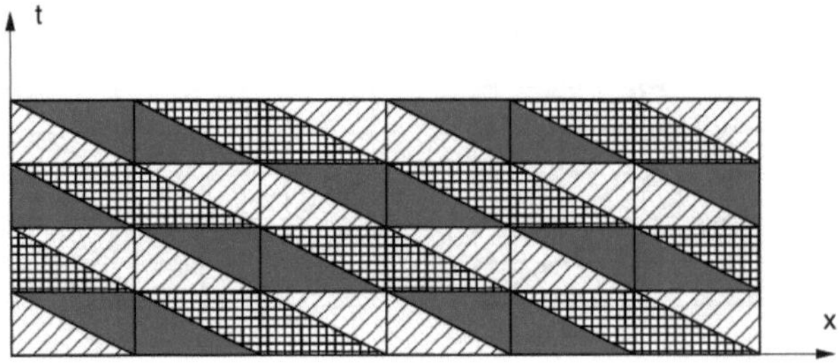

Fig. 6.27 Areas of activity of individual processors in a multiprocessor solution.

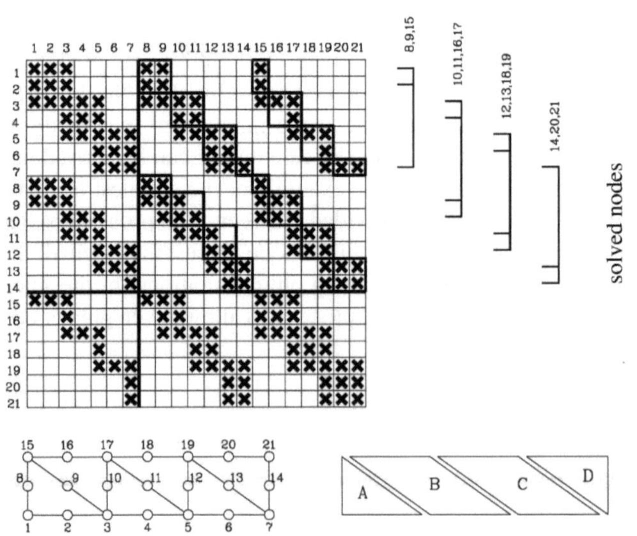

Fig. 6.28 Constructing and solving the system of equations for higher order space–time elements.

Another kind of division of the time-space is proposed in the papers [80, 114]. The nodes are arranged alternately every second time (Figure 6.29). The advantage of this approach is its lengthening of the integration step (although the stability of such a scheme has not yet been fully explored), isotropy, and above all decomposition of the system of equations. This helps reduce the number of iterative processes (such as at material nonlinearities) to each node, without having to manipulate entire arrays. A major drawback is the difficulty in practically applying such a division. Problems arise in covering even simple flat areas with such elements. At the edges and corners of the area we need to build three-dimensional objects of unusual shape, differing from the rest of the grid. The amount of work in the programming phase is so large that it can be profitable only for repeatedly performed structural calculations for huge meshes.

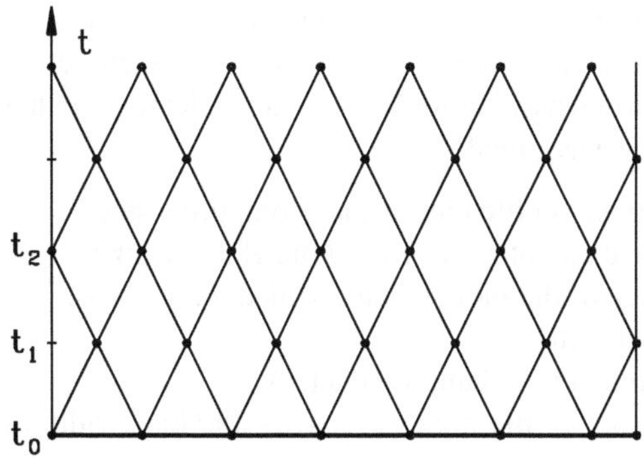

Fig. 6.29 Staggered arrangement of the element nodes.

6.6.2 *Numerical Efficiency*

Numerical efficiency is one of basic features that should be sought for when solving non-linear problems. Finding solutions in a single computational step consists of two phases: building the global matrix of coefficients, with prior modification of the geometric and material properties, and solving the algebraic equations. The computational cost of the process of integrating the equations of motion is significantly affected by the form of the matrix of inertia and the damping. A diagonal inertia matrix and numerical damping, described by a combination of stiffness and inertia matrix, allows shortening and simplifying the computations. The use of explicit integration schemes in time results in decoupling the system of equations. However, in some cases, consistent mass and damping matrices are required. In such cases, the need to solve the full system of equations raises the cost significantly.

The simplest assessment of the quality of the algorithm is to estimate the number of arithmetic operations necessary to carry out the calculation in one step. It can be assumed that the share of operations of multiplication M in the total number of arithmetic operations is constant. In the space–time element method the number of multiplications needed to solve the system of equations (not including the setting up the coefficients of the system) depends on the amount of memory used for collecting the coefficients. It is

$$M = 2sN(c+1) \tag{6.114}$$

if we engage $3.5sN(c+1)+1.5sN$ memory units, and it is

$$M = 3sN(c+1) \tag{6.115}$$

if we engage $1.5sN(c+1)+1.5sN$ memory units. By memory unit, we mean that portion of memory that stores a real number. By c we denote the number of nodes adjacent to a single node in the mesh, N denotes the total number of degrees of freedom, and s denotes the number of nodal degrees of freedom.

Comparing the full cost of computing one step in the space–time element method and in the finite element method combined with the method of central differences requires making certain assumptions. In the finite element method and in the central difference method we assumed

- creating a matrix of coefficients at each computing step,
- symmetry of the element matrices and the global matrix,
- regular band edge of the global matrix (filled with zeros),
- consistent inertia matrix,
- negligible damping (zero damping matrix),
- optimal numbering of the nodes to narrow the bandwidth of the matrix in the finite element method,
- perform multiplications by zero within the band,
- plane stress/strain elastic task as a model task adopted in estimation.

The comparison of the cost calculations is shown in Figure 6.30. We can compare the above estimates, (6.114) and (6.115), with [102]. The Newmark method requires $M = Nb^2/2n + N(4b+3)$ multiplications per one step of calculations (where n is the number of time steps and b is the width of the half band). The Trujillo method [102, 139] for a matrix stored in band form requires $M = N(4b+6)$ operations, and for storage in block form, $M = 2sN(c+1) + 10N$. A summary of the computational cost of each method is given in Table 6.4. It should be noted that the width of the matrix half band b is usually proportional to $N^{1/2}$ in plane tasks and to $N^{1/3}$ in three-dimensional tasks. In the comparison, the entire group of methods using diagonal matrices of inertia and damping were omitted. They give decoupled equations and the cost of solving is then low.

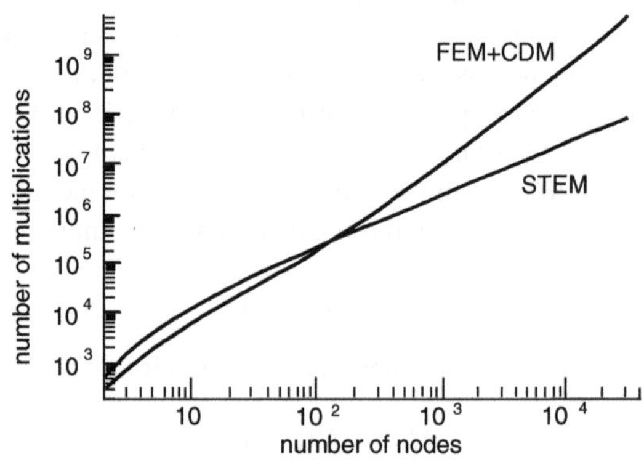

Fig. 6.30 Comparison of computational costs of the finite element method and the method of central differences with the space–time element method.

Table 6.4 Computational cost of numerical methods.

method	number of multiplications
Gauss m. (symmetric matrices)	$\frac{N^3}{6}+\frac{N^2}{4}+\frac{N}{12}$
Gauss m. (symmetric and band matrices)	$\frac{1}{2}b^2N$
STEM (higher memory)	$2sN(c+1)$
STEM (lower memory)	$3sN(c+1)$
Newmark m. [102]	$4bN+3N$
Trujillo m. (with half band) [102]	$4bN+6N$
Trujillo m. (with block notation) [102]	$2bN+10N$

Conclusions

- The computational cost of the space–time element method grows linearly with the number of grid nodes.
- The number of operations in the space–time element method does not depend on the method of numbering the nodes (mesh topology).
- In classical methods, the half band width b increases with the size of the task. The number of arithmetic operations per one time step is then proportional to $bN \approx N^{3/2}$.
- Estimates are rough and may vary considerably for different specific tasks.

6.7 Simplex Elements in the Displacement Description

In the following sections we will derive the stiffness, inertia, and damping matrices of the basic simplicial space–time finite elements. We will deal with a rod or string element, the element of a beam of medium thickness, the slab, and a plate of medium thickness and three-dimensional body. The way to build a matrix is simple. We use the well-known procedure from the finite element method. The only difference is the inclusion of time as an additional coordinate of the coordinate system in which we describe a discrete element.

6.7.1 Triangular Element of a Bar Vibrating Axially

We use a linear displacement distribution inside the triangular element

$$w(x,t) = a_1 x + a_2 t + a_3 \,, \tag{6.116}$$

where a_1, a_2, a_3 are constants dependent on the geometry of the element and the values of the nodal displacements. It can be written in another way in surface coordinates L_i

$$w = L_1 u_1 + L_2 u_2 + L_3 u_3 \,, \tag{6.117}$$

where

$$L_i = \frac{1}{2\Delta} \det \begin{bmatrix} x & t & 1 \\ x_k & t_k & 1 \\ x_l & t_l & 1 \end{bmatrix} \,, \tag{6.118}$$

where i, k, l and j, m, n are the permutations of the numbers of the vertices in the triangle, and Δ is the surface area of a triangle.
The shape functions $N_i(x,t)$ are directly expressed by coordinates L_i

$$\mathbf{N} = [L_1, L_2, L_3] \,. \tag{6.119}$$

The differential operator can be considered as a matrix containing a single element $\mathscr{D} = \partial/\partial x$. If we take the origin to be the centre of gravity of the triangle, the stiffness $\mathbf{K}$ and inertia $\mathbf{M}$ matrices are

$$K_{ij} = \frac{EA}{4\Delta}(t_k - t_l)(t_m - t_n) \,, \tag{6.120}$$

$$M_{ij} = -\frac{\rho A}{4\Delta}(x_k - x_l)(x_m - x_n) \,. \tag{6.121}$$

The external damping $\mathbf{Z}$ has the form

$$Z_{ij} = \frac{\eta_z}{4\Delta}(x_k t_l - x_l t_k)(x_n - x_m) \,, \tag{6.122}$$

and the internal, due to the double differentiation of (6.116), is equal to zero.

6.7.2 Space–Time Finite Element of the Beam of Moderate Height

In a beam of moderate height, we take into account the effects of shear. Both the deflection w and the rotation θ are expressed by independent linear functions

$$\begin{Bmatrix} w \\ \theta \end{Bmatrix} = \begin{Bmatrix} a_1 x + a_2 t + a_3 \\ b_1 x + b_2 t + b_3 \end{Bmatrix} \,. \tag{6.123}$$

The coefficients a_i and b_i depend on the geometry of the space–time element. In surface coordinates L_1, L_2, and L_3, we can write the dependencies of the displacements on the nodal values:

$$\begin{Bmatrix} w \\ \theta \end{Bmatrix} = \begin{Bmatrix} L_1 w_1 + L_2 w_2 + L_3 w_3 \\ L_1 \theta_1 + L_2 \theta_2 + L_3 \theta_3 \end{Bmatrix} , \tag{6.124}$$

where

$$L_i = \frac{1}{2\Delta} \begin{vmatrix} x & t & 1 \\ x_j & t_j & 1 \\ x_k & t_k & 1 \end{vmatrix} , \tag{6.125}$$

and Δ is the surface area of a triangular element. The coefficients a_i can then be determined:

$$\begin{aligned} a_1 &= \frac{1}{2\Delta}(t_{23}w_1 + t_{31}w_2 + t_{12}w_3) , \\ a_2 &= \frac{1}{2\Delta}(x_{32}w_1 + x_{13}w_2 + x_{21}w_3) , \\ a_3 &= \frac{1}{2\Delta}(x_2 t_3 w_1 + x_3 t_1 w_2 + x_1 t_2 w_3) , \end{aligned} \tag{6.126}$$

where $x_{ij} = x_i - x_j$, $t_{ij} = t_i - t_j$. The coefficients b_i are computed in the same way. They contain the parameters of rotation θ_k instead of transverse displacements w_k. Total decoupling of the deflections and rotations results in the shape function matrix becoming diagonal:

$$\mathbf{N}_i = \begin{bmatrix} L_i & 0 \\ 0 & L_i \end{bmatrix} , \quad i = 1,2,3 . \tag{6.127}$$

The strains ε are described by the average share angle β and curvature κ

$$\varepsilon = \begin{Bmatrix} \beta \\ \kappa \end{Bmatrix} = \begin{Bmatrix} \frac{\partial w}{\partial x} + \theta \\ \frac{\partial \theta}{\partial x} \end{Bmatrix} . \tag{6.128}$$

The differential operator $\mathscr{D}$ contains two derivatives in terms of the variable x and has the form

$$\mathscr{D} = \begin{bmatrix} \frac{\partial}{\partial x} & 1 \\ 0 & \frac{\partial}{\partial x} \end{bmatrix} . \tag{6.129}$$

The constitutive relationship is described by the matrix of elasticity $\mathbf{E}$

$$\sigma = \begin{Bmatrix} Q \\ M \end{Bmatrix} = \mathbf{E}\varepsilon = \begin{bmatrix} \frac{GA}{k} & 0 \\ 0 & EI \end{bmatrix} \varepsilon . \tag{6.130}$$

The next step is to determine the products, using the general formulas (6.15), (6.16), (6.19), and (6.20), and then integrate them in the area of the triangle. If we take the origin at the centre of gravity of the triangle, we can use the known rules

$$\int_{\Delta} \mathrm{d}x\mathrm{d}t = \Delta, \ \int_{\Delta} x\,\mathrm{d}x\mathrm{d}t = 0, \ \int_{\Delta} x^2\,\mathrm{d}x\mathrm{d}t = \frac{\Delta}{12}\Sigma x_i^2, \ \int_{\Delta} xt\,\mathrm{d}x\mathrm{d}t = \frac{\Delta}{12}\Sigma x_i t_i \, . \quad (6.131)$$

The resulting space–time stiffness matrix consists of 36 elements, in 2×2 frames, in each of the nodes. One of these frames $\mathbf{K}^*_{ij}$ is given below

$$\mathbf{K}^*_{ij} = \left[\begin{array}{c:c} \begin{array}{l} \frac{GA}{4k\Delta} t_{kl} t_{mn} - \\ -\frac{\rho A}{4\Delta} x_{kl} x_{mn} + \\ +\frac{\eta_z}{4\Delta} x_{mn}(x_k t_l - x_j t_k) \end{array} & \begin{array}{l} \frac{GA}{4k\Delta} t_{kl}(x_m t_n - x_n t_m) + \\ +\frac{\eta_z}{4\Delta} t_{kl} x_{mn} \end{array} \\ \hdashline \frac{GA}{4k\Delta} t_{mn}(x_k t_l - x_l t_k) & \begin{array}{l} \frac{EI}{4\Delta} t_{kl} t_{mn} - \frac{\rho I}{4\Delta} x_{kl} x_{mn} + \\ +\frac{\eta_w + \eta_z}{4\Delta}(x_k t_l - x_l t_k) x_{mn} + \\ +\frac{GA}{4k\Delta}\left[t_{kl} t_{mn} \frac{\Sigma x^2}{12} + x_{kl} x_{nm} \frac{\Sigma t^2}{12} + \right. \\ \left. +(t_{kl} x_{nm} + x_{lk} t_{mn}) \frac{\Sigma xt}{12} \right] + \\ \frac{GA}{4k\Delta}(x_k t_l - x_l t_k)(x_m t_n - x_n t_m) \end{array} \end{array}\right] . \quad (6.132)$$

Here we have used the following notation: $x_{ij} = x_i - x_j$, $t_{ij} = t_i - t_j$, $\Sigma x^2 = x_1^2 + x_2^2 + x_3^2$, $\Sigma t^2 = t_1^2 + t_2^2 + t_3^2$, $\Sigma xt = x_1 t_1 + x_2 t_2 + x_3 t_3$.

Figure 6.31 shows the amplitude of the displacements of the free end of a cantilever loaded with a concentrated force with a Heaviside distribution, in relation to the value determined analytically. With the increase in the length of the elements b compared to the height H of the cross section, the accuracy decreases. With the length of the element equal to the height of the element's cross-section, the amplitude error reaches 24%.

6.7.3 Tetrahedral Space–Time Element of a Plate

We assume linear interpolation functions to describe the distribution of the generalized displacements in the space–time element domain. The vertical displacement w and the rotations in both x and y directions, i.e., θ_x and θ_y, will be expressed in terms of nodal parameters. The following polynomials of the first order are used

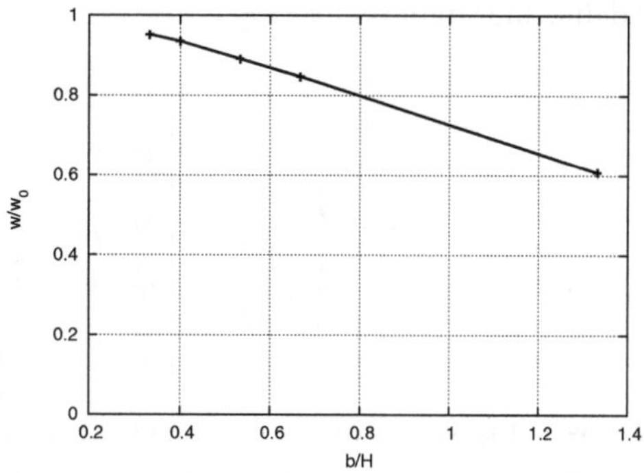

Fig. 6.31 The accuracy of the calculated displacements of the free end of the beam depending on the length of the spatial elements.

$$\begin{Bmatrix} w \\ \theta_x \\ \theta_y \end{Bmatrix} = \begin{Bmatrix} a_1x+b_1y+c_1t+d_1 \\ a_2x+b_2y+c_2t+d_2 \\ a_3x+b_3y+c_3t+d_3 \\ a_4x+b_4y+c_4t+d_4 \end{Bmatrix} . \tag{6.133}$$

The constants a_i, b_i, c_i, and d_i, $i = 1,2,3$ are determined from the relationships

$$\begin{Bmatrix} a_1 \\ b_1 \\ c_1 \\ d_1 \end{Bmatrix} = \mathbf{P}^{-1} \begin{Bmatrix} w_1 \\ w_2 \\ w_3 \\ w_4 \end{Bmatrix}, \quad \begin{Bmatrix} a_2 \\ b_2 \\ c_2 \\ d_2 \end{Bmatrix} = \mathbf{P}^{-1} \begin{Bmatrix} \theta_{x1} \\ \theta_{x2} \\ \theta_{x3} \\ \theta_{x4} \end{Bmatrix}, \quad \begin{Bmatrix} a_3 \\ b_3 \\ c_3 \\ d_3 \end{Bmatrix} = \mathbf{P}^{-1} \begin{Bmatrix} \theta_{y1} \\ \theta_{y2} \\ \theta_{y3} \\ \theta_{y4} \end{Bmatrix}, \tag{6.134}$$

where w_i, θ_{xi}, θ_{yi} are nodal displacements at node i. The matrix $\mathbf{P}$ then has the form

$$\mathbf{P} = \begin{bmatrix} x_1 & y_1 & t_1 & 1 \\ x_2 & y_2 & t_2 & 1 \\ x_3 & y_3 & t_3 & 1 \\ x_4 & y_4 & t_4 & 1 \end{bmatrix} . \tag{6.135}$$

If we denote by $\mathbf{r}_i$, $i = 1,\ldots,4$, the columns of the matrix $\mathbf{P}^{-1}$ and by $\mathbf{g}$ the vector of monomials

$$\mathbf{g} = [x, y, t, 1] , \tag{6.136}$$

then the shape functions of the tetrahedron are

$$\mathbf{N} = [\mathbf{N}_1, \mathbf{N}_2, \mathbf{N}_3, \mathbf{N}_4] , \quad \mathbf{N}_i = \mathbf{g}\,\mathbf{r}_i, \quad i = 1,\ldots,4. \tag{6.137}$$

We define the strain vector ε by

$$\varepsilon = \begin{Bmatrix} \kappa_x \\ \kappa_y \\ \kappa_{xy} \\ \beta_x \\ \beta_y \end{Bmatrix} = \begin{bmatrix} 0 & \frac{\partial}{\partial x} & 0 \\ 0 & 0 & \frac{\partial}{\partial x} \\ 0 & \frac{\partial}{\partial y} & \frac{\partial}{\partial x} \\ \frac{\partial}{\partial x} & 1 & 0 \\ \frac{\partial}{\partial y} & 0 & 1 \end{bmatrix} \begin{Bmatrix} w \\ \theta_x \\ \theta_y \end{Bmatrix} = \mathscr{D}\mathbf{w} . \tag{6.138}$$

Here, $\mathbf{w}$ is the vector of displacements. The stress vector σ contains the bending moments m_x, m_y, and m_{xy} and shear forces q_x and q_y per unit length:

$$\sigma = \begin{Bmatrix} m_x \\ m_y \\ m_{xy} \\ q_x \\ q_y \end{Bmatrix} . \tag{6.139}$$

The elasticity matrix,

$$\mathbf{E} = \begin{bmatrix} D & \nu D & 0 & 0 & 0 \\ & D & 0 & 0 & 0 \\ & & \frac{1-\nu}{2} D & 0 & 0 \\ & & & H & 0 \\ & & & & H \end{bmatrix} , \quad D = \frac{t^3}{12} \frac{E}{1-\nu^2} , \quad H = \frac{5}{6} G t , \tag{6.140}$$

where E is the Young modulus, ν is the Poisson coefficient, G is the shear modulus, and t is the thickness of the plate, allows us to write the stress–strain relation:

$$\sigma = \mathbf{E}\varepsilon . \tag{6.141}$$

The integration over the space–time tetrahedron is simplified if the origin of the coordinate system is placed at the centre of gravity of the element. We denote the entries of the matrix $\mathbf{P}^{-1}$ (6.135) by p_{ij}. Moreover, ρ is the mass density, η_w is the internal damping coefficient of the Kelvin–Voight model, η_z is the external damping coefficient, and V is the volume of the tetrahedral space–time element. Then the space–time stiffness matrix entries derived in terms of the displacements are as follows

$$
\begin{aligned}
K_{ij11} =& HV(p_{1i}p_{1j}+p_{2i}p_{2j})-\rho h p_{3i}p_{3j}\ ,\\
K_{ij12} =& HV\,p_{1i}p_{4j}+\eta_w V\,p_{1i}p_{3j}\ ,\\
K_{ij13} =& HV\,p_{2i}p_{4j}+\eta_w V\,p_{2i}p_{3j}\ ,\\
K_{ij21} =& HV\,p_{4i}p_{1j}\ ,\\
K_{ij22} =& DV\,p_{1i}p_{1j}+\frac{1-nu}{2}DV\,p_{2i}p_{2j}+\frac{HV}{20}p_{1i}p_{1j}\sum x_k^2+\frac{HV}{20}p_{2i}p_{2j}\sum y_k^2+\\
&+\frac{HV}{20}p_{3i}p_{3j}\sum t_k^2+\frac{HV}{20}(p_{1i}p_{2j}+p_{2i}p_{1j})\sum x_k y_k+\\
&+\frac{HV}{20}(p_{2i}p_{3j}+p_{3i}p_{2j})\sum y_k t_k+\frac{HV}{20}(p_{1i}p_{3j}+p_{3i}p_{1j})\sum x_k t_k+\\
&+HV\,p_{4i}p_{4j}-p_{3i}p_{3j}\frac{\rho h^3}{12}V+(\eta_w+\eta_z)V\,p_{4i}p_{3j}\ ,\\
K_{ij23} =& \nu DV\,p_{1i}p_{2j}+\frac{1-nu}{2}DV\,p_{2i}p_{1j}\ ,\\
K_{ij31} =& HV\,p_{4i}p_{2j}\ ,\\
K_{ij32} =& \nu DV\,p_{2i}p_{1j}+\frac{1-nu}{D}V2p_{1i}p_{2j}\ ,\\
K_{ij33} =& DV\,p_{2i}p_{1j}+\frac{1-nu}{2}DV\,p_{1i}p_{1j}+\frac{HV}{20}p_{1i}p_{1j}\sum x_k^2+\frac{HV}{20}p_{2i}p_{2j}\sum y_k^2+\\
&+\frac{HV}{20}p_{3i}p_{3j}\sum t_k^2+\frac{HV}{20}(p_{1i}p_{2j}+p_{2i}p_{1j})\sum x_k y_k+\\
&+\frac{HV}{20}(p_{2i}p_{3j}+p_{3i}p_{2j})\sum y_k t_k+\frac{HV}{20}(p_{1i}p_{3j}+p_{3i}p_{1j})\sum x_k t_k+\\
&+HV\,p_{4i}p_{4j}-p_{3i}p_{3j}\frac{\rho h^3}{12}V+(\eta_w+\eta_z)V\,p_{4i}p_{3j}\ .
\end{aligned}
\tag{6.142}
$$

The test problem depicted in Figure 6.32 results in amplitudes which are there compared with other results collected in Table 6.5.

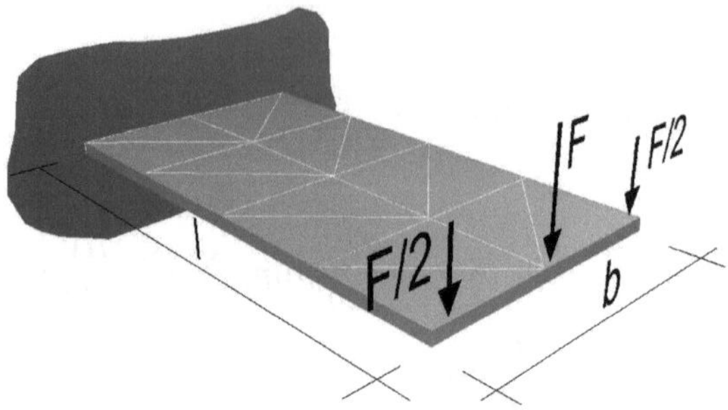

Fig. 6.32 Test problem — a cantilever plate.

Table 6.5 Summary results of calculations of test analysis of a disc.

mesh	No. of nodes	deflection w	period T	$\frac{Dw}{QL^2}$	$\frac{\omega}{\sqrt{D/\rho t L^4}}$
2×1	6	318.	109.	0.564	4.84
4×2	15	537.	133.	0.953	3.97
8×4	45	720.	157.	1.277	3.37
Other results for comparison*:					
— Ritz method					3.47
— experimental					3.42
— Plunkett'e experiments					3.50
— finite elements (modal analysis)					
— 2×1 (4 triangular elements)					3.39
— 4×2 (16 triangular elements)					3.44

*O.C. Zienkiewicz, *The Finite Element Method in Engineering Science*, 2nd edition, McGraw-Hill, 1971

6.8 Triangular Elements Expressed in Velocities

The final system of equations obtained using the velocity formulation has the same features as when the displacement formulation is used. We will try to construct a model of a triangular element of an axially vibrating rod and investigate its properties. The accuracy of the calculated amplitudes, the periods of the free vibrations, and the stability of solutions obtained by using these elements will be important when we evaluate the velocity simplex elements.

The virtual power equation (6.61) is used to build a recursive matrix equation. Below we will use triangular elements. In the beginning, let us take two complementary triangles, shown in Figure 6.33. The principle of virtual work (6.62) allows us to write

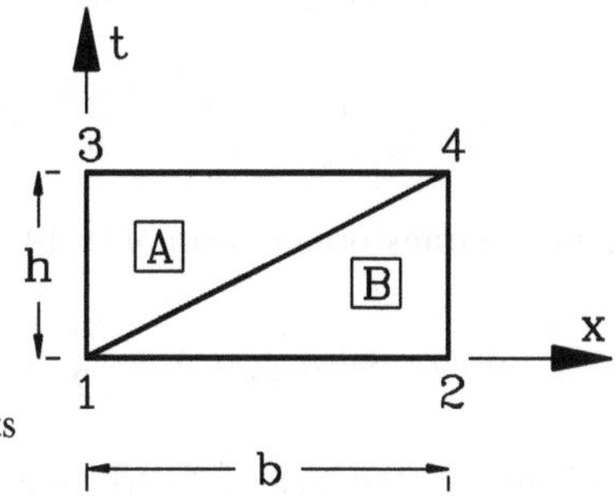

Fig. 6.33 Triangular system of elements described with velocities.

$$\int_{\Omega} (\mathbf{N}^*)^T \rho A \frac{\partial v}{\partial t} \mathrm{d}\Omega + \int_{\Omega} \frac{\partial (\mathbf{N}^*)^T}{\partial x} EA \frac{\partial}{\partial x} \left[\int_{t_0}^{t} v \mathrm{d}t \right] \mathrm{d}\Omega = 0 \,. \tag{6.143}$$

Here, $\mathbf{N}^*$ is the matrix of virtual shape functions, v is the velocity at the point (x,t), Ω is the domain of the triangle A or B, and E, ρ, A are the Young modulus, mass density and cross sectional area, respectively. In the element A the real velocity is described by

$$v_A(x,y) = \mathbf{N}_A(x,t)\, \mathbf{v} = \left[1 - \frac{t}{h}, \frac{x}{b}, -\frac{x}{b} + \frac{t}{h} \right] \mathbf{v} \,, \tag{6.144}$$

and the virtual functions in nodes 3 and 4 have the form

$$\mathbf{N}_A^*(x,t) = \left[-\frac{x}{b} + \frac{t}{h}, \frac{x}{b} \right] \,. \tag{6.145}$$

Inertia matrix is then determined from the relation

$$\mathbf{M}_A = \int_0^b \int_{\frac{hx}{b}}^{h} (\mathbf{N}_A)^T \rho A \frac{\partial}{\partial t} \mathbf{N}_A \,\mathrm{d}t\,\mathrm{d}x = \rho A \frac{b}{6} \begin{bmatrix} -1 & 0 & 1 \\ -1 & 0 & 1 \end{bmatrix} \,. \tag{6.146}$$

The real shape functions and the virtual function of the node 4 in the element B have the following forms

$$\mathbf{N}_B = \left[1 - \frac{x}{b}, \frac{x}{b} - \frac{t}{h}, \frac{t}{h} \right], \quad \mathbf{N}_B^* = \frac{t}{h} \,. \tag{6.147}$$

The mass matrix is thus

$$\mathbf{M}_B = \rho A \frac{b}{6} \left[0, -1, 1 \right] \,. \tag{6.148}$$

The stiffness matrices described in the second part of the equation (6.143) have the following form

$$\mathbf{K}_A = \int_0^b \int_{\frac{hx}{b}}^{h} \frac{\partial (\mathbf{N}_A^*)^T}{\partial x} EA \frac{\partial}{\partial x} \left[\int_0^t \mathbf{N}_A \mathrm{d}t \right] \mathrm{d}t\,\mathrm{d}x = EA \frac{h^2}{3b} \begin{bmatrix} 0 & -1 & 1 \\ 0 & 1 & -1 \end{bmatrix},$$
$$\mathbf{K}_B = \mathbf{0} \,. \tag{6.149}$$

The initial force vector in the element is

$$\mathbf{s} = EA\int_0^b \int_{\frac{hx}{b}}^h \frac{\partial(\mathbf{N}_A^*)^T}{\partial x}\,\frac{\partial w(t_0)}{\partial x}\,\mathrm{d}t\,\mathrm{d}x = EA\varepsilon_0\frac{h}{2}\begin{bmatrix}-1\\ 1\end{bmatrix}. \qquad (6.150)$$

We can use the transformed form of (6.143):

$$\frac{1}{4\Delta}(x_m t_n - x_n t_m)\sum_i x_{kj}\,v_i + \frac{t_{mn}}{2}\varepsilon_0 - \frac{1}{4\Delta}t_{mn}t_0\sum_i x_{jk}\,v_i = 0\,, \qquad (6.151)$$

where Δ is the surface area of the triangle, $t_{mn} = t_m - t_n$, $x_{kj} = x_k - x_j$, and t_0 is the time location of the node with the lowest coordinate t_i in the coordinate system with origin at the centre of gravity of the triangle.

The global arrays of the system allow us to write a recursive equation:

$$\left(\frac{1}{6}\left[\begin{array}{cc|cc}-1 & 0 & 1 & 0\\ -1 & -1 & 1 & 1\end{array}\right] + \frac{c^2h^2}{3b^2}\left[\begin{array}{cc|cc}0 & 0 & 1 & -1\\ 0 & 0 & -1 & 1\end{array}\right]\right)\mathbf{v} + \frac{h}{2b}c^2\varepsilon_0\begin{bmatrix}-1\\ 1\end{bmatrix} = \mathbf{0}\,. \qquad (6.152)$$

If we fix the right degree of freedom, we obtain an oscillating system with one degree of freedom, described by the system of equations

$$\begin{cases} v_{i+1} = \frac{(1-2\kappa^2)v_i + 3\varepsilon_i}{1+\kappa^2}\,, \quad \kappa = \frac{ch}{b}\,,\\ \varepsilon_{i+1} = \varepsilon_i - \left(\frac{2}{3}v_i + \frac{1}{3}v_{i+1}\right)\frac{h}{b}\,. \end{cases} \qquad (6.153)$$

The period of oscillations of the oscillator modelled by the system (6.153) (with unit values for E, S, ρ, b) is 3.628, whereas the exact value is $2\pi/\sqrt{3} \approx 3.6276$.

The derived matrices were tested by examples. The task was an axially vibrating rod, divided into 20 spatial elements. The meshes of type a and b are shown in Figure 6.34. In order to compare the results, the task was solved with the use of a mesh of rectangular elements, with parameter $\alpha = 1/2$. In all cases, the time step was $h = 0.01$. The displacements of the free end of the bar in time, as the system response, are shown in Figure 6.34. Comparing the graphs, one can see the correct behaviour of the given system. The amplitudes of the parasitic frequencies are small. The difference illustrated in Figures 6.34a and 6.34b results from the manner in which the initial conditions were introduced. In the case of a, the initial speed $v_0 = 1$ was applied to the extreme node, which belongs to only one triangle. In case b, the extreme node at $t = 0$ joins two triangular elements. Finally, in this way of loading, the grid b better models the propagation of pulses. The stiffness matrices in the case of a grid consisting of two triangles (e.g. Figure 6.33) can also be derived in a simpler manner. We must calculate the work of the internal forces on the virtual displacements in the domain $0 \le x \le b$, $0 \le t \le h$. For example, if we take the system shown in Figure 6.34a, the increment of deformations in the interval $[0,h]$, calculated at $x = b/3$, and thus at the centre of gravity of the left triangle, is

$$\Delta\varepsilon = \frac{2}{3}h\frac{v_2 - v_1}{b} + \frac{1}{3}h\frac{v_4 - v_3}{b}\,. \qquad (6.154)$$

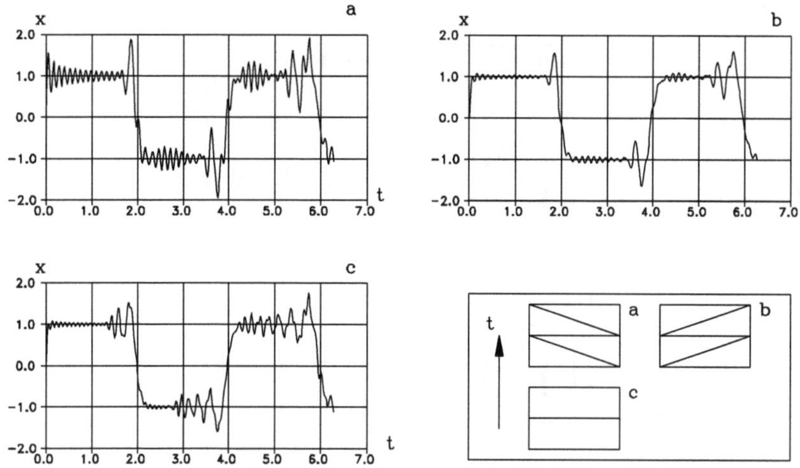

Fig. 6.34 Displacements in time of the free end of the rod with triangular and rectangular meshes (20 spatial elements).

Finally, we get the matrix

$$\frac{EAh^2}{6b}\left[\begin{array}{cc|cc} 2 & -2 & 1 & -1 \\ -2 & 2 & -1 & 1 \end{array}\right]. \tag{6.155}$$

The displacements at the end of the interval $[0,h]$ are determined by the formula $x_1 = x_0 + hv_0$. The system response does not depend on which time we calculate the strain increment or on how situated the time interval of length h along which the integration is carried out. Sample diagrams with the corresponding stiffness matrices are given in Table 6.6. The first set of options there is conditionally stable. Others are unconditionally stable due to the time step. Assuming the inertia matrix given in (6.146) and (6.148), summed in (6.152), and the stiffness matrix of the system according to the first system in Table 6.6, we obtain triangular matrices of coefficients (right submatrix, associated with $\mathbf{v}_{i+1}$).

Triangular elements are ideal for refining and coarsening the grid over time. We can also modify the mesh in selected spatial areas and decrease the time step (Figure 6.35). A subintegration in time carried out in selected areas enables us, for example, to easilydescribe the contact. In Figure 6.35b, two joints are introduced: t_c is the entry of the point into contact and t_f is the end of the contact interval. In practice, the whole space–time layer, i.e., all the nodes placed at t_{i+1} and the supplementary nodes (sub–division), are treated in the same stage of calculations, in one matrix equation.

Table 6.6 Stiffness matrices in the different schemes.

scheme	stiffness matrix
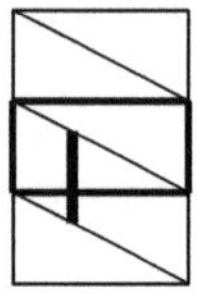	$\frac{EAh^2}{6b}\left[\begin{array}{rr\|rr} 3 & -3 & 0 & 0 \\ -3 & 3 & 0 & 0 \end{array}\right]$
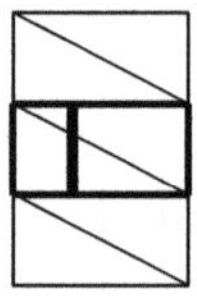	$\frac{EAh^2}{6b}\left[\begin{array}{rr\|rr} 2 & -2 & 1 & -1 \\ -2 & 2 & -1 & 1 \end{array}\right]$
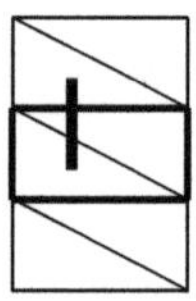	$\frac{EAh^2}{6b}\left[\begin{array}{rr\|rr} 1 & -1 & 2 & -2 \\ -1 & 1 & -2 & 2 \end{array}\right]$
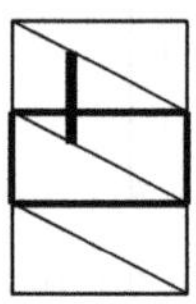	$\frac{EAh^2}{6b}\left[\begin{array}{rr\|rr} 0 & 0 & 3 & -3 \\ 0 & 0 & -3 & 3 \end{array}\right]$

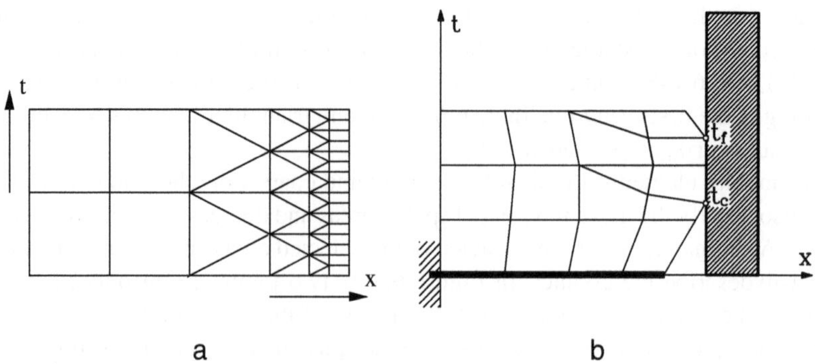

Fig. 6.35 Examples of grid refinement in time-space using triangular elements: a) general idea, b) scheme used for contact problems.

Chapter 7
Space–Time Finite Elements and a Moving Load

Non-classical problems are usually poorly treated by classical and commonly known solution methods. Time dependent problems, especially vibrations, described by partial differential equations are classically treated with the finite element method in space and the family of Newmark methods in time. Such time integration methods were described in Chapter 5. We discussed in the introduction to Chapter 6 the disadvantages of such an approach and the necessity for a more general treatment of phenomena in space and time. The space–time finite element method extends the finite element approximation of the differential equation over the time domain. The main advantage in our moving mass problems concerns its facility in treating the partial derivatives obtained from the chain rule applied to the acceleration of the inertial particle in a moving coordinate system, equations (3.119) or (3.121).

First we will consider a string subjected to a moving mass accompanying a moving point force. We will derive the matrices that concern only the mass particle. The rest of the structure, i.e., the string and its space–time elements, were described in Section 6.3.2. Then a Bernoulli–Euler beam will be considered. In this case, the resulting matrices are relatively complex, although their derivation is simple. The Timoshenko beam exhibits, in turn, simple matrices of the elements carrying the inertial particle. The plate space–time element will show the approach to two-dimensional structures. In the next chapter, the string, a Bernoulli–Euler beam, and a Timoshenko beam will also be considered with the use of the Newmark time integration method. In this case we will propose mass matrices which would be applied to a general structure. This is an advantageous solution since most computer codes involve only the Newmark method.

In this chapter, the velocity formulation will be used since it, with its two-level scheme, is a more efficient approach. The displacement formulation is a three-level scheme and is not sufficiently advantageous for our purpose.

C.I. Bajer and B. Dyniewicz: Numerical Analysis of Vibrations of Structures, LNACM 65, pp. 181–221.
springerlink.com

7.1 Space–Time Finite Element of a String

We will here treat the matrices derived from various terms of the differential equation called the inertia matrix, the damping matrix, and the stiffness matrix. When multiplied by the nodal velocities, they give the vertical transverse forces, the Coriolis forces, and the centrifugal forces, respectively. We will also give the algorithm for the solution of the vibrating string subjected to a moving mass.

7.1.1 Discretization of the String Element Carrying a Moving Mass

The last term $\delta(x-\mathrm{v}t)m\mathrm{d}^2w(\mathrm{v}t,t)/\mathrm{d}t^2$ in the equation of motion, (3.1), describes an inertial moving mass. $\mathrm{d}^2w(\mathrm{v}t,t)/\mathrm{d}t^2$ is the vertical acceleration of the moving mass and at the same time the acceleration of the point of the string at which the mass is temporarily placed (it is $x = x_0 + \mathrm{v}t$). The acceleration of the mass $\mathrm{d}^2w(\mathrm{v}t,t)/\mathrm{d}t^2$ moving with a constant velocity v, according to the Renaudot formula (which in fact is the chain rule for differentiation), results in three terms:

$$\frac{\mathrm{d}^2w(\mathrm{v}t,t)}{\mathrm{d}t^2} = \left.\frac{\partial^2 w(x,t)}{\partial t^2}\right|_{x=\mathrm{v}t} + 2\mathrm{v}\left.\frac{\partial^2 w(x,t)}{\partial x\partial t}\right|_{x=\mathrm{v}t} + \mathrm{v}^2\left.\frac{\partial^2 w(x,t)}{\partial x^2}\right|_{x=\mathrm{v}t} . \tag{7.1}$$

Thus we can separate the transverse acceleration, the Coriolis acceleration, and the centrifugal acceleration. This is the so-called Renaudot notation for the constant speed v.

In our space–time finite element method we formulate equations in terms of velocities. The mass acceleration $\mathrm{d}^2w(\mathrm{v}t,t)/\mathrm{d}t^2$ is expressed in terms of velocities as well

$$\frac{\mathrm{d}^2w(\mathrm{v}t,t)}{\mathrm{d}t^2} = \frac{\mathrm{d}v(\mathrm{v}t,t)}{\mathrm{d}t} = \left.\frac{\partial v(x,t)}{\partial t}\right|_{x=\mathrm{v}t} + \mathrm{v}\left.\frac{\partial v(x,t)}{\partial x}\right|_{x=\mathrm{v}t} + \mathrm{v}^2\frac{\partial^2 w_0}{\partial x^2} . \tag{7.2}$$

The first term on the right-hand side states the real inertia (when multiplied by m), the second term (multiplied by m) expresses forces similar to damping forces and the third term (also multiplied by m) expresses forces similar to stiffness forces.

In the final stage three resulting matrices are responsible for transverse inertia (the matrix has the form of the inertia matrix), damping forces (the matrix multiplied by the velocity vector has a form similar to the Coriolis forces) and stiffness (potential) forces (the matrix, if multiplied by the velocity vector, has a form similar to the centrifugal forces). The third matrix appears as the result of initial displacements in the time interval.

Let us now follow this idea and treat numerically the right-hand side inertial term of (7.2). The same mathematical steps as in the case of pure string enables us to integrate the inertial term

$$\int_0^h \int_0^b v^* \delta(x-\mathrm{v}t)m\frac{\mathrm{d}^2w(x_0+\mathrm{v}t,t)}{\mathrm{d}t^2}\,\mathrm{d}x\,\mathrm{d}t . \tag{7.3}$$

We use the same linear interpolation of the velocity (6.64). The virtual velocity v^*:

$$v^*(x,t) = \mathbf{N}^*\,\dot{\mathbf{q}}_p = \delta(t-\alpha h)\left[1-\frac{x}{b},\;\frac{x}{b}\right]\dot{\mathbf{q}}_p\,. \tag{7.4}$$

Consequent integration results in two matrices: the moving mass inertia matrix $\mathbf{M}_m$

$$\mathbf{M}_m = \frac{m}{h}\left[\begin{array}{cc|cc} -(1-\kappa)^2 & -\kappa(1-\kappa) & (1-\kappa)^2 & \kappa(1-\kappa) \\ -\kappa(1-\kappa) & -\kappa^2 & \kappa(1-\kappa) & \kappa^2 \end{array}\right], \tag{7.5}$$

where $\kappa = (x_0 + \mathrm{v}\alpha h)/b$, x_0 is a starting position of the mass in the space–time element (at $t = t_0$) (see Figure 7.1), and the moving mass damping matrix $\mathbf{C}_m$

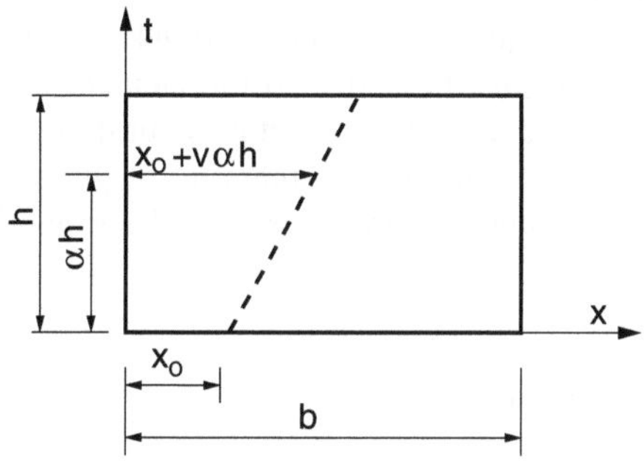

Fig. 7.1 Mass path in the space–time finite element.

$$\mathbf{C}_m = \frac{m\mathrm{v}}{b}\left[\begin{array}{cc|cc} -(1-\kappa)(1-\beta) & (1-\kappa)(1-\beta) & -(1-\kappa)\beta & (1-\kappa)\beta \\ -\kappa(1-\beta) & \kappa(1-\beta) & -\kappa\beta & \kappa\beta \end{array}\right]. \tag{7.6}$$

Let us now consider the contribution of $w(x,0)$ in (7.2). We integrate by parts the virtual work

$$\mathrm{v}^2\int_0^h\int_0^b v^*\frac{\partial^2 w(x,0)}{\partial x^2}\,\mathrm{d}x\mathrm{d}t = -\mathrm{v}^2\int_0^h\int_0^b \frac{\partial v^*}{\partial x}\frac{\partial w(x,0)}{\partial x}\,\mathrm{d}x\mathrm{d}t\,. \tag{7.7}$$

Since displacements of the left and right node of the element are expressed by $w_L = w_L^0 + h[\beta v_1 + (1-\beta)v_3]$ and $w_R = w_R^0 + h[\beta v_2 + (1-\beta)v_4]$, we can derive the required $\partial w_0/\partial x$

$$\frac{\partial w_0}{\partial x} = \frac{w_R - w_L}{b} = \frac{w_R^0 - w_L^0}{b} + \frac{h}{b}[-\beta v_1 + \beta v_2 - (1-\beta)v_3 + (1-\beta)v_4]\,. \tag{7.8}$$

The numbering of the nodes is presented in Figure 7.1. The matrix $\mathbf{K}_m$ is the stiffness mass matrix

$$\mathbf{K}_m = \frac{hm\mathrm{v}^2}{b^2}\left[\begin{array}{cc|cc} \beta & -\beta & 1-\beta & -(1-\beta) \\ -\beta & \beta & -(1-\beta) & 1-\beta \end{array}\right], \tag{7.9}$$

The term $(w_R^0 - w_L^0)/b$ in (7.8) multiplied by mv^2/b results in initial nodal forces $\mathbf{e}_m$ in the space–time layer.

7.1.1.1 Technical Remarks

Even in the simplest case of uni-dimensional structures, we must be careful as to in what time the respective terms should be posed. This issue arises for the inter-element zones. Let us look at Figure 7.2. In step 1 we proceed from nodes 1 and 2 to nodes 3 and 4. After solving the system of equations for the displacements at the final time in the time layer, we must compute the nodal forces introduced by the mass, aside from the nodal forces resulting from the classical space–time analysis of the string. In practice, we must add the term $-mv^2/b \cdot \varepsilon_0$ to the respective element, corresponding with the degree of freedom of nodes 1 and 3 in the vector $\mathbf{e}_m$, located on the left-hand side of the equilibrium equation (6.84)–(6.88). The same term with a positive sign $mv^2/b \cdot \varepsilon_0$ must be added to the element corresponding with the degree of freedom of nodes 2 and 4 in the vector $\mathbf{e}_m$. In step 2 we compute the nodal forces based on nodes 6 and 7 and we put them in the places that correspond with the degrees of freedom of nodes 4 and 5 or of nodes 6 and 7. This procedure is presented in Algorithm 11.

7.1.2 Numerical Results

The numerical results obtained with the proposed space–time approach can be compared with the semi-analytical solution. Moreover, the spring–mass finite element solution can also be plotted. In our tests, the string was discretized by a set of 200 finite elements. The time step h was equal to $b/40$v. This means that the mass passes

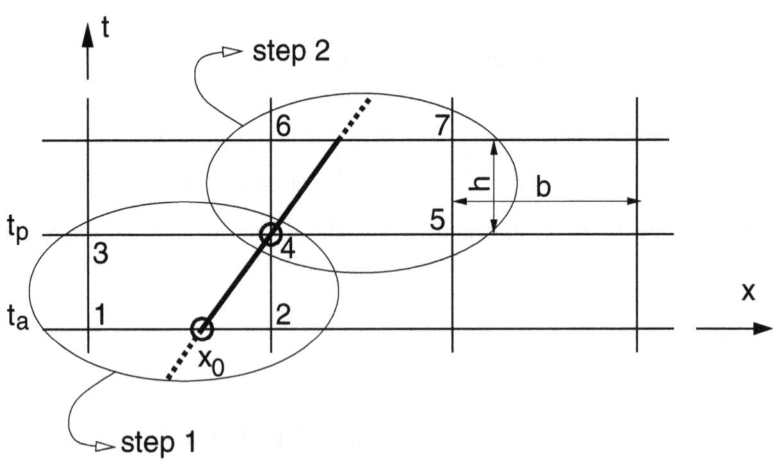

Fig. 7.2 Inter-element zones.

Algorithm 11. The solution for the vibrating string with concentrated moving mass.

1. Set the problem data: length L, number of spatial elements n_e, number of nodes $n_p = n_e + 1$, element length $b = L/n_e$, mass density ρ, tension N, concentrated moving mass m, speed v, point force P, time step h.
2. Compute global space–time matrices of the string: left-hand side square matrix $\mathbf{A}$ and right-hand side matrix $\mathbf{B}$, both of dimension $n_p \times n_p$.
3. Initialize the nodal force vector $\mathbf{e} = \mathbf{0}$.
4. Initialize time of the process: $t = 0$.
5. Increment time $t := t + h$ (which is the final time in time layer).

   - Determine the element i_{el} on which the load travels:
     $i_{el} = \mathrm{INT}(\mathrm{v}t/b - eps) + 1, eps = 0.0001$,
     ε is a small number, significantly smaller than $\mathrm{v}h/b$ (for example $\mathrm{v}h/b/100$).
   - Determine the load position on the element at the beginning of the time step x_0:
     $x_0 = (t-h)\mathrm{v} - (i_{el} - 1)b$.
   - Compute $\kappa = (x_0 + \mathrm{v}\alpha h)/b$.
   - Compute matrices $\mathbf{M}_m$, $\mathbf{C}_m$, and $\mathbf{K}_m$ of the mass placed on the element. These are $n_p \times 2n_p$ matrices:
     $\mathbf{M}_m = [\mathbf{M}_m^A|\mathbf{M}_m^B]$, $\mathbf{C}_m = [\mathbf{C}_m^A|\mathbf{C}_m^B]$, $\mathbf{K}_m = [\mathbf{K}_m^A|\mathbf{K}_m^B]$.
   - Add the mass matrices to the global string matrices: $\mathbf{A} := \mathbf{A} + \mathbf{M}_m^A + \mathbf{C}_m^A + \mathbf{K}_m^A$, $\mathbf{B} := \mathbf{B} + \mathbf{M}_m^B + \mathbf{C}_m^B + \mathbf{K}_m^B$.
   - Set the external load vector $\mathbf{Q}$, which has two non-zero components, at the places i_{el} and $i_{el} + 1$:
     $Q_{i_{el}} = P \cdot (-\mathrm{v}t/b + i_{el})$, $Q_{i_{el}+1} = P \cdot (\mathrm{v}t/b - i_{el} + 1)$.
   - Compute the right-hand side vector of the system of algebraic equations: $\mathbf{f} = \mathbf{Q} - \mathbf{e}$.
   - Solve the system of equations for velocities $\mathbf{v}_{i+1}$:
     $\mathbf{A}\mathbf{v}_i + \mathbf{B}\mathbf{v}_{i+1} = \mathbf{Q} - \mathbf{e}$.
   - Compute the displacements:
     $\mathbf{w}_{i+1} = \mathbf{w}_i + ((1-\beta)\mathbf{v}_i + \beta\mathbf{v}_{i+1})h$, $\beta = 1 - \alpha$.
   - Compute two elements of the vector $\mathbf{e}$:
     $e_{i_{el}} = -m\mathrm{v}^2/b \cdot \varepsilon_0$, $e_{i_{el}+1} = m\mathrm{v}^2/b \cdot \varepsilon_0$, $\varepsilon_0 = (w_{i_{el}+1} - w_{i_{el}})/b$.

6. Repeat the time loop until the end of the simulation time.

---

from joint to joint in 40 time steps. We assume $m = 1.0$, $\rho A = 1.0$, $l = 1.0$, $N = 1$, and $P = -1$. The results obtained by the space–time finite element method are presented in Figure 7.3. We notice that at lower speeds, up to 0.3–0.4c, the coincidence with the semi-analytical results is almost perfect. We observe the convergence of the results to the semi-analytical solution with decreasing time steps and increasing numbers of spatial elements. Unfortunately, the convergence is slow. At higher speeds, the total time of the simulation is shorter and only a lesser number of time steps is required to reach the end support during the mass motion. All

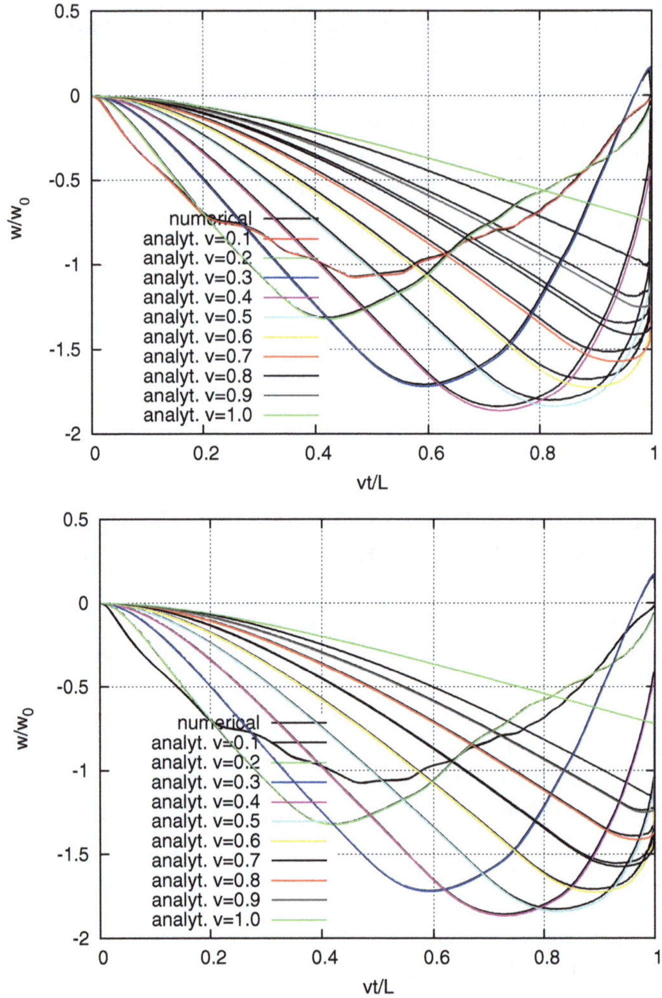

Fig. 7.3 Displacements under a moving mass using the space–time finite element solution for $\alpha = 0.5$ (upper) and $\alpha = 1.0$ (lower), compared with the semi-analytical solution at different speeds v ($c = 1$).

the important features of the resulting curves, especially the high gradients of displacement near the end support, are then represented with lower accuracy. The error analysis of this method allows us to say that in the general case the error is equal to $h^2(1/2-\alpha)+\mathscr{O}h^3$. In the particular case $\alpha = 0.5$, the error is $1/12h^3+\mathscr{O}h^4$. However, in this case the time integration scheme of the space–time finite element method is conditionally stable. The gap between the numerical and semi-analytical results is visible in both diagrams. The plot for $\alpha = 1/2$ is visually better in the higher speed range.

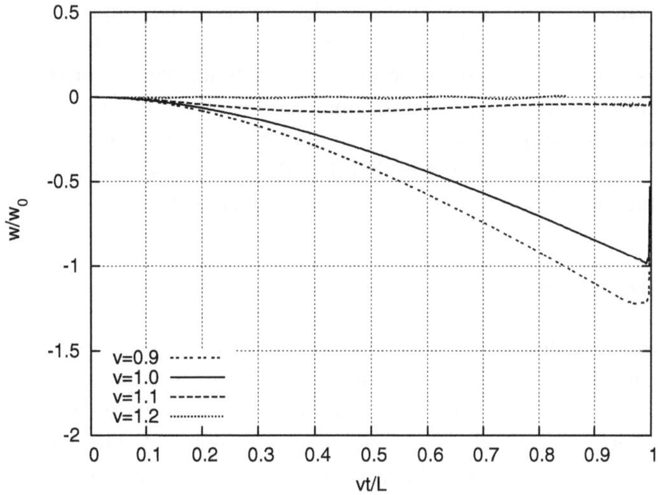

Fig. 7.4 Displacements under a moving mass for v equal to $0.9c$, $1.0c$, $1.1c$, and $1.2c$.

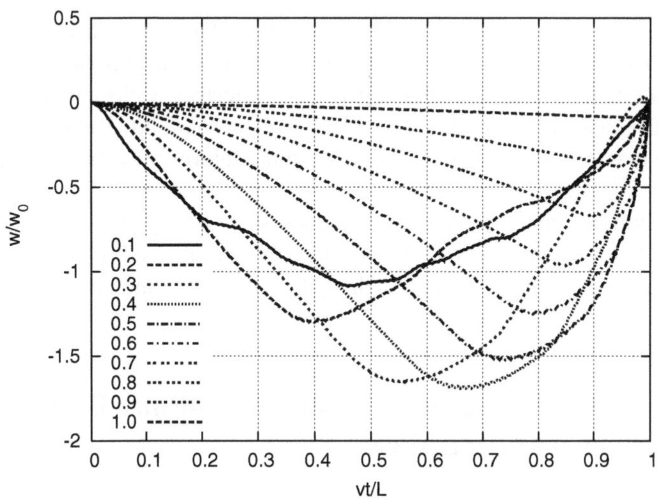

Fig. 7.5 Finite element solution—displacements of a string under an oscillator for speeds $\mathsf{v} = 0.1c$–$1.0c$.

Higher velocities can also be considered. Figure 7.4 presents the displacements in time of the particle for $0.9 \le \mathsf{v}/c \le 1.2$. We notice a good coincidence of the plot with the expected zero line. Further examples prove the efficiency and accuracy of this approach. One can plot the displacements of selected fixed points of the string. In such a case, there results a very good coincidence with the semi-analytical approach.

We can compare our results with the displacements of the contact point of a string under a travelling oscillator (Figure 7.5). Note the significant difference, especially for the higher speed range, between the semi-analytical results or the space–time finite element approach and the curves obtained for the oscillator. A pure mass can not be replaced in computations by an oscillator for problems with a large moving mass influence, i.e., for velocities higher than 0.3–0.5 times the wave speed and for a ration of the moving mass to the string mass higher than 0.2.

7.1.3 Conclusions

In this section, we dealt with the numerical analysis of string vibrations under a moving inertial load. We derived a matrix formula for the time integration procedure using the space–time finite element method. The solutions presented in the literature are derived from classical time integration schemes. The published results are acceptable for low speeds of the travelling mass. In such a case, the errors in the formulations do not contribute visible differences to the results. In common practice, a massless force acting on a string in the form of an oscillator is applied. Such results are gross underestimates, and for a velocity higher than $0.2\,c$–$0.3\,c$ they cannot be relied on.

The approach presented in this book can be applied to the whole range of speeds, up to the wave speed itself: $\mathrm{v} = c$. The precision of the results is high. In the case of a speed higher than the wave speed, the particle's trajectory is close to the theoretical zero line. The discontinuities at $x = l$ which have been exhibited and proved analytically for a massless string are easily visible in the figures presenting the numerical trajectories.

The method presented in this section can be successfully applied to other structures subjected to an inertial load: beams, frames, and plates. Moreover, the space–time finite element approach can be adapted to classical time integration schemes (Newmark, etc.).

7.2 Space–Time Elements for a Bernoulli–Euler Beam Carrying a Moving Mass

A discrete beam element, in both the classical finite element method and space–time finite element method, is more complicated than A string element. The derivation of the matrices by using Dirac delta virtual time functions are conceptually difficult. The product of a Dirac-type virtual function and a Dirac distribution of the mass in space, with the argument varying in time, causes mathematical problems. In this section we use hat-shaped virtual functions in our analysis. They are simple to analyse and have a lower error rate. The value of this function is constant in time and its respective derivatives and double integrals can be computed relatively simply.

The mathematical steps will be performed here in another way than for the string element carrying a mass. The beam element results in larger matrices with significantly more complicated expressions. In the following, we will consider mathematically the first element of the inertia matrix only. All remaining elements of $\mathbf{M}_m$ can be computed matricially. Other matrices, i.e., $\mathbf{C}_m$, $\mathbf{K}_m$, and $\mathbf{e}_m$ in (6.77), are obtained in the same way. The matrices of the element carrying the mass differ in each time step since the position of the mass particle varies in time. Thus the global matrices must be established for each time step.

We remember that a hat-shaped virtual time function v^* is constant in time and in the case of the Bernoulli–Euler beam it has the following form:

$$v^*(x) = \left[1 - 3\frac{x^2}{b^2} + 2\frac{x^3}{b^3},\ x - 2\frac{x^2}{b} + \frac{x^3}{b^2},\ 3\frac{x^2}{b^2} - 2\frac{x^3}{b^3},\ -\frac{x^2}{b} + \frac{x^3}{b^2}\right]\mathbf{v}\,. \tag{7.10}$$

We recognize here the well-known shape functions that describe displacements (or velocities) in terms of the nodal displacements and the nodal rotations. The same interpolation formulas are used as for real spatial shape functions. In the case of the Euler beam, the nodal displacements are closely related to the rotation angles, hence the velocity approximation in the finite element is given by the following formula:

$$v(x,t) = N_1 v_1 + N_2\dot{\theta}_1 + N_3 v_2 + N_4\dot{\theta}_2 + N_5 v_3 + N_6\dot{\theta}_3 + N_7 v_4 + N_8\dot{\theta}_4\,. \tag{7.11}$$

As in (7.11), we use a linear interpolation in time and third order polynomials relative to space:

$$\begin{aligned}
N_1 &= \left(1 - 3\tfrac{x^2}{b^2} + 2\tfrac{x^3}{b^3}\right)\left(1 - \tfrac{t}{h}\right), & N_5 &= \left(1 - 3\tfrac{x^2}{b^2} + 2\tfrac{x^3}{b^3}\right)\tfrac{t}{h}\,,\\
N_2 &= \left(x - 2\tfrac{x^2}{b} + \tfrac{x^3}{b^2}\right)\left(1 - \tfrac{t}{h}\right), & N_6 &= \left(x - 2\tfrac{x^2}{b} + \tfrac{x^3}{b^2}\right)\tfrac{t}{h}\,,\\
N_3 &= \left(3\tfrac{x^2}{b^2} - 2\tfrac{x^3}{b^3}\right)\left(1 - \tfrac{t}{h}\right), & N_7 &= \left(3\tfrac{x^2}{b^2} - 2\tfrac{x^3}{b^3}\right)\tfrac{t}{h}\,,\\
N_4 &= \left(-\tfrac{x^2}{b} + \tfrac{x^3}{b^2}\right)\left(1 - \tfrac{t}{h}\right), & N_8 &= \left(-\tfrac{x^2}{b} + \tfrac{x^3}{b^2}\right)\tfrac{t}{h}\,.
\end{aligned} \tag{7.12}$$

The assumed shape functions allow us to derive all terms of the Renaudot formula (7.1). Then the elements of the matrix $\mathbf{M}_m$ can be computed. We present here only the analysis of the first element $(\cdot)_{11}$ of the inertia matrix

$$\begin{aligned}
(\mathbf{M}_m)_{11} &= -\frac{m}{h}\int_0^h\int_0^b \delta(x - x_0 - \mathrm{v}t)\left(1 - 3\frac{x^2}{b^2} + 2\frac{x^3}{b^3}\right)^2\left(1 - \frac{t}{h}\right)\mathrm{d}x\mathrm{d}t = \\
&= -\frac{m}{h}\int_0^h\int_0^b \left[1 - 3\frac{(x_0 + \mathrm{v}t)^2}{b^2} + 2\frac{(x_0 + \mathrm{v}t)^3}{b^3}\right]^2\left(1 - \frac{t}{h}\right)\mathrm{d}x\mathrm{d}t\,.
\end{aligned} \tag{7.13}$$

Taking into account the integration limits, we have the following:

$$
\begin{aligned}
(\mathbf{M}_m)_{11} = & -\frac{m}{560b^6}\Big[560b^6\left(4\kappa^6 - 12\kappa^5 + 9\kappa^4 + 4\kappa^3 - 6\kappa^2 + 1\right) + \\
& + 280b^4\text{v}^2h^2\left(10\kappa^4 - 20\,\kappa^3 + 9\kappa^2 + 2\kappa - 1\right) + \\
& + 21b^2\text{v}^4h^4\left(20\kappa^2 - 20\kappa + 3\right) + 5\text{v}^6h^h\Big]\;,
\end{aligned}
\tag{7.14}
$$

where

$$
\kappa = \frac{x_0 + \text{v}h/2}{b}\;. \tag{7.15}
$$

The complete matrices carrying mass particles $\mathbf{M}_m$, $\mathbf{C}_m$, $\mathbf{K}_m$, and $\mathbf{e}_m$ are large. They are listed in Tables 7.1–7.7. We must remember that the first three matrices join velocity vectors at two successive times and are composed of two square submatrices, left and right. They have dimensions $n \times 2n$, where n is the total number of degrees of freedom of the structure. The matrix $\mathbf{e}_m$ has dimension $n \times n$. All matrices are established for $m = 1$, so they must be multiplied by the actual value of m. We have also introduced $\xi = \text{v}h/b$ and κ given by (7.15).

7.2.1 Numerical Results

The numerical results for the displacements in time of the Bernoulli–Euler simply-supported beam are presented in Figure 7.6. The following parameter values were employed: $E = 1.0, A = 1.0, I = 0.01, l = 1.0, \rho = 1.0, m = 1.0$, and $P = -1.0$. The following boundary conditions were assumed in this example: $w(0,t) = w(l,t) = 0$. Additionally, natural boundary conditions were supplied by element interpolation functions: $w''(0,t) = w''(l,t) = 0$. The vertical displacements are related to the static deflection ($w_0 = Pl^3/48/E/I$) of the middle of the span under the force placed at $x = l/2$. We note the perfect coincidence of the numerical analysis with the hat-shape virtual functions and semi-analytical curves. The Dirac-shape virtual functions result in a small error. A significant decrease in the time step reduces the difference, however, and all three curves coincide.

In the next example, a cantilever beam was subjected to a travelling inertial load. The data were taken from the previous example. The boundary conditions are $w(0,t) = w'(0,t) = 0$. Additionally, natural boundary conditions were supplied by element interpolation functions: $w''(l,t) = w'''(l,t) = 0$. Figure 7.7a shows the deflection of the point following the mass and Figure 7.7b is the deflection of the free end of the beam. The displacements are related to the static deflection of the free end of the beam under a force placed at $x = l$.

Table 7.1 Matrix $\mathbf{M}_m$—Bernoulli–Euler beam (left-hand square)

$-4\kappa^6+12\kappa^5-$ $\kappa^4(5\xi^2+9)+$ $2\kappa^3(5\xi^2-2)-$ $3\kappa^2(\xi^4+6\xi^2-8)/4+$ $\kappa\xi^2(3\xi^2-4)/4-$ $(5\xi^6+63\xi^4-280\xi^2+560)/560$	$-2hv\kappa^6/\xi+7hv\kappa^5/\xi-$ $hv\kappa^4(5\xi^2+16)/(2\xi)+$ $hv\kappa^3(35\xi^2+12)/(6\xi)-$ $hv\kappa^2(3\xi^4+32\xi^2-16)/(8\xi)+$ $hv\kappa(7\xi^4+8\xi^2-16)/(16\xi)-$ $hv\xi(15\xi^4+336\xi^2-560)/3360$	$4\kappa^6-12\kappa^5+$ $\kappa^4(5\xi^2+9)+$ $2\kappa^3(1-5\xi^2)+$ $3\kappa^2(\xi^4+6\xi^2-4)/4+$ $\kappa\xi^2(2-3\xi^2)/4+$ $\xi^2(5\xi^4+63\xi^2-140)/560$	$-2hv\kappa^6/\xi+5hv\kappa^5/\xi-$ $hv\kappa^4(5\xi^2+6)/(2\xi)+$ $hv\kappa^3(25\xi^2-6)/(6\xi)-$ $hv\kappa^2(3\xi^4+12\xi^2-8)/(8\xi)+$ $hv\kappa\xi(5\xi^2-4)/16-$ $hv\xi(15\xi^4+126\xi^2-280)/3360$
$-2hv\kappa^6/\xi+7hv\kappa^5/\xi-$ $hv\kappa^4(5\xi^2+16)/(2\xi)+$ $hv\kappa^3(35\xi^2+12)/(6\xi)-$ $hv\kappa^2(3\xi^4+32\xi^2-16)/(8\xi)+$ $hv\kappa(7\xi^4+8\xi^2-16)/(16\xi)-$ $hv\xi(15\xi^4+336\xi^2-560)/3360$	$-h^2v^2\kappa^6/\xi^2+4h^2v^2\kappa^5/\xi^2-$ $h^2v^2\kappa^4(5\xi^2+24)/(4\xi^2)+$ $2h^2v^2\kappa^3(5\xi^2+6)/(3\xi^2)-$ $h^2v^2\kappa^2(3\xi^4+48\xi^2+16)/(16\xi^2)+$ $h^2v^2\kappa(\xi^2+4)/4-$ $h^2v^2(15\xi^4+504\xi^2+560)/6720$	$2hv\kappa^6/\xi-7hv\kappa^5/\xi+$ $hv\kappa^4(5\xi^2+16)/(2\xi)-$ $hv\kappa^3(35\xi^2+18)/(6\xi)+$ $hv\kappa^2\xi(3\xi^2+32)/8-$ $hv\kappa\xi(7\xi^2+12)/16+$ $hv\xi^3(5\xi^2+112)/1120$	$-h^2v^2\kappa^6/\xi^2+3h^2v^2\kappa^5/\xi^2-$ $h^2v^2\kappa^4(5\xi^2+12)/(4\xi^2)+$ $h^2v^2\kappa^3(5\xi^2+2)/(2\xi^2)-$ $3h^2v^2\kappa^2(\xi^2+8)/16+$ $h^2v^2\kappa(3\xi^2+4)/16-$ $h^2v^2\xi^2(5\xi^2+84)/2240$
$4\kappa^6-12\kappa^5+$ $\kappa^4(5\xi^2+9)+$ $2\kappa^3(1-5\xi^2)+$ $3\kappa^2(\xi^4+6\xi^2-4)/4+$ $\kappa\xi^2(2-3\xi^2)/4+$ $\xi^2(5\xi^4+63\xi^2-140)/560$	$2hv\kappa^6/\xi-7hv\kappa^5/\xi+$ $hv\kappa^4(5\xi^2+16)/(2\xi)-$ $hv\kappa^3(35\xi^2+18)/(6\xi)+$ $hv\kappa^2\xi(3\xi^2+32)/8-$ $hv\kappa\xi(7\xi^2+12)/16+$ $hv\xi^3(5\xi^2+112)/1120$	$-4\kappa^6+12\kappa^5-$ $\kappa^4(5\xi^2+9)+$ $10\kappa^3\xi^2-$ $3\kappa^2\xi^2(\xi^2+6)/4+$ $3\kappa\xi^4/4-$ $\xi^4(5\xi^2+63)/560$	$2hv\kappa^6/\xi-5hv\kappa^5/\xi+$ $hv\kappa^4(5\xi^2+6)/(2\xi)-$ $25hv\kappa^3\xi/6+$ $3hv\kappa^2\xi(\xi^2+4)/8-$ $5hv\kappa\xi^3/16+$ $hv\xi^3(5\xi^2+42)/1120$
$-2hv\kappa^6/\xi+5hv\kappa^5/\xi-$ $hv\kappa^4(5\xi^2+6)/(2\xi)+$ $hv\kappa^3(25\xi^2-6)/(6\xi)-$ $hv\kappa^2(3\xi^4+12\xi^2-8)/(8\xi)+$ $hv\kappa\xi(5\xi^2-4)/16-$ $hv\xi(15\xi^4+126\xi^2-280)/3360$	$-h^2v^2\kappa^6/\xi^2+3h^2v^2\kappa^5/\xi^2-$ $h^2v^2\kappa^4(5\xi^2+12)/(4\xi^2)+$ $h^2v^2\kappa^3(5\xi^2+2)/(2\xi^2)-$ $3h^2v^2\kappa^2(\xi^2+8)/16+$ $h^2v^2\kappa(3\xi^2+4)/16-$ $h^2v^2\xi^2(5\xi^2+84)/2240$	$2hv\kappa^6/\xi-5hv\kappa^5/\xi+$ $hv\kappa^4(5\xi^2+6)/(2\xi)-$ $25hv\kappa^3\xi/6+$ $3hv\kappa^2\xi(\xi^2+4)/8-$ $5hv\kappa\xi^3/16+$ $hv\xi^3(5\xi^2+42)/1120$	$-h^2v^2\kappa^6/\xi^2+2h^2v^2\kappa^5/\xi^2-$ $h^2v^2\kappa^4(5\xi^2+4)/(4\xi^2)+$ $5h^2v^2\kappa^3/3-$ $h^2v^2\kappa^2(3\xi^2+8)/16+$ $h^2v^2\kappa\xi^2/8-$ $h^2v^2\xi^2(5\xi^2+28)/2240$

Table 7.2 Matrix $\mathbf{M}_m$—Bernoulli–Euler beam (right-hand square)

$4\kappa^6-12\kappa^5+$ $\kappa^4(5\xi^2+9)+$ $2\kappa^3(2-5\xi^2)+$ $3\kappa^2(\xi^4+6\xi^2-8)/4+$ $\kappa\xi^2(4-3\xi^2)/4+$ $(5\xi^6+63\xi^4-280\xi^2+560)/560$	$2hv\kappa^6/\xi-7hv\kappa^5/\xi+$ $hv\kappa^4(5\xi^2+16)/(2\xi)-$ $hv\kappa^3(35\xi^2+12)/(6\xi)+$ $hv\kappa^2(3\xi^4+32\xi^2-16)/(8\xi)-$ $hv\kappa(7\xi^4+8\xi^2-16)/(16\xi)+$ $hv\xi(15\xi^4+336\xi^2-560)/3360$	$-4\kappa^6+12\kappa^5-$ $\kappa^4(5\xi^2+9)+$ $2\kappa^3(5\xi^2-1)-$ $3\kappa^2(\xi^4+6\xi^2-4)/4+$ $\kappa\xi^2(3\xi^2-2)/4-$ $\xi^2(5\xi^4+63\xi^2-140)/560$	$2hv\kappa^6/\xi-5hv\kappa^5/\xi+$ $hv\kappa^4(5\xi^2+6)/(2\xi)+$ $hv\kappa^3(6-25\xi^2)/(6\xi)+$ $hv\kappa^2(3\xi^4+12\xi^2-8)/(8\xi)+$ $hv\kappa\xi(4-5\xi^2)/16+$ $hv\xi(15\xi^4+126\xi^2-280)/3360$
$2hv\kappa^6/\xi-7hv\kappa^5/\xi+$ $hv\kappa^4(5\xi^2+16)/(2\xi)-$ $hv\kappa^3(35\xi^2+12)/(6\xi)+$ $hv\kappa^2(3\xi^4+32\xi^2-16)/(8\xi)-$ $hv\kappa(7\xi^4+8\xi^2-16)/(16\xi)+$ $hv\xi(15\xi^4+336\xi^2-560)/3360$	$h^2v^2\kappa^6/\xi^2-4h^2v^2\kappa^5/\xi^2+$ $h^2v^2\kappa^4(5\xi^2+24)/(4\xi^2)-$ $2h^2v^2\kappa^3(5\xi^2+6)/(3\xi^2)+$ $h^2v^2\kappa^2(3\xi^4+48\xi^2+16)/(16\xi^2)-$ $h^2v^2\kappa(\xi^2+4)/4+$ $h^2v^2(15\xi^4+504\xi^2+560)/6720$	$-2hv\kappa^6/\xi+7hv\kappa^5/\xi-$ $hv\kappa^4(5\xi^2+16)/(2\xi)+$ $hv\kappa^3(35\xi^2+18)/(6\xi)-$ $hv\kappa^2\xi(3\xi^2+32)/8+$ $hv\kappa\xi(7\xi^2+12)/16-$ $hv\xi^3(5\xi^2+112)/1120$	$h^2v^2\kappa^6/\xi^2-3h^2v^2\kappa^5/\xi^2+$ $h^2v^2\kappa^4(5\xi^2+12)/(4\xi^2)-$ $h^2v^2\kappa^3(5\xi^2+2)/(2\xi^2)+$ $3h^2v^2\kappa^2(\xi^2+8)/16-$ $h^2v^2\kappa(3\xi^2+4)/16+$ $h^2v^2\xi^2(5\xi^2+84)/2240$
$-4\kappa^6+12\kappa^5-$ $\kappa^4(5\xi^2+9)+$ $2\kappa^3(5\xi^2-1)-$ $3\kappa^2(\xi^4+6\xi^2-4)/4+$ $\kappa\xi^2(3\xi^2-2)/4-$ $\xi^2(5\xi^4+63\xi^2-140)/560$	$-2hv\kappa^6/\xi+7hv\kappa^5/\xi-$ $hv\kappa^4(5\xi^2+16)/(2\xi)+$ $hv\kappa^3(35\xi^2+18)/(6\xi)-$ $hv\kappa^2\xi(3\xi^2+32)/8+$ $hv\kappa\xi(7\xi^2+12)/16-$ $hv\xi^3(5\xi^2+112)/1120$	$4\kappa^6-12\kappa^5+$ $\kappa^4(5\xi^2+9)-$ $10\kappa^3\xi^2+$ $3\kappa^2\xi^2(\xi^2+6)/4-$ $3\kappa\xi^4/4+$ $\xi^4(5\xi^2+63)/560$	$-2hv\kappa^6/\xi+5hv\kappa^5/\xi-$ $hv\kappa^4(5\xi^2+6)/(2\xi)+$ $25hv\kappa^3\xi/6-$ $3hv\kappa^2\xi(\xi^2+4)/8+$ $5hv\kappa\xi^3/16-$ $hv\xi^3(5\xi^2+42)/1120$
$2hv\kappa^6/\xi-5hv\kappa^5/\xi+$ $hv\kappa^4(5\xi^2+6)/(2\xi)+$ $hv\kappa^3(6-25\xi^2)/(6\xi)+$ $hv\kappa^2(3\xi^4+12\xi^2-8)/(8\xi)+$ $hv\kappa\xi(4-5\xi^2)/16+$ $hv\xi(15\xi^4+126\xi^2-280)/3360$	$h^2v^2\kappa^6/\xi^2-3h^2v^2\kappa^5/\xi^2+$ $h^2v^2\kappa^4(5\xi^2+12)/(4\xi^2)-$ $h^2v^2\kappa^3(5\xi^2+2)/(2\xi^2)+$ $3h^2v^2\kappa^2(\xi^2+8)/16-$ $h^2v^2\kappa(3\xi^2+4)/16+$ $h^2v^2\xi^2(5\xi^2+84)/2240$	$-2hv\kappa^6/\xi+5hv\kappa^5/\xi-$ $hv\kappa^4(5\xi^2+6)/(2\xi)+$ $25hv\kappa^3\xi/6-$ $3hv\kappa^2\xi(\xi^2+4)/8+$ $5hv\kappa\xi^3/16-$ $hv\xi^3(5\xi^2+42)/1120$	$h^2v^2\kappa^6/\xi^2-2h^2v^2\kappa^5/\xi^2+$ $h^2v^2\kappa^4(5\xi^2+4)/(4\xi^2)-$ $5h^2v^2\kappa^3/3+$ $h^2v^2\kappa^2(3\xi^2+8)/16-$ $h^2v^2\kappa\xi^2/8+$ $h^2v^2\xi^2(5\xi^2+28)/2240$

Table 7.3 Matrix $\mathbf{C}_m$—Bernoulli–Euler beam (left-hand square)

$12\xi\kappa^5-10\xi\kappa^4(\xi+3)+$ $2\xi\kappa^3(5\xi^2+10\xi+9)-$ $3\xi\kappa^2(\xi^3+5\xi^2+3\xi-2)+$ $\xi\kappa(3\xi^4+12\xi^3+$ $18\xi^2-8\xi-24)/4-$ $\xi^2(15\xi^4+105\xi^3+126\xi^2-$ $-140\xi-280)/280$	$6h\nu\kappa^5-h\nu\kappa^4(5\xi+17)+$ $h\nu\kappa^3(15\xi^2+34\xi+42)/3-$ $h\xi\nu\kappa^2(3\xi^2+17\xi+14)/2+$ $h\nu\kappa(15\xi^4+68\xi^3+140\xi^2-160)/40-$ $h\nu(45\xi^5+357\xi^4+588\xi^3-$ $-1120\xi-1680)/1680$	$-12\xi\kappa^5+10\xi\kappa^4(\xi+3)-$ $2\xi\kappa^3(5\xi^2+10\xi+9)+$ $3\xi\kappa^2(\xi^3+5\xi^2+3\xi-2)-$ $\xi\kappa(3\xi^4+12\xi^3+18\xi^2-8\xi-$ $24)/4+\xi^2(15\xi^4+105\xi^3+$ $+126\xi^2-140\xi-280)/280$	$6h\nu\kappa^5-h\nu\kappa^4(5\xi+13)+$ $h\nu\kappa^3(15\xi^2+26\xi+18)/3-$ $h\nu\kappa^2(3\xi^3+13\xi^2+6\xi-6)/2+$ $h\nu\kappa(15\xi^4+52\xi^3+60\xi^2-40\xi-$ $-80)/40-h\xi\nu(45\xi^4+273\xi^3+$ $+252\xi^2-420\xi-560)/1680$
$6h\nu\kappa^5-h\nu\kappa^4(5\xi+18)+$ $h\nu\kappa^3(5\xi^2+12\xi+18)-$ $3h\nu\kappa^2(\xi^3+6\xi^2+6\xi+4)/2+$ $h\xi\nu\kappa(15\xi^3+72\xi^2+180\xi+$ $+80)/40-h\xi^2\nu(15\xi^3+126\xi^2+$ $+252\xi+280)/560$	$3h^2\nu^2\kappa^5/\xi-5h^2\nu^2\kappa^4(\xi+4)/(2\xi)+$ $h^2\nu^2\kappa^3(15\xi^2+40\xi+72)/(6\xi)-$ $h^2\nu^2\kappa^2(3\xi^3+20\xi^2+24\xi+24)/(4\xi)+$ $h^2\nu^2\kappa(3\xi^4+16\xi^3+48\xi^2+$ $+32\xi+16)/(16\xi)-$ $h^2\nu^2(45\xi^4+420\xi^3+1008\xi^2+$ $+1680\xi+560)/3360$	$-6h\nu\kappa^5+h\nu\kappa^4(5\xi+18)-$ $h\nu\kappa^3(5\xi^2+12\xi+18)+$ $3h\nu\kappa^2(\xi^3+6\xi^2+6\xi+4)/2-$ $h\xi\nu\kappa(15\xi^3+72\xi^2+180\xi+$ $+80)/40+h\xi^2\nu(15\xi^3+126\xi^2+$ $+252\xi+280)/560$	$3h^2\nu^2\kappa^5/\xi-h^2\nu^2\kappa^4(5\xi+16)/(2\xi)+$ $h^2\nu^2\kappa^3(15\xi^2+32\xi+42)/(6\xi)-$ $h^2\nu^2\kappa^2(3\xi^3+16\xi^2+14\xi+8)/(4\xi)+$ $h^2\nu^2\kappa(45\xi^3+192\xi^2+420\xi+$ $+160)/240-h^2\xi\nu^2(45\xi^3+336\xi^2+$ $+588\xi+560)/3360$
$-12\xi\kappa^5+10\xi\kappa^4(\xi+3)-$ $2\xi\kappa^3(5\xi^2+10\xi+9)+$ $3\xi^2\kappa^2(\xi^2+5\xi+3)-$ $3\xi^3\kappa(\xi^2+4\xi+6)/4+$ $3\xi^4(5\xi^2+35\xi+42)/280$	$-6h\nu\kappa^5+h\nu\kappa^4(5\xi+17)-$ $h\nu\kappa^3(15\xi^2+34\xi+42)/3+$ $h\nu\kappa^2(3\xi^3+17\xi^2+14\xi+6)/2-$ $h\xi\nu\kappa(15\xi^3+68\xi^2+140\xi+40)/40+$ $h\xi^2\nu(15\xi^3+119\xi^2+196\xi+140)/560$	$12\xi\kappa^5-10\xi\kappa^4(\xi+3)+$ $2\xi\kappa^3(5\xi^2+10\xi+9)-$ $3\xi^2\kappa^2(\xi^2+5\xi+3)+$ $3\xi^3\kappa(\xi^2+4\xi+6)/4-$ $3\xi^4(5\xi^2+35\xi+42)/280$	$-6h\nu\kappa^5+h\nu\kappa^4(5\xi+13)-$ $h\nu\kappa^3(15\xi^2+26\xi+18)/3+$ $h\xi\nu\kappa^2(3\xi^2+13\xi+6)/2-$ $h\xi^2\nu\kappa(15\xi^2+52\xi+60)/40+$ $h\xi^3\nu(15\xi^2+91\xi+84)/560$
$6h\nu\kappa^5-h\nu\kappa^4(5\xi+12)+$ $h\nu\kappa^3(5\xi^2+8\xi+6)-$ $3h\xi\nu\kappa^2(\xi^2+4\xi+2)/2+$ $3h\xi^2\nu\kappa(5\xi^2+16\xi+20)/40-$ $3h\xi^3\nu(5\xi^2+28\xi+28)/560$	$3h^2\nu^2\kappa^5/\xi-h^2\nu^2\kappa^4(5\xi+14)/(2\xi)+$ $h^2\nu^2\kappa^3(15\xi^2+28\xi+30)/(6\xi)-$ $h^2\nu^2\kappa^2(3\xi^3+14\xi^2+10\xi+4)/(4\xi)+$ $h^2\nu^2\kappa(45\xi^3+168\xi^2+300\xi+80)/240-$ $h^2\xi\nu^2(45\xi^3+294\xi^2+420\xi+280)/3360$	$-6h\nu\kappa^5+h\nu\kappa^4(5\xi+12)-$ $h\nu\kappa^3(5\xi^2+8\xi+6)+$ $3h\xi\nu\kappa^2(\xi^2+4\xi+2)/2-$ $3h\xi^2\nu\kappa(5\xi^2+16\xi+20)/40+$ $3h\xi^3\nu(5\xi^2+28\xi+28)/560$	$3h^2\nu^2\kappa^5/\xi-5h^2\nu^2\kappa^4(\xi+2)/(2\xi)+$ $h^2\nu^2\kappa^3(15\xi^2+20\xi+12)/(6\xi)-$ $h^2\nu^2\kappa^2(3\xi^2+10\xi+4)/4+$ $+h^2\xi\nu^2\kappa(3\xi^2+8\xi+8)/16-$ $h^2\xi^2\nu^2(15\xi^2+70\xi+56)/1120$

Table 7.4 Matrix $\mathbf{C}_m$—Bernoulli–Euler beam (right-hand square)

$12\xi\kappa^5+10\xi\kappa^4(\xi-3)+2\xi\kappa^3(5\xi^2-10\xi+9)+3\xi\kappa^2(\xi^3-5\xi^2+3\xi+2)+\xi\kappa(3\xi^4-12\xi^3+18\xi^2+8\xi-24)/4+\xi^2(15\xi^4-105\xi^3+126\xi^2++140\xi-280)/280$	$6h\nu\kappa^5+h\nu\kappa^4(5\xi-17)+h\nu\kappa^3(15\xi^2-34\xi+42)/3+h\xi\nu\kappa^2(3\xi^2-17\xi+14)/2+h\nu\kappa(15\xi^4-68\xi^3+140\xi^2-160)/40+h\nu(45\xi^5-357\xi^4+588\xi^3-1120\xi++1680)/1680$	$-12\xi\kappa^5+10\xi\kappa^4(3-\xi)-2\xi\kappa^3(5\xi^2-10\xi+9)-3\xi\kappa^2(\xi^3-5\xi^2+3\xi+2)-\xi\kappa(3\xi^4-12\xi^3+18\xi^2+8\xi-24)/4-\xi^2(15\xi^4-105\xi^3+126\xi^2+140\xi-280)/280$	$6h\nu\kappa^5+h\nu\kappa^4(5\xi-13)+h\nu\kappa^3(15\xi^2-26\xi+18)/3+h\nu\kappa^2(3\xi^3-13\xi^2+6\xi+6)/2+h\nu\kappa(15\xi^4-52\xi^3+60\xi^2+40\xi-80)/40+h\xi\nu(45\xi^4-273\xi^3+252\xi^2+420\xi-560)/1680$
$6h\nu\kappa^5+h\nu\kappa^4(5\xi-18)+h\nu\kappa^3(5\xi^2-12\xi+18)+3h\nu\kappa^2(\xi^3-6\xi^2+6\xi-4)/2+h\xi\nu\kappa(15\xi^3-72\xi^2+180\xi-80)/40+h\xi^2\nu(15\xi^3-126\xi^2++252\xi-280)/560$	$3h^2\nu^2\kappa^5/\xi+5h^2\nu^2\kappa^4(\xi-4)/(2\xi)+h^2\nu^2\kappa^3(15\xi^2-40\xi+72)/(6\xi)+h^2\nu^2\kappa^2(3\xi^3-20\xi^2+24\xi-24)/(4\xi)+h^2\nu^2\kappa(3\xi^4-16\xi^3+48\xi^2-32\xi+16)/(16\xi)+h^2\nu^2(45\xi^4-420\xi^3++1008\xi^2-1680\xi+560)/3360$	$-6h\nu\kappa^5+h\nu\kappa^4(18-5\xi)-h\nu\kappa^3(5\xi^2-12\xi+18)-3h\nu\kappa^2(\xi^3-6\xi^2+6\xi-4)/2-h\xi\nu\kappa(15\xi^3-72\xi^2+180\xi-80)/40-h\xi^2\nu(15\xi^3--126\xi^2+252\xi-280)/560$	$3h^2\nu^2\kappa^5/\xi+h^2\nu^2\kappa^4(5\xi-16)//(2\xi)+h^2\nu^2\kappa^3(15\xi^2-32\xi++42)/(6\xi)+h^2\nu^2\kappa^2\cdot\cdot(3\xi^3-16\xi^2+14\xi-8)/(4\xi)+h^2\nu^2\kappa(45\xi^3-192\xi^2+420\xi--160)/240+h^2\xi\nu^2(45\xi^3--336\xi^2+588\xi-560)/3360$
$-12\xi\kappa^5+10\xi\kappa^4(3-\xi)-2\xi\kappa^3(5\xi^2-10\xi+9)-3\xi^2\kappa^2(\xi^2-5\xi+3)-3\xi^3\kappa(\xi^2-4\xi+6)/4-3\xi^4(5\xi^2-35\xi+42)/280$	$-6h\nu\kappa^5+h\nu\kappa^4(17-5\xi)-h\nu\kappa^3(15\xi^2-34\xi+42)/3-h\nu\kappa^2(3\xi^3-17\xi^2+14\xi-6)/2-h\xi\nu\kappa(15\xi^3-68\xi^2+140\xi-40)/40-h\xi^2\nu(15\xi^3-119\xi^2+196\xi-140)/560$	$12\xi\kappa^5+10\xi\kappa^4(\xi-3)+2\xi\kappa^3(5\xi^2-10\xi+9)+3\xi^2\kappa^2(\xi^2-5\xi+3)+3\xi^3\kappa(\xi^2-4\xi+6)/4+3\xi^4(5\xi^2-35\xi+42)/280$	$-6h\nu\kappa^5+h\nu\kappa^4(13-5\xi)-h\nu\kappa^3(15\xi^2-26\xi+18)/3-h\xi\nu\kappa^2(3\xi^2-13\xi+6)/2-h\xi^2\nu\kappa(15\xi^2-52\xi+60)/40-h\xi^3\nu(15\xi^2-91\xi+84)/560$
$6h\nu\kappa^5+h\nu\kappa^4(5\xi-12)+h\nu\kappa^3(5\xi^2-8\xi+6)+3h\xi\nu\kappa^2(\xi^2-4\xi+2)/2+3h\xi^2\nu\kappa(5\xi^2-16\xi+20)/40+3h\xi^3\nu(5\xi^2-28\xi+28)/560$	$3h^2\nu^2\kappa^5/\xi+h^2\nu^2\kappa^4(5\xi-14)/(2\xi)+h^2\nu^2\kappa^3(15\xi^2-28\xi+30)/(6\xi)+h^2\nu^2\kappa^2(3\xi^3-14\xi^2+10\xi-4)/(4\xi)+h^2\nu^2\kappa(45\xi^3-168\xi^2+300\xi-80)/240+h^2\xi\nu^2(45\xi^3-294\xi^2+420\xi-280)/3360$	$-6h\nu\kappa^5+h\nu\kappa^4(12-5\xi)-h\nu\kappa^3(5\xi^2-8\xi+6)-3h\xi\nu\kappa^2(\xi^2-4\xi+2)/2-3h\xi^2\nu\kappa(5\xi^2-16\xi+20)/40-3h\xi^3\nu(5\xi^2-28\xi+28)/560$	$3h^2\nu^2\kappa^5/\xi+5h^2\nu^2\kappa^4(\xi-2)/(2\xi)+h^2\nu^2\kappa^3(15\xi^2-20\xi+12)/(6\xi)+h^2\nu^2\kappa^2(3\xi^2-10\xi+4)/4+h^2\xi\nu^2\kappa(3\xi^2-8\xi+8)/16+h^2\xi^2\nu^2(15\xi^2-70\xi+56)/1120$

Table 7.5 Matrix $\mathbf{K}_m$—Bernoulli–Euler beam (left-hand square)

$8\xi^2\kappa^4+4\xi^2\kappa^3(\xi-4)+$ $6\xi^2\kappa^2(3\xi^2-5\xi+5)/5+$ $\xi^2\kappa(6\xi^3-36\xi^2+15\xi+40)/10+$ $\xi^2(12\xi^4-42\xi^3+63\xi^2+$ $+70\xi-280)/140$	$4h\xi\nu\kappa^4+$ $2h\xi\nu\kappa^3(3\xi-13)/3+$ $h\xi\nu\kappa^2(36\xi^2-65\xi+80)/20+$ $h\xi\nu\kappa(6\xi^3-39\xi^2+20\xi+40)/20+$ $h\xi\nu(72\xi^4-273\xi^3+504\xi^2+$ $420\xi-2240)/1680$	$-8\xi^2\kappa^4+$ $4\xi^2\kappa^3(4-\xi)-$ $6\xi^2\kappa^2(3\xi^2-5\xi+5)/5-$ $\xi^2\kappa(6\xi^3-36\xi^2+15\xi+40)/10-$ $\xi^2(12\xi^4-42\xi^3+63\xi^2+$ $+70\xi-280)/140$	$4h\xi\nu\kappa^4+$ $2h\xi\nu\kappa^3(3\xi-11)/3+$ $h\xi\nu\kappa^2(36\xi^2-55\xi+40)/20+$ $h\xi\nu\kappa(6\xi^3-33\xi^2+10\xi+40)/20+$ $h\xi\nu(72\xi^4-231\xi^3+$ $252\xi^2+420\xi-1120)/1680$
$4h\xi\nu\kappa^4+$ $2h\xi\nu\kappa^3(\xi-5)+$ $h\xi\nu\kappa^2(36\xi^2-75\xi+160)/20+$ $h\xi\nu\kappa(6\xi^3-45\xi^2+40\xi-40)/20+$ $h\xi^2\nu(24\xi^3-105\xi^2+336\xi-$ $-140)/560$	$2h^2\nu^2\kappa^4+h^2\nu^2\kappa^3(3\xi-16)/3+$ $h^2\nu^2\kappa^2(27\xi^2-60\xi+140)/30+$ $h^2\nu^2\kappa(9\xi^3-72\xi^2+70\xi-80)/60+$ $h^2\xi\nu^2(9\xi^3-42\xi^2+147\xi-70)/420$	$-4h\xi\nu\kappa^4+$ $2h\xi\nu\kappa^3(5-\xi)-$ $h\xi\nu\kappa^2(36\xi^2-75\xi+160)/20-$ $h\xi\nu\kappa(6\xi^3-45\xi^2+40\xi-40)/20-$ $h\xi^2\nu(24\xi^3-105\xi^2+336\xi-140)/560$	$2h^2\nu^2\kappa^4+$ $h^2\nu^2\kappa^3(3\xi-14)/3+$ $h^2\nu^2\kappa^2(54\xi^2-105\xi+200)/60+$ $h^2\nu^2\kappa(9\xi^3-63\xi^2+50\xi-40)/60+$ $h^2\xi\nu^2(36\xi^3-147\xi^2+420\xi-$ $-140)/1680$
$-8\xi^2\kappa^4+$ $4\xi^2\kappa^3(4-\xi)-$ $6\xi^2\kappa^2(3\xi^2-5\xi+5)/5-$ $3\xi^3\kappa(2\xi^2-12\xi+5)/10-$ $3\xi^4(4\xi^2-14\xi+21)/140$	$-4h\xi\nu\kappa^4+$ $2h\xi\nu\kappa^3(13-3\xi)/3-$ $h\xi\nu\kappa^2(36\xi^2-65\xi+80)/20-$ $h\xi^2\nu\kappa(6\xi^2-39\xi+20)/20-$ $h\xi^3\nu(24\xi^2-91\xi+168)/560$	$8\xi^2\kappa^4+$ $4\xi^2\kappa^3(\xi-4)+$ $6\xi^2\kappa^2(3\xi^2-5\xi+5)/5+$ $3\xi^3\kappa(2\xi^2-12\xi+5)/10+$ $3\xi^4(4\xi^2-14\xi+21)/140$	$-4h\xi\nu\kappa^4+$ $2h\xi\nu\kappa^3(11-3\xi)/3-$ $h\xi\nu\kappa^2(36\xi^2-55\xi+40)/20-$ $h\xi^2\nu\kappa(6\xi^2-33\xi+10)/20-$ $h\xi^3\nu(24\xi^2-77\xi+84)/560$
$4h\xi\nu\kappa^4+$ $2h\xi\nu\kappa^3(\xi-3)+$ $h\xi\nu\kappa^2(36\xi^2-45\xi+40)/20+$ $h\xi^2\nu\kappa(6\xi^2-27\xi+10)/20+$ $3h\xi^3\nu(8\xi^2-21\xi+28)/560$	$2h^2\nu^2\kappa^4+$ $h^2\nu^2\kappa^3(3\xi-10)/3+$ $h^2\nu^2\kappa^2(54\xi^2-75\xi+80)/60+$ $h^2\xi\nu^2\kappa(9\xi^2-45\xi+20)/60+$ $h^2\xi^2\nu^2(12\xi^2-35\xi+56)/560$	$-4h\xi\nu\kappa^4+$ $2h\xi\nu\kappa^3(3-\xi)-$ $h\xi\nu\kappa^2(36\xi^2-45\xi+40)/20-$ $h\xi^2\nu\kappa(6\xi^2-27\xi+10)/20-$ $3h\xi^3\nu(8\xi^2-21\xi+28)/560$	$2h^2\nu^2\kappa^4+$ $h^2\nu^2\kappa^3(3\xi-8)/3+$ $h^2\nu^2\kappa^2(27\xi^2-30\xi+20)/30+$ $h^2\xi\nu^2\kappa(9\xi^2-36\xi+10)/60+$ $h^2\xi^2\nu^2(3\xi^2-7\xi+7)/140$

Table 7.6 Matrix $\mathbf{K}_m$—Bernoulli–Euler beam (right-hand square)

$4\xi^2\kappa^4+4\xi^2\kappa^3(\xi-2)+$ $3\xi^2\kappa^2(4\xi^2-10\xi+5)/5+$ $\xi^2\kappa(6\xi^3-24\xi^2+15\xi+20)/10+$ $\xi^2(9\xi^4-42\xi^3+42\xi^2+70\xi-$ $-140)/140$	$2h\xi\nu\kappa^4+h\xi\nu\kappa^3(6\xi-13)/3+$ $h\xi\nu\kappa^2(24\xi^2-65\xi+40)/20+$ $h\xi\nu\kappa(3\xi^3-13\xi^2+10\xi+10)/10+$ $h\xi\nu(54\xi^4-273\xi^3+336\xi^2+$ $+420\xi-1120)/1680$	$-4\xi^2\kappa^4+4\xi^2\kappa^3(2-\xi)-$ $3\xi^2\kappa^2(4\xi^2-10\xi+5)/5-$ $\xi^2\kappa(6\xi^3-24\xi^2+15\xi+20)/10-$ $\xi^2(9\xi^4-42\xi^3+$ $42\xi^2+70\xi-140)/140$	$2h\xi\nu\kappa^4+h\xi\nu\kappa^3(6\xi-11)/3+$ $h\xi\nu\kappa^2(24\xi^2-55\xi+20)/20+$ $h\xi\nu\kappa(3\xi^3-11\xi^2+5\xi+10)/10+$ $h\xi\nu(54\xi^4-231\xi^3+$ $168\xi^2+420\xi-560)/1680$
$2h\xi\nu\kappa^4+h\xi\nu\kappa^3(2\xi-5)+$ $h\xi\nu\kappa^2(24\xi^2-75\xi+80)/20+$ $h\xi\nu\kappa(3\xi^3-15\xi^2+20\xi-10)/10+$ $h\xi^2\nu(18\xi^3-105\xi^2+224\xi-$ $-140)/560$	$h^2\nu^2\kappa^4+h^2\nu^2\kappa^3(3\xi-8)/3+$ $h^2\nu^2\kappa^2(9\xi^2-30\xi+35)/15+$ $h^2\nu^2\kappa(9\xi^3-48\xi^2+70\xi-40)/60+$ $h^2\xi\nu^2(27\xi^3-168\xi^2+392\xi-$ $-280)/1680$	$-2h\xi\nu\kappa^4+h\xi\nu\kappa^3(5-2\xi)-$ $h\xi\nu\kappa^2(24\xi^2-75\xi+80)/20-$ $h\xi\nu\kappa(3\xi^3-15\xi^2+20\xi-10)/10-$ $h\xi^2\nu(18\xi^3-105\xi^2+224\xi-$ $-140)/560$	$h^2\nu^2\kappa^4+h^2\nu^2\kappa^3(3\xi-7)/3+$ $h^2\nu^2\kappa^2(36\xi^2-105\xi+100)/60+$ $h^2\nu^2\kappa(9\xi^3-42\xi^2+$ $50\xi-20)/60+$ $h^2\xi\nu^2(27\xi^3-147\xi^2+280\xi-$ $-140)/1680$
$-4\xi^2\kappa^4+4\xi^2\kappa^3(2-\xi)-$ $3\xi^2\kappa^2(4\xi^2-10\xi+5)/5-$ $3\xi^3\kappa(2\xi^2-8\xi+5)/10-$ $3\xi^4(3\xi^2-14\xi+14)/140$	$-2h\xi\nu\kappa^4+h\xi\nu\kappa^3(13-6\xi)/3-$ $h\xi\nu\kappa^2(24\xi^2-65\xi+40)/20-$ $h\xi^2\nu\kappa(3\xi^2-13\xi+10)/10-$ $h\xi^3\nu(18\xi^2-91\xi+112)/560$	$4\xi^2\kappa^4+4\xi^2\kappa^3(\xi-2)+$ $3\xi^2\kappa^2(4\xi^2-10\xi+5)/5+$ $3\xi^3\kappa(2\xi^2-8\xi+5)/10+$ $3\xi^4(3\xi^2-14\xi+14)/140$	$-2h\xi\nu\kappa^4+h\xi\nu\kappa^3(11-6\xi)/3-$ $h\xi\nu\kappa^2(24\xi^2-55\xi+20)/20-$ $h\xi^2\nu\kappa(3\xi^2-11\xi+5)/10-$ $h\xi^3\nu(18\xi^2-77\xi+56)/560$
$2h\xi\nu\kappa^4+h\xi\nu\kappa^3(2\xi-3)+$ $h\xi\nu\kappa^2(24\xi^2-45\xi+20)/20+$ $h\xi^2\nu\kappa(3\xi^2-9\xi+5)/10+$ $h\xi^3\nu(18\xi^2-63\xi+56)/560$	$h^2\nu^2\kappa^4+h^2\nu^2\kappa^3(3\xi-5)/3+$ $h^2\nu^2\kappa^2(36\xi^2-75\xi+40)/60+$ $h^2\xi\nu^2\kappa(9\xi^2-30\xi+20)/60+$ $h^2\xi^2\nu^2(27\xi^2-105\xi+112)/1680$	$-2h\xi\nu\kappa^4+h\xi\nu\kappa^3(3-2\xi)-$ $h\xi\nu\kappa^2(24\xi^2-45\xi+20)/20-$ $h\xi^2\nu\kappa(3\xi^2-9\xi+5)/10-$ $h\xi^3\nu(18\xi^2-63\xi+56)/560$	$h^2\nu^2\kappa^4+h^2\nu^2\kappa^3(3\xi-4)/3+$ $h^2\nu^2\kappa^2(9\xi^2-15\xi+5)/15+$ $h^2\xi\nu^2\kappa(9\xi^2-24\xi+10)/60+$ $h^2\xi^2\nu^2(27\xi^2-84\xi+56)/1680$

Table 7.7 Matrix $\mathbf{E}_m$—Bernoulli–Euler beam

$$
\left[\begin{array}{llll}
\begin{array}{l} 24\xi^2\kappa^4/h- \\ 48\xi^2\kappa^3/h+ \\ 6\xi^2\kappa^2(2\xi^2+3)/h+ \\ 12\xi^2\kappa(1-\xi^2)/h+ \\ 3\xi^2(\xi^4+5\xi^2-20)/(10h) \end{array} &
\begin{array}{l} 12\xi\nu\kappa^4- \\ 26\xi\nu\kappa^3+ \\ 6\xi\nu\kappa^2(\xi^2+2)+ \\ \xi\nu\kappa(12-13\xi^2)/2+ \\ \xi\nu(3\xi^4+20\xi^2-80)/20 \end{array} &
\begin{array}{l} -24\xi^2\kappa^4/h+ \\ 48\xi^2\kappa^3/h- \\ 6\xi^2\kappa^2(2\xi^2+3)/h+ \\ 12\xi^2\kappa(\xi^2-1)/h- \\ 3\xi^2(\xi^4+5\xi^2-20)/(10h) \end{array} &
\begin{array}{l} 12\xi\nu\kappa^4- \\ 22\xi\nu\kappa^3+ \\ 6\xi\nu\kappa^2(\xi^2+1)+ \\ \xi\nu\kappa(12-11\xi^2)/2+ \\ \xi\nu(3\xi^4+10\xi^2-40)/20 \end{array} \\
\\
\begin{array}{l} 12\xi\nu\kappa^4- \\ 30\xi\nu\kappa^3+ \\ 6\xi\nu\kappa^2(\xi^2+4)- \\ 3\xi\nu\kappa(5\xi^2+4)/2+ \\ \xi^3\nu(3\xi^2+40)/20 \end{array} &
\begin{array}{l} 6h\nu^2\kappa^4- \\ 16h\nu^2\kappa^3+ \\ h\nu^2\kappa^2(3\xi^2+14)- \\ 4h\nu^2\kappa(\xi^2+1)+ \\ h\xi^2\nu^2(9\xi^2+140)/120 \end{array} &
\begin{array}{l} -12\xi\nu\kappa^4+ \\ 30\xi\nu\kappa^3- \\ 6\xi\nu\kappa^2(\xi^2+4)+ \\ 3\xi\nu\kappa(5\xi^2+4)/2- \\ \xi^3\nu(3\xi^2+40)/20 \end{array} &
\begin{array}{l} 6h\nu^2\kappa^4- \\ 14h\nu^2\kappa^3+ \\ h\nu^2\kappa^2(3\xi^2+10)- \\ h\nu^2\kappa(7\xi^2+4)/2+ \\ h\xi^2\nu^2(9\xi^2+100)/120 \end{array} \\
\\
\begin{array}{l} -24\xi^2\kappa^4/h+ \\ 48\xi^2\kappa^3/h- \\ 6\xi^2\kappa^2(2\xi^2+3)/h+ \\ 12\xi^4\kappa/h- \\ 3\xi^4(\xi^2+5)/(10h) \end{array} &
\begin{array}{l} -12\xi\nu\kappa^4+ \\ 26\xi\nu\kappa^3- \\ 6\xi\nu\kappa^2(\xi^2+2)+ \\ 13\xi^3\nu\kappa/2- \\ \xi^3\nu(3\xi^2+20)/20 \end{array} &
\begin{array}{l} 24\xi^2\kappa^4/h- \\ 48\xi^2\kappa^3/h+ \\ 6\xi^2\kappa^2(2\xi^2+3)/h- \\ 12\xi^4\kappa/h+ \\ 3\xi^4(\xi^2+5)/(10h) \end{array} &
\begin{array}{l} -12\xi\nu\kappa^4+ \\ 22\xi\nu\kappa^3- \\ 6\xi\nu\kappa^2(\xi^2+1)+ \\ 11\xi^3\nu\kappa/2- \\ \xi^3\nu(3\xi^2+10)/20 \end{array} \\
\\
\begin{array}{l} 12\xi\nu\kappa^4- \\ 18\xi\nu\kappa^3+ \\ 6\xi\nu\kappa^2(\xi^2+1)- \\ 9\xi^3\nu\kappa/2+ \\ \xi^3\nu(3\xi^2+10)/20 \end{array} &
\begin{array}{l} 6h\nu^2\kappa^4- \\ 10h\nu^2\kappa^3+ \\ h\nu^2\kappa^2(3\xi^2+4)- \\ 5h\xi^2\nu^2\kappa/2+ \\ h\xi^2\nu^2(9\xi^2+40)/120 \end{array} &
\begin{array}{l} -12\xi\nu\kappa^4+ \\ 18\xi\nu\kappa^3- \\ 6\xi\nu\kappa^2(\xi^2+1)+ \\ 9\xi^3\nu\kappa/2- \\ \xi^3\nu(3\xi^2+10)/20 \end{array} &
\begin{array}{l} 6h\nu^2\kappa^4- \\ 8h\nu^2\kappa^3+ \\ h\nu^2\kappa^2(3\xi^2+2)- \\ 2h\xi^2\nu^2\kappa+ \\ h\xi^2\nu^2(9\xi^2+20)/120 \end{array}
\end{array}\right]
$$

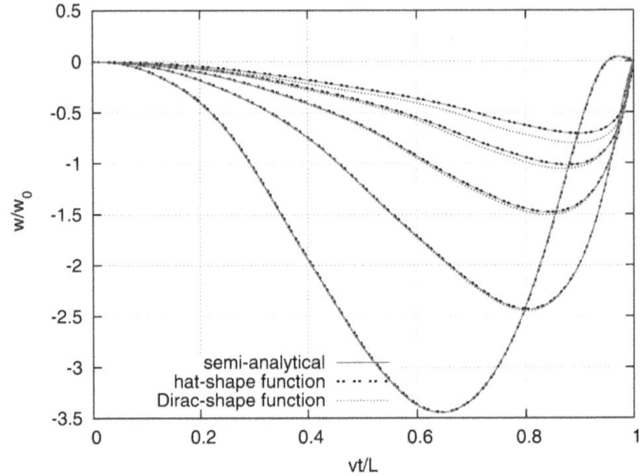

Fig. 7.6 Displacements of a Bernoulli–Euler simply-supported beam under a moving mass at speeds $\mathsf{v} = 0.1c, 0.2c, \ldots, 0.5c$ (space–time and semi-analytical results); $\mathsf{v}_{cr} = 0.314$.

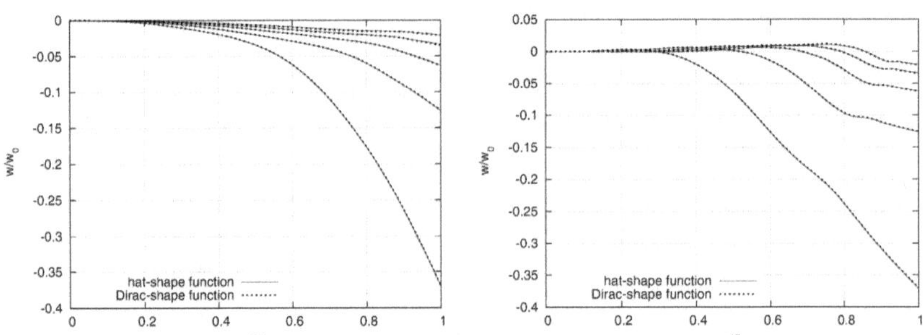

Fig. 7.7 Displacements under a mass moving on the Bernoulli–Euler cantilever beam (a) and displacements of the free end (b) at speeds $\mathsf{v} = 0.1c, 0.2c, \ldots, 0.5c$, with hat-shape and Dirac-shape virtual functions.

7.3 Space–Time Element of Timoshenko Beam Carrying a Moving Mass

Applying this numerical scheme to a moving inertial point requires modifying the global matrices describing our step-by-step scheme. We must change not only the inertia matrix, but also the matrices that express the damping and stiffness.

The vertical point acceleration of the moving inertial particle is computed from the displacement of the contact point determined on the supporting structure $w(\mathsf{v}t, t)$. We must apply the chain rule (for the time derivative) twice. The result is called the Renaudot formulation:

$$\frac{\mathrm{d}^2 w(\mathrm{v}t,t)}{\mathrm{d}t^2} = \left.\frac{\partial^2 w(x,t)}{\partial t^2}\right|_{x=\mathrm{v}t} + 2\mathrm{v}\left.\frac{\partial^2 w(x,t)}{\partial x \partial t}\right|_{x=\mathrm{v}t} + \mathrm{v}^2 \left.\frac{\partial^2 w(x,t)}{\partial x^2}\right|_{x=\mathrm{v}t} . \tag{7.16}$$

The coordinate increment of the time, dt, is, in consequence, the increment of the spatial coordinates x as well as of the time coordinates t. As a result, we have three terms: the vertical acceleration, the Coriolis acceleration, and the centrifugal acceleration.

The equation for the virtual power in the space–time region $\Omega = \{(x,t)\colon 0 \le x \le b,\ 0 \le t \le h\}$ (Figure 7.8) describing a moving material point can be written in the form

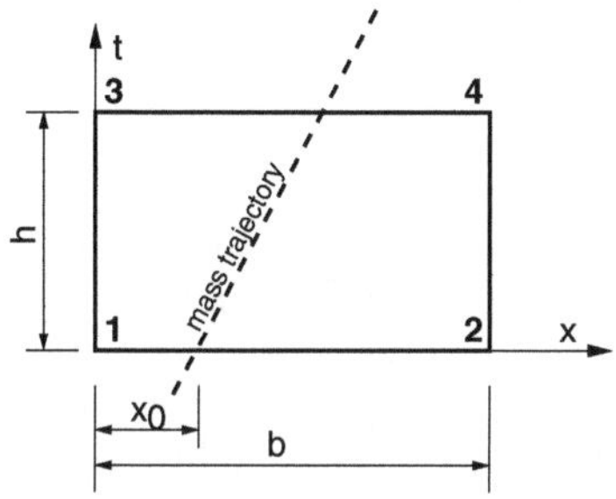

Fig. 7.8 Space–time element.

$$\Pi_m = \int_0^h \int_0^b v^*(x,t)\,\delta(x - x_0 - \mathrm{v}t)\, m \frac{\mathrm{d}^2 w(\mathrm{v}t,t)}{\mathrm{d}t^2} \mathrm{d}x\mathrm{d}t . \tag{7.17}$$

We assume a linear distribution of the nodal velocity in x and t

$$v(x,t) = \sum_{i=1}^{4} N_i(x,t)\, v_i . \tag{7.18}$$

In the domain Ω, the shape function $\mathbf{N} = [N_1, \ldots, N_4]$ has the form

$$\mathbf{N} = \left[\frac{1}{bh}(x-b)(t-h)\,,\ -\frac{1}{bh}x(t-h)\,,\ -\frac{1}{bh}(x-b)t\,,\ \frac{1}{bh}xt\right] . \tag{7.19}$$

In this case, it is not possible, in interpolating the nodal velocity, to determine all the required term in (7.16). Due to the distribution in (7.17), it is not possible to reduce the order of the derivative by integrating by parts. The Renaudot formula can be written in the equivalent form

$$\frac{\mathrm{d}^2 w(\mathrm{v}t,t)}{\mathrm{d}t^2} = \left.\frac{\partial v(x,t)}{\partial t}\right|_{x=\mathrm{v}t} + \mathrm{v}\left.\frac{\partial v(x,t)}{\partial x}\right|_{x=\mathrm{v}t} + \mathrm{v}\frac{\mathrm{d}}{\mathrm{d}t}\left[\left.\frac{\partial w(x,t)}{\partial x}\right|_{x=\mathrm{v}t}\right] . \tag{7.20}$$

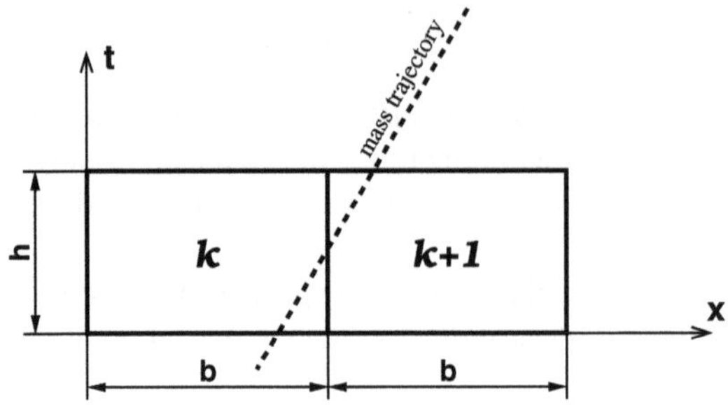

Fig. 7.9 The transition mass between elements.

Using (7.18) and (7.19), we obtain

$$\frac{\partial v(x,t)}{\partial t} = -\frac{1}{h}\left(1-\frac{x}{b}\right)v_1 - \frac{1}{h}\frac{x}{b}v_2 + \frac{1}{h}\left(1-\frac{x}{b}\right)v_3 + \frac{1}{h}\frac{x}{b}v_4\,, \tag{7.21}$$

$$\frac{\partial v(x,t)}{\partial x} = -\frac{1}{b}\left(1-\frac{t}{h}\right)v_1 + \frac{1}{b}\left(1-\frac{t}{h}\right)v_2 - \frac{1}{b}\frac{t}{h}v_3 + \frac{1}{b}\frac{t}{h}v_4\,. \tag{7.22}$$

In the case of the third part of the equation (7.20), we use the backward difference formula. In this case we have

$$\frac{\mathrm{d}}{\mathrm{d}t}\left[\left.\frac{\partial w(x,t)}{\partial x}\right|_{x=vt}\right] = \frac{1}{h}\left[\left.\frac{\partial w(x,t)}{\partial x}\right|_{x=vt}\right]^{t+h} - \frac{1}{h}\left[\left.\frac{\partial w(x,t)}{\partial x}\right|_{x=vt}\right]^{t}\,. \tag{7.23}$$

The upper indices indicate the time at which the respective terms are defined. At the time of transition of the moving load between the elements k and $k+1$, Figure 7.9, the current displacements are calculated in the element $k+1$:

$$\left[\left.\frac{\partial w(x,t)}{\partial x}\right|_{x=vt}\right]^{t+h} = \frac{1}{b}\left(w_4^{k+1} - w_3^{k+1}\right)\,. \tag{7.24}$$

However, the initial displacement in the element k is

$$\left[\left.\frac{\partial w(x,t)}{\partial x}\right|_{x=vt}\right]^{t} = \frac{1}{b}\left(w_2^{k} - w_1^{k}\right)\,. \tag{7.25}$$

Using (6.54), (7.24), and (7.25), the finite difference scheme (7.23) can be written as follows

$$\begin{aligned}\frac{\mathrm{d}}{\mathrm{d}t}\left[\left.\frac{\partial w(x,t)}{\partial x}\right|_{x=vt}\right] &= \frac{1}{bh}\left(w_2^{k+1} - w_2^{k} - w_1^{k+1} + w_1^{k}\right) + \\ &+ \frac{1}{b}\left[-\beta v_1^{k+1} + \beta v_2^{k+1} - (1-\beta)v_3^{k+1} + (1-\beta)v_4^{k+1}\right]\,.\end{aligned} \tag{7.26}$$

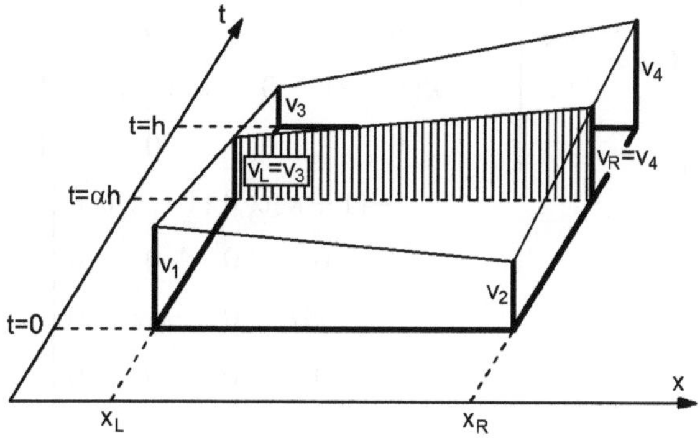

Fig. 7.10 Dirac delta virtual function.

The exact solution is obtained by $\beta = 1 - \alpha$ [9]. Therefore, we can write

$$\frac{\mathrm{d}}{\mathrm{d}t}\left[\left.\frac{\partial w(x,t)}{\partial x}\right|_{x=vt}\right] = \frac{1}{bh}\left(w_2^{k+1} - w_2^k - w_1^{k+1} + w_1^k\right) + \\ + \frac{1}{b}\left[-(1-\alpha)v_1^{k+1} + (1-\alpha)v_2^{k+1} - \alpha v_3^{k+1} + \alpha v_4^{k+1}\right] . \tag{7.27}$$

In order to calculate the virtual power describing a moving inertial point (7.17), we use the virtual impulse function, Figure 7.10,

$$v^*(x,t) = \delta(t-\alpha h)\left[\left(1-\frac{x}{b}\right)v_3 + \frac{x}{b}v_4\right] . \tag{7.28}$$

We can control the properties of the procedure with the parameter α. As a result of minimizing the virtual value of power Π_m, we obtain the matrices describing the inertial moving load:

$$\mathbf{M}_m = \frac{m}{h}\begin{bmatrix} -(1-\kappa)^2 & 0 & -\kappa(1-\kappa) & 0 & (1-\kappa)^2 & 0 & \kappa(1-\kappa) & 0 \\ 0 & 0 & 0 & 0 & 0 & 0 & 0 & 0 \\ -\kappa(1-\kappa) & 0 & -\kappa^2 & 0 & \kappa(1-\kappa) & 0 & \kappa^2 & 0 \\ 0 & 0 & 0 & 0 & 0 & 0 & 0 & 0 \end{bmatrix} , \tag{7.29}$$

$$\mathbf{C}_m = \frac{2m\mathrm{v}}{b}\left[\begin{array}{cccccccc} -(1-\kappa)(1-\alpha) & 0 & (1-\kappa)(1-\alpha) & 0 & -(1-\kappa)\alpha & 0 & (1-\kappa)\alpha & 0 \\ 0 & 0 & 0 & 0 & 0 & 0 & 0 & 0 \\ -\kappa(1-\alpha) & 0 & \kappa(1-\alpha) & 0 & -\kappa\alpha & 0 & \kappa\alpha & 0 \\ 0 & 0 & 0 & 0 & 0 & 0 & 0 & 0 \end{array}\right], \tag{7.30}$$

and

$$\mathbf{e}_m = \frac{m\mathrm{v}}{bh}\left[\begin{array}{c} (1-\kappa)\left(w_2^{k+1} - w_2^k - w_1^{k+1} + w_1^k\right) \\ 0 \\ \kappa\left(w_2^{k+1} - w_2^k - w_1^{k+1} + w_1^k\right) \\ 0 \end{array}\right], \tag{7.31}$$

with the coefficient

$$\kappa = \frac{x_0 + \mathrm{v}\alpha h}{b}, \quad 0 < \kappa \leq 1. \tag{7.32}$$

On the basis of the actual position of the moving load matrix (7.29) and (7.30) fall into the appropriate cells of the global inertia and damping matrices describing the structure under consideration. An important component is the vector of nodal forces (7.31), which is calculated based on the displacements from the current and the

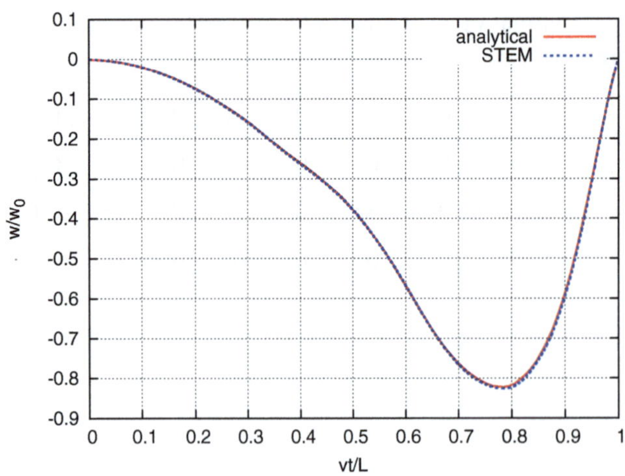

Fig. 7.11 Space–time finite element solution—displacements of the Timoshenko beam under a moving load.

previous step and increases the vector on the right-hand side describing the gravitational moving load.

The numerical results are depicted in Figure 7.11. The data were taken from [86], the speed being $\mathsf{v} = 85.55$ m/s. Algorithm 12 presents the computational steps.

Algorithm 12. Space–time element method applied to a Timoshenko beam.

1. Define geometric and material data, velocity v, length of the element b, time step h, etc.
2. Compute element matrices of repeated structural elements and formulate global coefficient matrix.
3. Initialize loop parameters: number of elements carrying the mass $iel_{rem} = 1$.
4. Perform computations of a single time step:
   - Number of the current step $it = 0$, current time of the beginning of the time interval $t = 0$.
   - Number of the element carrying the load $iel = \mathrm{INT}(t \cdot \mathsf{v}/b - eps) + 1$, $eps = 0.0001$.
   - Position of the mass on the element $x_0 = t \cdot \mathsf{v} - (iel - 1) \cdot b$, parameter $\kappa = (x_0 + \mathsf{v}\alpha h)/b$.
   - Compute mass space–time element and load vector.
   - Compute the nodal force vector of the mass element:
     $\mathbf{e}_m = \frac{m\mathsf{v}}{bh}(w_r^{ielmem} - w_l^{ielmem} - w_r^{ielmem} + w_l^{ielmem})$ (lower index is left of right node of the element, upper index is number of element; notice that $iel = ielmem$ always except in the passage from element to element.
   - Formulate the system of algebraic equations.
   - Solve the system of equations for velocities $\mathbf{v}$.
   - Compute displacements at time $t+h$.
   - Compute nodal forces provided by the mass:
     $(\mathbf{B}+\mathbf{B}_m)\mathbf{v}_{i+1} = \mathbf{F} - \mathbf{e} - \mathbf{e}_m - (\mathbf{A}+\mathbf{A}_m)\mathbf{v}_i$.
   - Remember the number of element $iel_{rem} = iel$.
   - Shift displacements and velocities: $\mathbf{w}_i \leftarrow \mathbf{w}_{i+1}$, $\mathbf{v}_i \leftarrow \mathbf{v}_{i+1}$.
5. Increment time step.

7.3.1 Conclusions

The presented matrices (7.29)–(7.31) describe the moving mass in the space–time formulation in generality. They can be applied to all type of structures. The numerical results perfectly coincide with the semi-analytical solution within a wide range of velocities of the mass. We applied non-dimensional speeds v up to 0.5, which corresponds with 0.3 times the critical speed. The critical speed means the speed of the force travelling in a cyclic way through a beam, when the vertical deflection increases to infinity. In the case of a moving mass, the critical speed has a considerably lower value, and in our example we come close to it.

The classical finite element approach does not allow us to obtain satisfactory results. They fail in the case of a string and exhibit very large errors in the case of beams. The typically applied methods of time integration, for example the Newmark method, fail since the moving inertial term cannot be considered in a continuous way in the time interval. Complex analysis could be performed with the space–time approach. Various virtual functions in time can be applied. They result in a solution scheme having a different accuracy. This is clearly demonstrated by the mass trajectory plots. At the final stage of the motion, the trajectories exhibit jumps (see [48]). In Figure 3.8 we showed the convergence of the semi-analytical solution with an increasing number of terms in the series. The jump of the trajectory should also be sufficiently accurate in numerical representations. The jumps in every case of numerical analyses are poorly portrayed by the numerical solutions. In our problem, in the higher speed range, the jumps are visible in the solutions with a sufficiently small error. A shorter time step increases the accuracy.

Figure 7.3 demonstrates that the hat-shaped virtual function results in better convergence. The choice of the virtual time step having the shape of a hat, instead of the Dirac delta type of this function, significantly improves the quality of the solution. Both cases, i.e., the Dirac delta virtual function with $\alpha = 1/2$ and the hat-shaped virtual function, exhibit theoretically the same estimated error of the method. However, the two sets of results differ. We can say that the error is contributed by and accumulated with different speeds because of other stages of the solution scheme: the velocity computation or displacement restitution.

The solutions given above are efficient in discrete vibration analysis with a travelling mass. Although this problem deals with a mass moving with a constant speed, the same mathematical procedure can be used to derive the characteristic matrices in the case of a mass moving with a varying speed. In this case, only the additional terms describing the influence of the acceleration along the structure need be taken into account. In this section, the horizontal acceleration of the mass is equal to zero.

The perfect coincidence with the semi-analytical solutions proves the efficiency of the space–time approach. The solution method can be easily implemented in code designed for the classical finite element method.

7.4 Space–Time Finite Plate Element Carrying a Moving Mass

In engineering practice, plate structures are as common as beams. In general, bending elements resist most types of load. In most cases, plates carry a static or dynamic load placed at fixed points, for example supporting vibrating machines, placed on an elastic foundation, or ceilings covering gaps. In some cases, the plates are subjected to moving loads: road plates, airfield plates, plates under subway tracks, etc.

7.4.1 Thin Plate

First we will consider a thin plate. The appropriate finite element formulation can be found, for example, in [153]. Let us consider a $2a \times 2b$ rectangular plate (Figure

7.12). In a rectangular area, we assume the displacement is distributed according to the polynomial

$$w(x,y) = [1,x,y,x^2,xy,y^2,x^3,x^2y,xy^2,y^3,x^3y,xy^3]\,\mathbf{c}\,. \tag{7.33}$$

The coefficients $c_1,\ldots,c_{12}$ collected in the vector $\mathbf{c}$ are determined in terms of the nodal generalized displacements, i.e., the vertical displacements w_i and rotations θ_{xi} and θ_{yi}, $i = 1,2,3,4$, and describes the number of a node.

The final static stiffness matrix $\mathbf{k}_e$ is the sum of four matrices:

$$\mathbf{k}_e = \frac{1}{60ab}\mathbf{L}\,(D_x\mathbf{k}_1 + D_y\mathbf{k}_2 + D_1\mathbf{k}_3 + D_{xy}\mathbf{k}_4)\,\mathbf{L}\,, \tag{7.34}$$

where

$$\mathbf{k}_1 = \frac{b^2}{a^2}\begin{bmatrix}
60 & 30 & 0 & 30 & 15 & 0 & -60 & 30 & 0 & -30 & 15 & 0\\
30 & 20 & 0 & 15 & 10 & 0 & -30 & 10 & 0 & -15 & 5 & 0\\
0 & 0 & 0 & 0 & 0 & 0 & 0 & 0 & 0 & 0 & 0 & 0\\
30 & 15 & 0 & 60 & 30 & 0 & -30 & 15 & 0 & -60 & 30 & 0\\
15 & 10 & 0 & 30 & 20 & 0 & -15 & 5 & 0 & -30 & 10 & 0\\
0 & 0 & 0 & 0 & 0 & 0 & 0 & 0 & 0 & 0 & 0 & 0\\
-60 & -30 & 0 & -30 & -15 & 0 & 60 & -30 & 0 & 30 & -15 & 0\\
30 & 10 & 0 & 15 & 5 & 0 & -30 & 20 & 0 & -15 & 10 & 0\\
0 & 0 & 0 & 0 & 0 & 0 & 0 & 0 & 0 & 0 & 0 & 0\\
-30 & -15 & 0 & -60 & -30 & 0 & 30 & -15 & 0 & 60 & -30 & 0\\
15 & 5 & 0 & 30 & 10 & 0 & -15 & 10 & 0 & -30 & 20 & 0\\
0 & 0 & 0 & 0 & 0 & 0 & 0 & 0 & 0 & 0 & 0 & 0
\end{bmatrix}, \tag{7.35}$$

$$\mathbf{k}_2 = \frac{a^2}{b^2}\begin{bmatrix}
60 & 0 & -30 & -60 & 0 & -30 & 30 & 0 & -15 & -30 & 0 & -15\\
0 & 0 & 0 & 0 & 0 & 0 & 0 & 0 & 0 & 0 & 0 & 0\\
-30 & 0 & 20 & 30 & 0 & 10 & -15 & 0 & 10 & 15 & 0 & 5\\
-60 & 0 & 30 & 60 & 0 & 30 & -30 & 0 & 15 & 30 & 0 & 15\\
0 & 0 & 0 & 0 & 0 & 0 & 0 & 0 & 0 & 0 & 0 & 0\\
-30 & 0 & 10 & 30 & 0 & 20 & -15 & 0 & 5 & 15 & 0 & 10\\
30 & 0 & -15 & -30 & 0 & -15 & 60 & 0 & -30 & -60 & 0 & -3\\
0 & 0 & 0 & 0 & 0 & 0 & 0 & 0 & 0 & 0 & 0 & 0\\
-15 & 0 & 10 & 15 & 0 & 5 & -30 & 0 & 20 & 30 & 0 & 10\\
-30 & 0 & 15 & 30 & 0 & 15 & -60 & 0 & 30 & 60 & 0 & 30\\
0 & 0 & 0 & 0 & 0 & 0 & 0 & 0 & 0 & 0 & 0 & 0\\
-15 & 0 & 5 & 15 & 0 & 10 & -30 & 0 & 10 & 30 & 0 & 20
\end{bmatrix}, \tag{7.36}$$

$$\mathbf{k}_3 = \begin{bmatrix} 30 & 15 & -15 & -30 & -15 & 0 & -30 & 0 & 15 & 30 & 0 & 0 \\ 15 & 0 & -15 & -15 & 0 & 0 & 0 & 0 & 0 & 0 & 0 & 0 \\ -15 & -15 & 0 & 0 & 0 & 0 & 15 & 0 & 0 & 0 & 0 & 0 \\ -30 & -15 & 0 & 30 & 15 & 15 & 30 & 0 & 0 & -30 & 0 & -15 \\ -15 & 0 & 0 & 15 & 0 & 15 & 0 & 0 & 0 & 0 & 0 & 0 \\ 0 & 0 & 0 & 15 & 15 & 0 & 0 & 0 & 0 & -15 & 0 & 0 \\ -30 & 0 & 15 & 30 & 0 & 0 & 30 & -15 & -15 & -30 & 15 & 0 \\ 0 & 0 & 0 & 0 & 0 & 0 & -15 & 0 & 15 & 15 & 0 & 0 \\ 15 & 0 & 0 & 0 & 0 & 0 & -15 & 15 & 0 & 0 & 0 & 0 \\ 30 & 0 & 0 & -30 & 0 & -15 & -30 & 15 & 0 & 30 & -15 & 15 \\ 0 & 0 & 0 & 0 & 0 & 0 & 15 & 0 & 0 & -15 & 0 & -15 \\ 0 & 0 & 0 & -15 & 0 & 0 & 0 & 0 & 0 & 15 & -15 & 0 \end{bmatrix}, \tag{7.37}$$

$$\mathbf{k}_4 = \begin{bmatrix} 84 & 6 & -6 & -84 & -6 & -6 & -84 & 6 & 6 & 84 & -6 & 6 \\ 6 & 8 & 0 & -6 & -8 & 0 & -6 & -2 & 0 & 6 & 2 & 0 \\ -6 & 0 & 8 & 6 & 0 & -2 & 6 & 0 & -8 & -6 & 0 & 2 \\ -84 & -6 & 6 & 84 & 6 & 6 & 84 & -6 & -6 & -84 & 6 & -6 \\ -6 & -8 & 0 & 6 & 8 & 0 & 6 & 2 & 0 & -6 & -2 & 0 \\ -6 & 0 & -2 & 6 & 0 & 8 & 6 & 0 & 2 & -6 & 0 & -8 \\ -84 & -6 & 6 & 84 & 6 & 6 & 84 & -6 & -6 & -84 & 6 & -6 \\ 6 & -2 & 0 & -6 & 2 & 0 & -6 & 8 & 0 & 6 & -8 & 0 \\ 6 & 0 & -8 & -6 & 0 & 2 & -6 & 0 & 8 & 6 & 0 & -2 \\ 84 & 6 & -6 & -84 & -6 & -6 & -84 & 6 & 6 & 84 & -6 & 6 \\ -6 & 2 & 0 & 6 & -2 & 0 & 6 & -8 & 0 & -6 & 8 & 0 \\ 6 & 0 & 2 & -6 & 0 & -8 & -6 & 0 & -2 & 6 & 0 & 8 \end{bmatrix}, \tag{7.38}$$

$$\mathbf{L} = \mathrm{diag}[1, 2a, 2b, 1, 2a, 2b, 1, 2a, 2b, 1, 2a, 2b]\,. \tag{7.39}$$

The node numbering is taken as in Figure 7.12. The coefficients in the elasticity matrix $\mathbf{D}$ are shown below

$$\mathbf{D} = \begin{bmatrix} D_x & D_1 & 0 \\ D_1 & D_y & 0 \\ 0 & 0 & D_{xy} \end{bmatrix}. \tag{7.40}$$

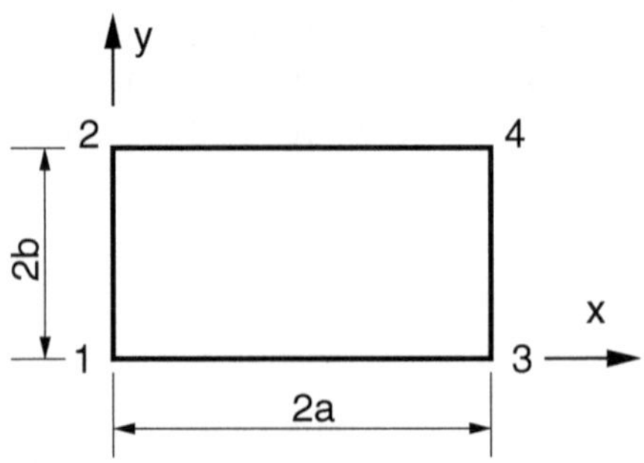

Fig. 7.12 Node numbering in a plate element.

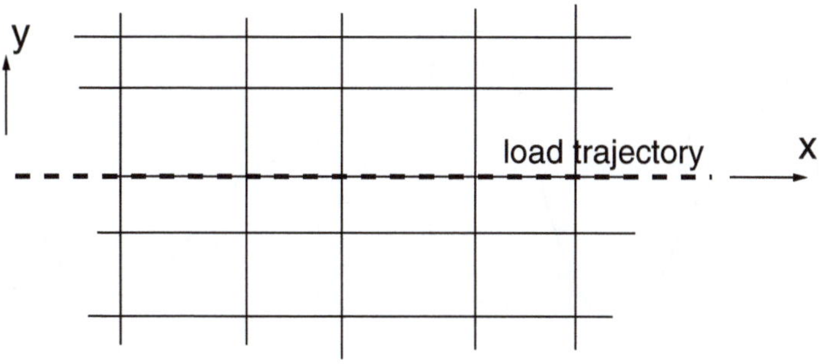

Fig. 7.13 Trajectory of the load along element edges.

In isotropic material, we have $D_x = D_y = D$, $D_1 = \nu D$, and $D_{xy} = (1-\nu)/2 \cdot D$, where D is the plate stiffness: $D = Et^3/12/(1-\nu^2)$.

Let us assume a concentrated load composed of a point with mass m and a force equivalent to the force of gravity, mg. For simplicity, the trajectory of the load coincides with the edge line of the finite elements (Figure 7.13). We use thin plate elements in the simulation of plate vibrations under a mass which moves along the symmetry axis of the plate. The data are: thickness $t = 40$ cm, dimensions $l_x = l_y = 12$ m, Young modulus $E = 30$ MPa, Poissone coefficient $\nu = 0.2$, mass density $\rho = 2.4$ g/cm^3. The moving load is composed of the mass $m = 10^4$ kg and the related force $P = 9.81 \cdot 10^4$ N.

The next figures (7.14, 7.15, and 7.16) are obtained for the thicknesses $t = 1, 40$, and 100 cm. Comparisons of the displacements of the contact point and the centre of the plate are depicted in Figures 7.17 and 7.18. The excellent coincidence is exhibited. The numerical analysis of the thicker plate subjected to an inertial load moving with the lower speed also coincides with the analytical results, see Figures 7.19 and 7.20.

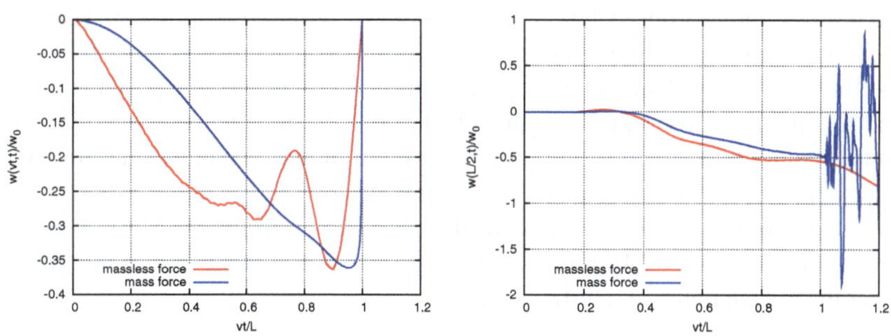

Fig. 7.14 Vertical displacements at the contact point and at the middle of the plate (thickness=10 cm, $v = 360$ km/h).

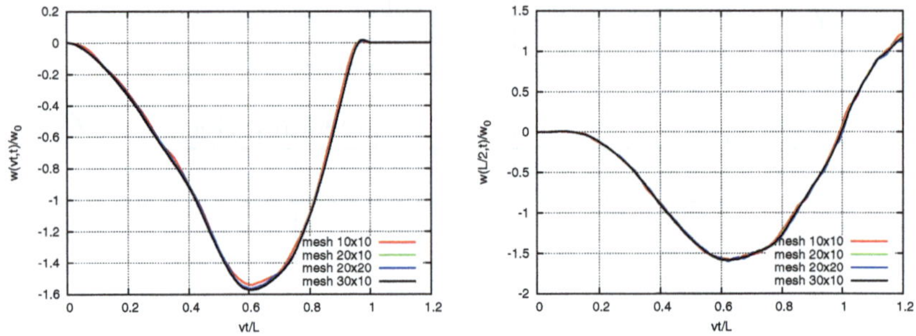

Fig. 7.15 Vertical displacements at the contact point and at the middle of the plate (thickness=40 cm, v = 360 km/h).

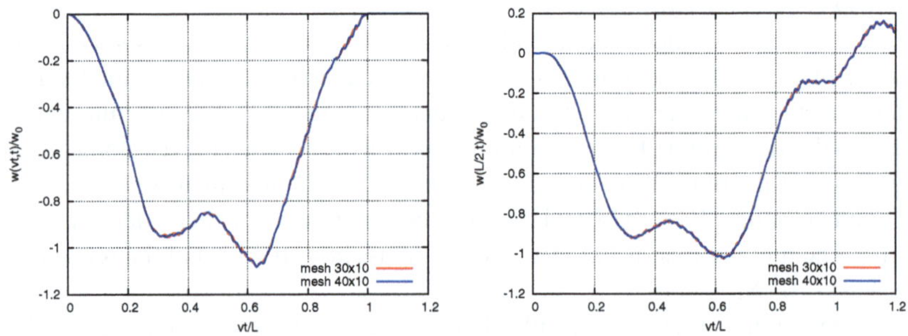

Fig. 7.16 Vertical displacements at the contact point and at the middle of the plate (thickness=100 cm, v = 360 km/h).

Now let us look at the acceleration of the mass. The plots can be obtained by differentiating the velocity or by double differentiating of the displacements. In both cases, high value peaks appear at the interfaces of the elements. They are induced by nodal forces that act only when the load passes the edge of an element. The values of the nodal forces are high if the number of elements in the mesh is low. If the mesh is sufficiently fine and the number of elements tends to infinity, the peak magnitudes tend to smooth to low values. Figures 7.21 and 7.22 show the accelerations at the contact point and at the mid-point. A large jump in the acceleration at the contact point is noticeable near the final edge of the plate.

We can compare these results with the acceleration in the massless force case, see Figures 7.23 and 7.24. In this case, the accelerations at the exit of the plate are lower than when under an inertial load.

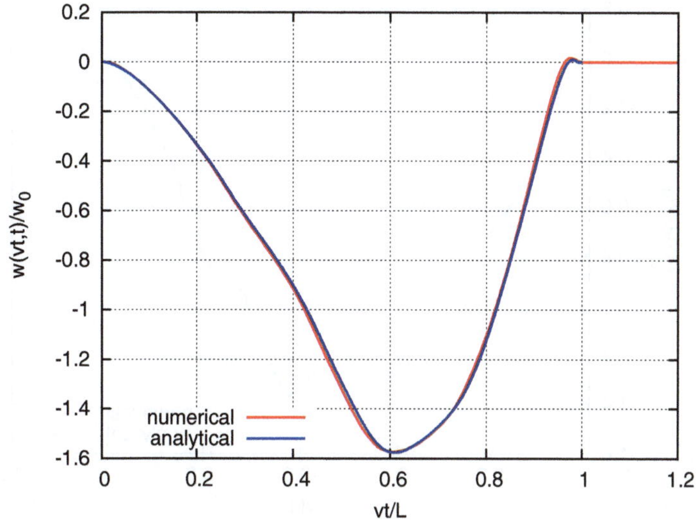

Fig. 7.17 Vertical displacements at the contact point (thickness=40 cm, v = 360 km/h).

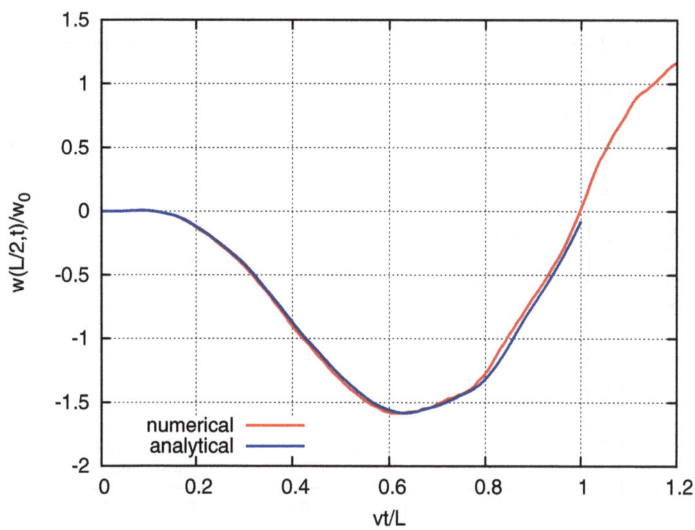

Fig. 7.18 Vertical displacements at the centre of the plate (thickness=40 cm, v = 360 km/h).

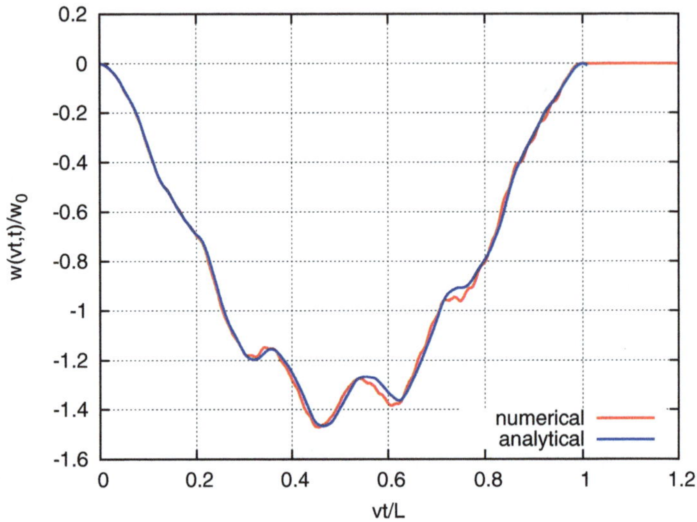

Fig. 7.19 Vertical displacements at the contact point of the plate (thickness=100 cm, v = 160 km/h).

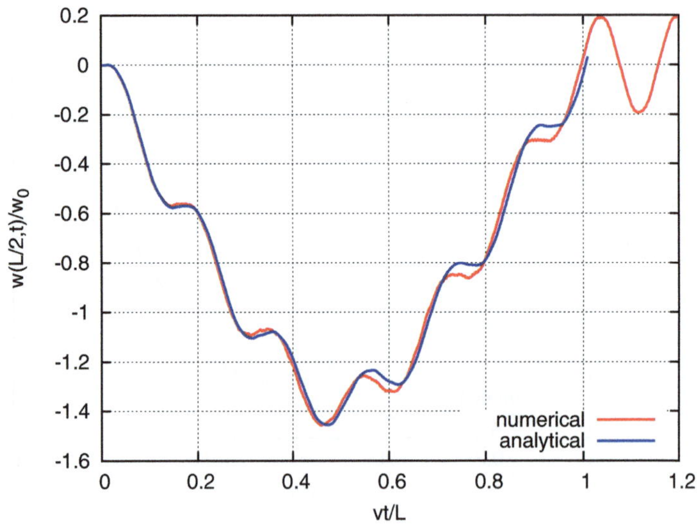

Fig. 7.20 Vertical displacements at the middle of the plate (thickness=100 cm, v = 160 km/h).

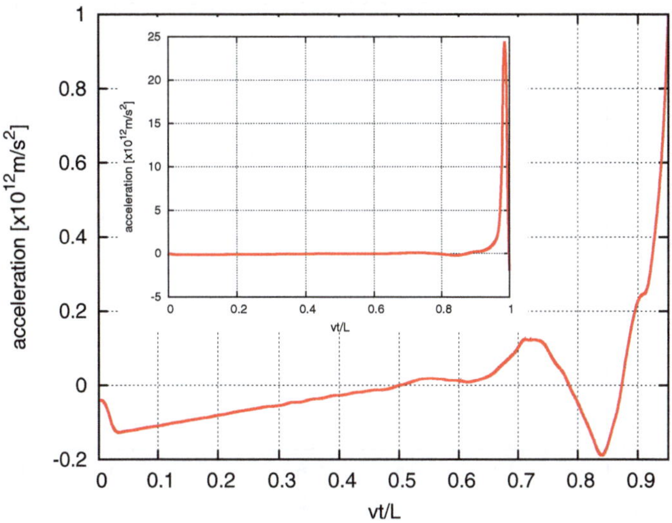

Fig. 7.21 Vertical accelerations at the contact point of the plate (thickness=10 cm, v = 360 km/h) is the inertial load.

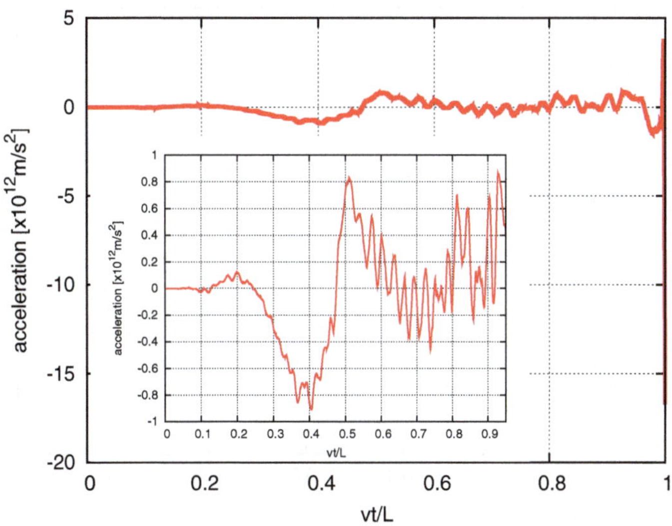

Fig. 7.22 Vertical accelerations at the mid-point of the plate (thickness=10 cm, v = 360 km/h) is the inertial load.

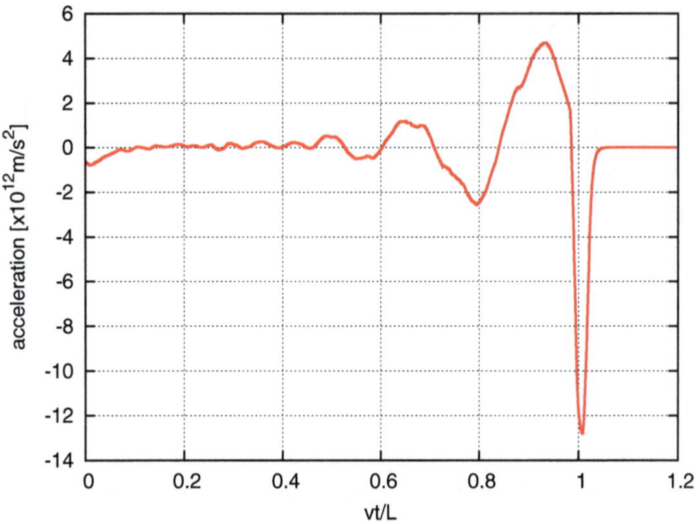

Fig. 7.23 Vertical accelerations at the contact point of the plate (thickness=10 cm, v = 360 km/h) is the massless force.

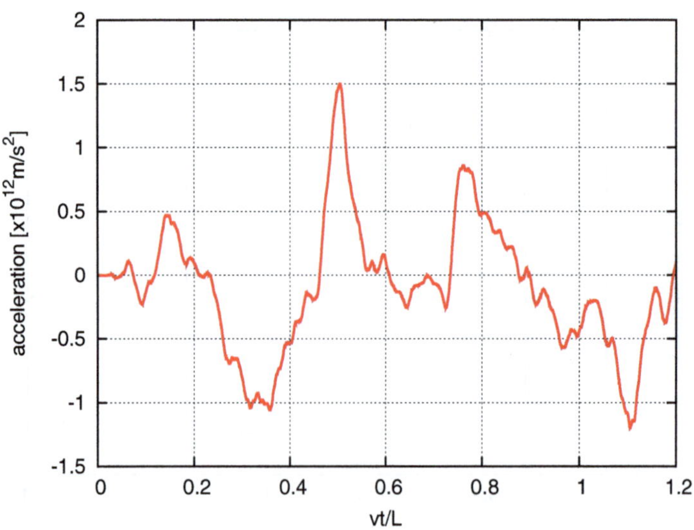

Fig. 7.24 Vertical accelerations at the mid-point of the plate (thickness=10 cm, v = 360 km/h) is the massless force.

7.4.2 Thick Plate

We consider here the Mindlin model of a plate. We use the formulation given for example in [39, 113]. We assume a linear distribution of both displacements and rotations along the element, according to the interpolation functions

$$N(\xi,\eta) = \frac{1}{4}(1+\xi\,\xi_i)(1+\eta\,\eta_i)\,. \tag{7.41}$$

Here, ξ and η are local dimensionless coordinates in the square reference element given by $-1 \le \xi \le 1$ and $-1 \le \eta \le 1$. The nodal coordinates ξ_i, η_i, $i = 1,\ldots,4$ therefore assume the values ± 1.

The strain–stress relations are given in Chapter 6.7.3, equations (6.138–6.141). Here we split the strain vector into the part governing bending $[\kappa_x, \kappa_y, \kappa_{xy}]$ and that governing sharing $[\beta_x, \beta_y]$.

The bending strain matrix $\mathbf{B}^f = [\mathbf{B}_1^f \,|\, \mathbf{B}_2^f \,|\, \mathbf{B}_3^f \,|\, \mathbf{B}_4^f]$ has the following components:

$$\mathbf{B}_i^f = \begin{bmatrix} 0 & 0 & -\partial N_i/\partial x \\ 0 & \partial N_i/\partial y & 0 \\ 0 & \partial N_i/\partial x & -\partial N_i/\partial y \end{bmatrix}, \quad i = 1,\ldots,4\,. \tag{7.42}$$

The shear strain matrix $\mathbf{B}^s = [\mathbf{B}_1^s \,|\ldots|\, \mathbf{B}_4^s]$ equals

$$\mathbf{B}_i^s = \begin{bmatrix} \partial N_i/\partial x & 0 & N_i \\ \partial N_i/\partial y & -N_i & 0 \end{bmatrix}, \quad i = 1,\ldots,4\,. \tag{7.43}$$

The derivation of (7.41), assuming that $x = \xi a$ and $y = \eta b$, via the chain rule (differentiating with respect to x and y) and then integrating over the element domain $-a \le x \le a$, $-b \le y \le b$, results in the final stiffness matrices. We integrate the bending part $\mathbf{B}^f$ analytically or numerically using 2×2 Gauss points, while the shear component $\mathbf{B}^s$ uses only one Gauss point. Thus we avoid the locking phenomenon. The final matrix is obtained as the sum $\mathbf{k} = \mathbf{k}^f + \mathbf{k}^s$, where

$$\mathbf{k}^f = \int_{-a}^{a}\int_{-b}^{b} \frac{t^3}{12} \mathbf{B}^{fT}\, \mathbf{D}^f \mathbf{B}^f \,\mathrm{d}x\mathrm{d}y\,, \tag{7.44}$$

$$\mathbf{k}^s = \int_{-a}^{a}\int_{-b}^{b} \kappa t \mathbf{B}^{sT}\, \mathbf{D}^s \mathbf{B}^s \,\mathrm{d}x\mathrm{d}y\,. \tag{7.45}$$

The matrices $\mathbf{D}^s$ and $\mathbf{D}^f$ can be taken from (6.140) as submatrices

$$\mathbf{D}^s = Gt \cdot \mathbf{I}_2\,. \tag{7.46}$$

Here, G is the shear modulus, $\kappa = 5/6$ is the shape factor, and t is the thickness of the plate. The matrix $\mathbf{D}^f$ has the form

$$\mathbf{D}^f = \begin{bmatrix} D & \nu D & 0 \\ \nu D & D & 0 \\ 0 & 0 & \frac{1-\nu}{2}D \end{bmatrix}, \quad D = \frac{t^3}{12} \cdot \frac{E}{1-\nu^2}\,. \tag{7.47}$$

Here, E is the Young modulus and ν is the Poisson coefficient. The stiffness matrices describing the bending and shear part have the final form:

$$\mathbf{K}_b = \frac{Et^3}{192ab} \begin{bmatrix} 0 & 0 & 0 \\ 0 & \frac{\xi_i\xi_j b}{2a(\nu+1)} + \frac{\eta_i\eta_j a}{b(1-\nu^2)} & \frac{2\eta_i\xi_j\nu+\xi_i\eta_j(1-\nu)}{2(\nu^2-1)} \\ 0 & \frac{\eta_i\xi_j(\nu-1)-2\xi_i\eta_j\nu}{2(1-\nu^2)} & \frac{\xi_i\xi_j b}{a(1-\nu^2)} + \frac{\eta_i\eta_j a}{2b(\nu+1)} \end{bmatrix}, \tag{7.48}$$

$$\mathbf{K}_s = \frac{Eab}{2(1+\nu)} \begin{bmatrix} \frac{\xi_i\xi_j}{16a^2} + \frac{\eta_i\eta_j}{16b^2} & -\frac{\eta_i}{16b} & \frac{\xi_i}{16a} \\ -\frac{\eta_i}{16b} & \frac{1}{16} & 0 \\ \frac{\xi_i}{16a} & 0 & \frac{1}{16} \end{bmatrix}. \tag{7.49}$$

The static problem was analysed first to determine the rate of subdivision required for adequate accuracy. The square plate simply supported at all four edges is subjected in its centre to a point force. The deflection of the loaded point is related to the static deflection of the thin plate. Figure 7.25 shows that even a coarse mesh gives sufficiently accurate results. Subdivision into 10–40 elements in both directions was used in the consequent dynamic tests.

The results are presented in Figures 7.26, 7.27, and 7.28. Three different thicknesses were used: 10 cm, 40 cm, and 100 cm. The plate was square, with edges

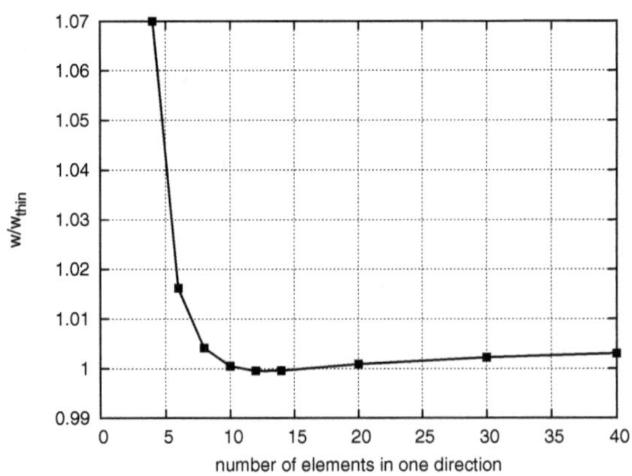

Fig. 7.25 The accuracy of the deflection of the centre of a Mindlin plate compared to the deflection of a thin plate.

of 12 m. The inertial load was compared with a massless load. The displacements depicted are of the follower point and of the centre of the plate. In the case of the thickness 10 cm, differences between both curves are visible, especially at the higher speed v = 360 km/h. The case of thickness equal 40 cm, both at low and high speed, exhibits a smooth response in the case of an inertial load. Moreover, the mass trajectory at v = 180 km/h oscillates with a higher amplitude than for the case of a pure massless load. This phenomenon is confirmed at the higher speed v = 360 km/h. The plate of thickness 100 cm is more rigid and its vibrations have higher frequency. The computer program listed in Appendix A.3 was used for these computations.

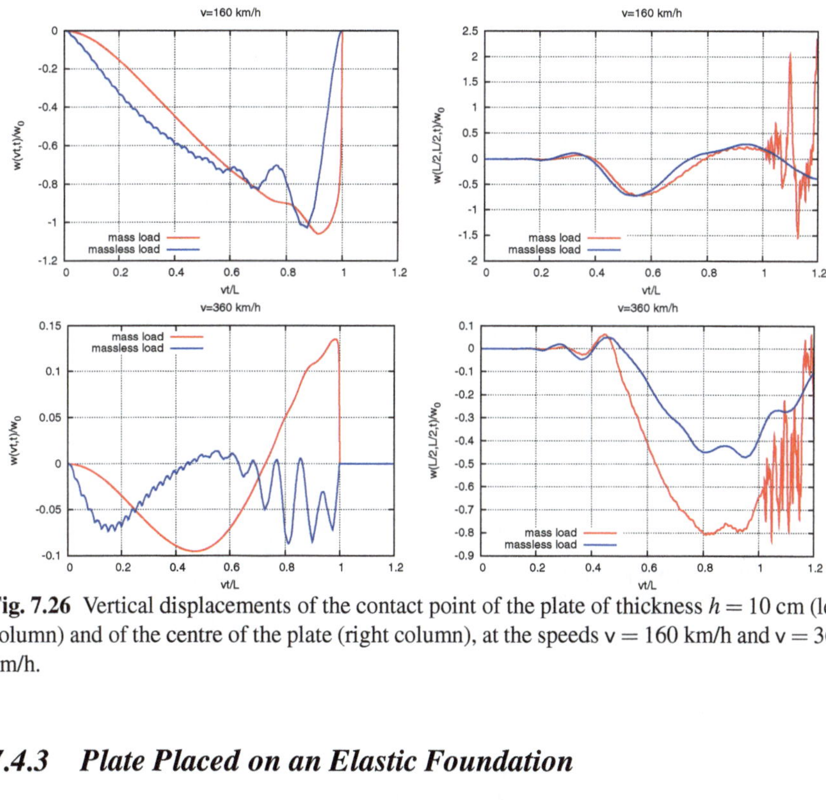

Fig. 7.26 Vertical displacements of the contact point of the plate of thickness $h = 10$ cm (left column) and of the centre of the plate (right column), at the speeds v = 160 km/h and v = 360 km/h.

7.4.3 Plate Placed on an Elastic Foundation

The Winkler type elastic foundation can be simply added to the elemental plate matrices. In the simplest case, we add the respective coefficients k_z multiplied by the fourth part of the area of the spatial finite element to the diagonal coefficients of the stiffness matrix. In a more complex approach, we should integrate the displacement shape functions with the coefficient k_z to obtain the consequent stiffness matrix instead of a granular one. In this case, the influence of the foundation is expressed also by off-diagonal coefficients. However, in our problem such a complex modelling does not influence the results significantly. Below we will perform the computations with the simplest, lumped effect of the foundation.

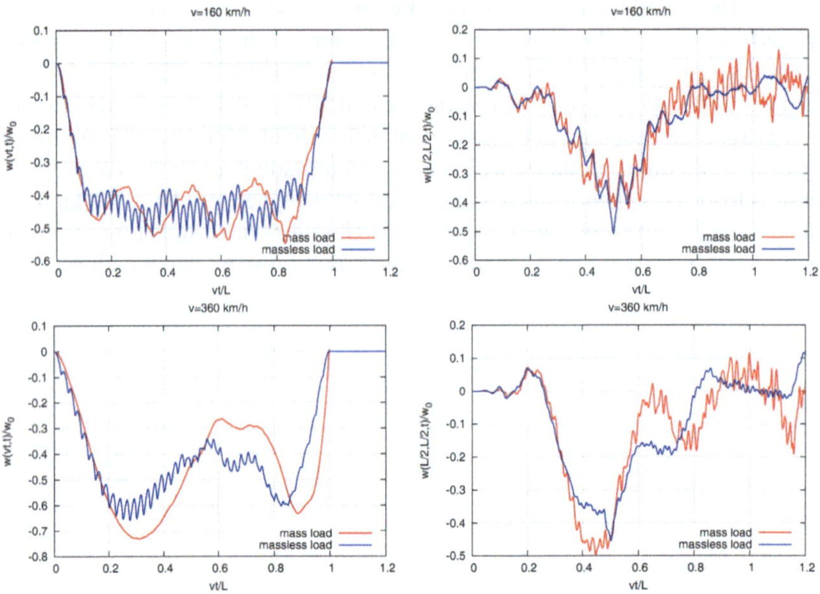

Fig. 7.27 Vertical displacements of the contact point of the plate of thickness $h = 40$ cm (left column) and of the centre of the plate (right column), at the speeds v = 160 km/h and v = 360 km/h.

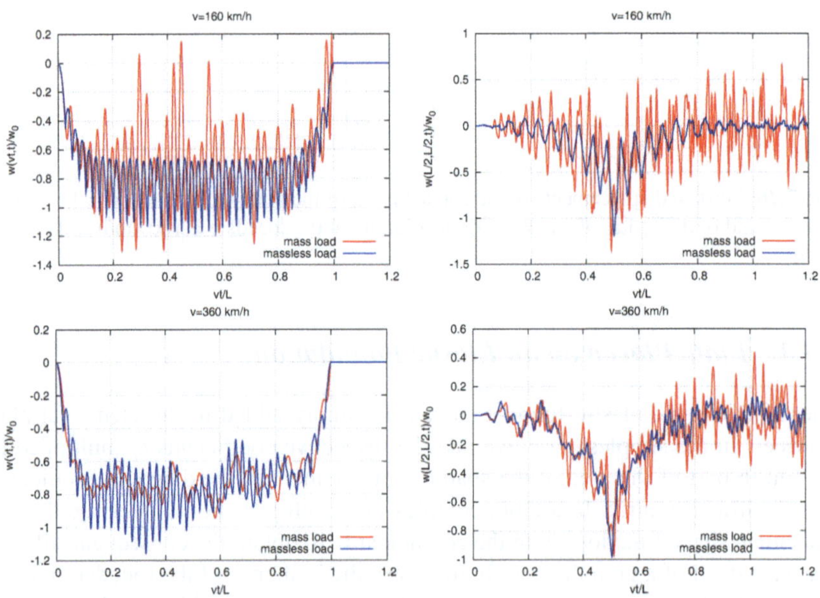

Fig. 7.28 Vertical displacements of the contact point of the plate of thickness $h = 100$ cm (left column) and of the centre of the plate (right column), at the speeds v = 160 km/h and v = 360 km/h.

We can also say that aside from the elasticity or visco–elastic effect of the foundation, we can also consider precisely its inertia. However, the final effect is similar to a direct increase of the mass density of the plate. That is why we will neglect the inertia of the foundation.

Let us look at the results of simulations. We consider a square plate of dimensions 12 m × 12 m, discretized by a mesh with 40 × 20 elements. The data is taken as in the previous chapters. The stiffness of the foundation is low ($k_z = 10^5$ Pa). We compare the deflections obtained with the thin plate model and with the thick plate model in the case of slow (v = 160 km/h) and fast (v = 360 km/h) mass passage velocities. Two thicknesses of the plate are considered: 10 cm in Figure 7.29 and 100 cm in Figure 7.30. In the case of a small thickness of the plate and a low velocity of the travelling mass, both curves practically coincide. This is clearly visible in the follower contact point. The centre of the plate exhibits a good agreement between both plate models, although the thick plate vibrates with higher frequency more intensively. The Mindlin plate model is more rigid and it is clearly visible for higher velocity (Figure 7.29, lower row of plots). The thicker plates ($t = 100$ cm), no matter in which model, vibrate with higher frequency, especially the thin plate.

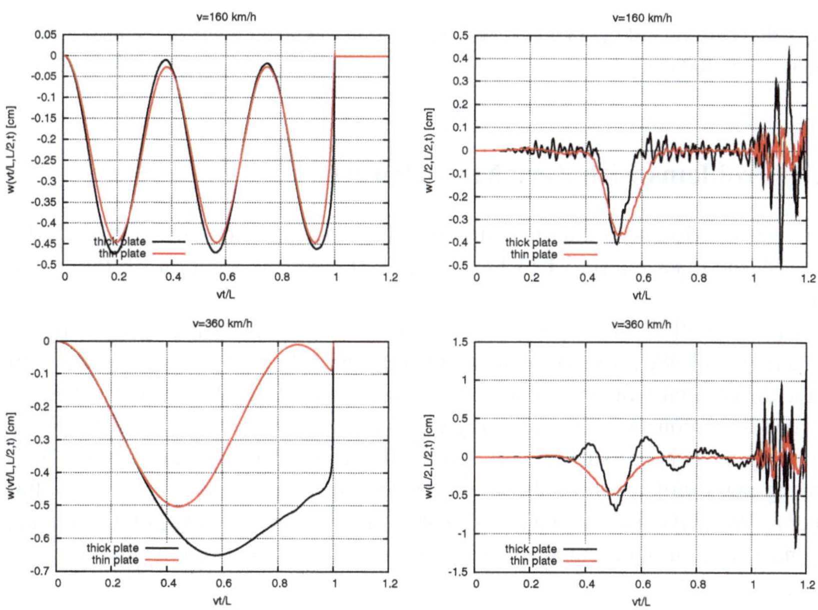

Fig. 7.29 Deflection of the moving contact point (left) and the middle of the plate (right) for thickness $t = 10$ cm.

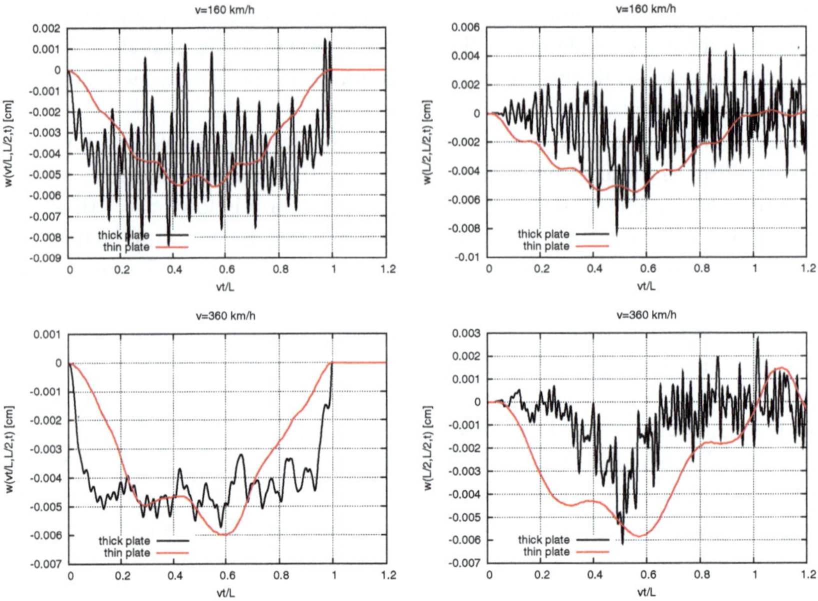

Fig. 7.30 Deflection of the moving contact point (left) and the middle of the plate (right) for thickness $t = 100$ cm.

7.5 Problems with Zero Mass Density

Now we will discuss a massless structure subjected to a moving inertial load. On page 34, the mass matrix **M** (3.14) with $\rho = 0$ gives the solution for the massless string.

Here we will assume zero mass density of the string in the space–time finite element model. We must recall Section 6.4.3. The only case for the time integration of the differential motion equation is possible with $\alpha = 1$. We must remember that this condition concerns the stationary geometry of the structure. The structure with a moving mass is not such a case. Our further examples will demonstrate that the moving mass violates the stability limitation for the massless space–time solution scheme. We will consider two cases: first with an inertial moving point load and the second with a moving massless point force. In both cases we consider a massless string.

Figure 7.31 presents the mass trajectories computed with the space–time element method. We assumed the parameter of the method $\alpha = 1$. The plot resembles the plot obtained for the inertial string subjected to a moving mass accompanied with the force. The shape of curves depends on the force to mass ratio. In our example this ratio is equal one. The figure perfectly corresponds with Figure 2.1 obtained with analytical computations.

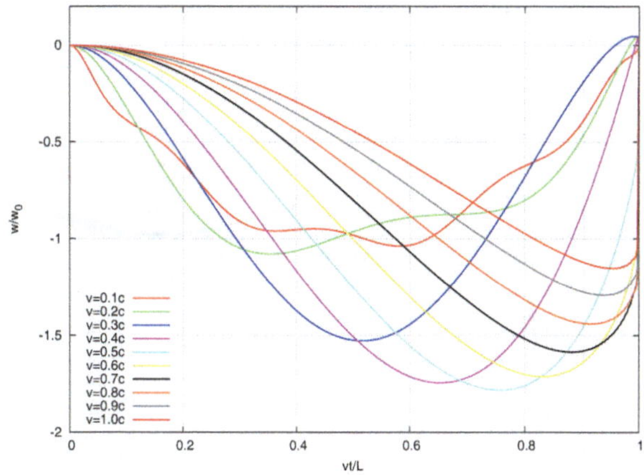

Fig. 7.31 Displacements under a moving mass: space–time finite element solution for massless string with $\alpha = 1$.

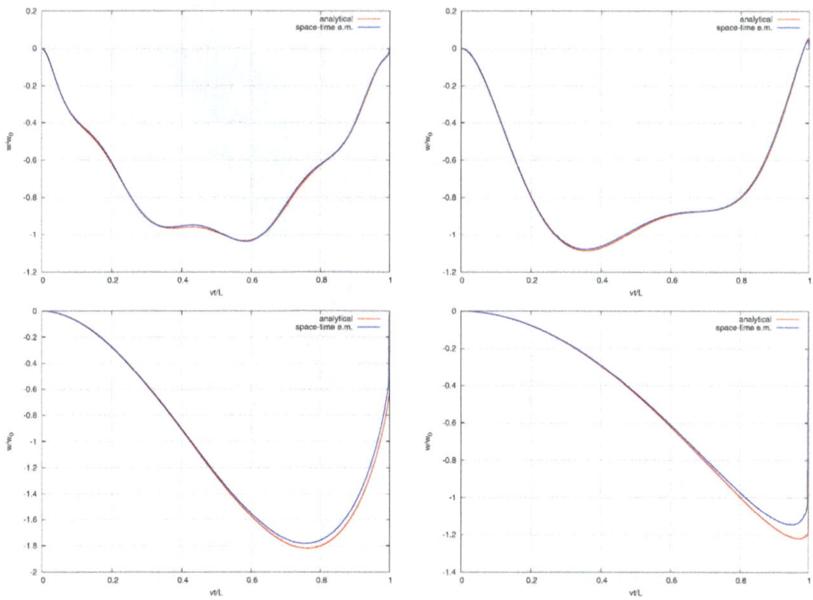

Fig. 7.32 Displacements under a moving mass—space–time finite element solution compared with the analytical results.

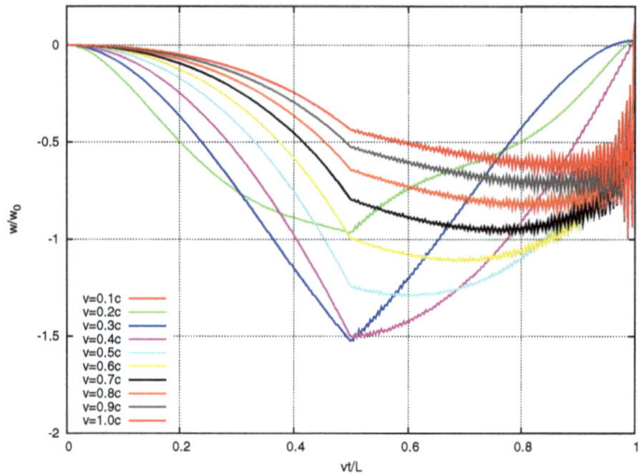

Fig. 7.33 Displacements at the midpoint—space–time finite element solution for a massless string with $\alpha = 0.9$.

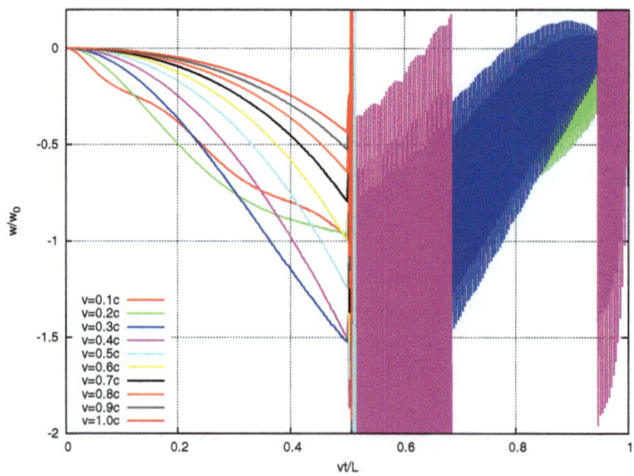

Fig. 7.34 Displacements at the midpoint—space–time finite element solution for a massless string with $\alpha = 1.0$.

In Figure 7.32, we compare the displacement trajectories of a massless string under a moving mass computed with the space–time element method at different speeds, with the analytical results. We took four values of the velocity: 0.1, 0.2, 0.5, and 1.0 times the wave speed evaluated for the inertial string in the previous chapters. The coincidence is practically perfect. We also compare the plots obtained with the space–time element method with the analytical curves.

Other test examples were computed with $\alpha = 0.9$ (Figure 7.33) and $\alpha = 1.0$ (Figure 7.34).

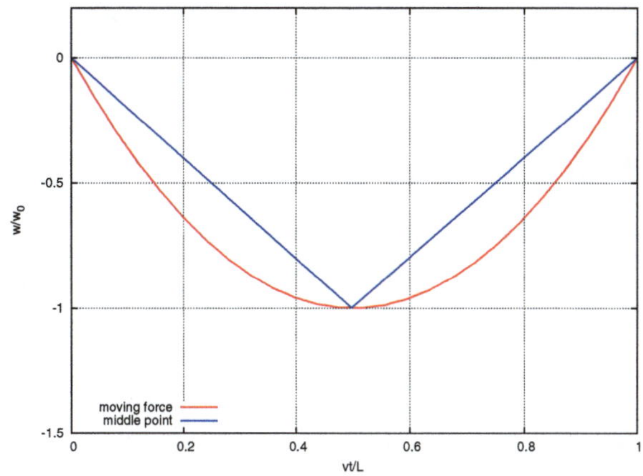

Fig. 7.35 Displacements in the case of a massless system ($m = 0$ and $\rho = 0$)—space–time finite element solution with $\alpha = 1.0$

The totally massless system, with zero mass density and zero moving mass, subjected to only massless forces, gives results as in the static solution, see Figure 7.35.

Chapter 8
The Newmark Method and a Moving Inertial Load

The Newmark method (see Section 5.5) is considered here as a representative example of a wide family of time integration methods. It is attractive since most of computational procedures in structural dynamics are based on this numerical scheme.

8.1 The Newmark Method in Moving Mass Problems

We must emphasize here that the matrices derived contribute only the point mass effects. They must be simply added to the classical matrices elaborated for a structure, i.e., for a string or a beam. The full discrete motion equation is

$$(\mathbf{M}+\mathbf{M}_m)\ddot{\mathbf{w}}^{i+1} + (\mathbf{C}+\mathbf{C}_m)\dot{\mathbf{w}}^{i+1} + (\mathbf{K}+\mathbf{K}_m)\mathbf{w}^{i+1} = \mathbf{F}^{i+1} + \mathbf{e}_m^i \,, \tag{8.1}$$

where $\mathbf{M}$ is the inertia matrix of the structure, $\mathbf{M}_m$ is the moving mass matrix, added only to the inertia matrix of the element on which it travels. The same occurs in the case of the damping matrix of the structure $\mathbf{C}$ and the point mass $\mathbf{C}_m$, and in the case of the stiffness matrix of the structure $\mathbf{K}$ and the point mass $\mathbf{K}_m$. The vector $\mathbf{F}^{i+1}$ is the vector of external forces established at time t_{i+1} and $\mathbf{e}_m^i$ is the right-hand side vector resulting from the the mass inertia term, established at the beginning of the time interval $[t_i, t_{i+1}]$. We will concentrate our attention on the mass influence only, thus we will derive the matrices $\mathbf{M}_m$, $\mathbf{C}_m$, $\mathbf{K}_m$, and $\mathbf{e}_m^i$ in the following equation

$$\mathbf{M}_m\ddot{\mathbf{w}}^{i+1} + \mathbf{C}_m\dot{\mathbf{w}}^{i+1} + \mathbf{K}_m\mathbf{w}^{i+1} = \mathbf{e}_m^i \,, \tag{8.2}$$

where the vector of nodal displacements in the case of beams is $\mathbf{w} = [w_1, \psi_1, w_2, \psi_2]$, and that in the case of a string is $\mathbf{w} = [w_1, w_2]$.

The matrices of the finite element that carry the inertial particle are composed, as a sum, of two sets: the matrices describing the element of the structure and the matrices that incorporate the mass influence. Since the elemental matrices are well known, below we will consider only the influence of the mass.

The solution of this problem concerns a mass particle moving on a general finite element. This can be applied to all types of structures: strings, beams, plates, shells,

C.I. Bajer and B. Dyniewicz: Numerical Analysis of Vibrations of Structures, LNACM 65, pp. 223–240.
springerlink.com

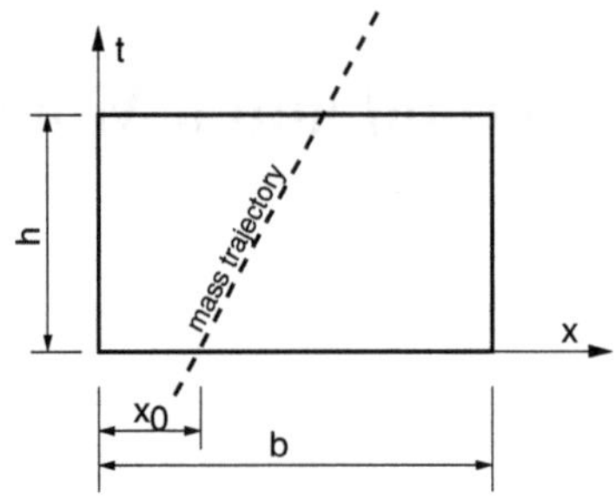

Fig. 8.1 The mass trajectory in space and time.

etc. Below we will derive the resulting matrices which will then be applied and tested with a string, an Euler beam, and a Timoshenko beam.

Let us consider a finite element of length b of the edge of the mass trajectory. The mass particle m passes through the finite element with velocity v in the time interval h, starting at the point $x = x_0$ (Figure 8.1). The equation of virtual work which describes the motion of the inertial particle is

$$\Pi_m = \int_0^b w^*(x)\delta(x - x_0 - \mathrm{v}t)\, m \frac{\mathrm{d}^2 w(\mathrm{v}t,t)}{\mathrm{d}t^2}\mathrm{d}x\,, \tag{8.3}$$

where the position of the moving point can be described by the function $x = \mathrm{v}t$. The virtual displacement function w^* is given by

$$w^*(x) = \left[1 - \frac{x}{b}, \frac{x}{b}\right] \mathbf{w}\,. \tag{8.4}$$

We take first-order polynomials as the shape functions describing the interpolation of the displacements:

$$w(x,t) = \left(1 - \frac{x}{b}\right) w_1(t) + \frac{x}{b} w_2(t)\,. \tag{8.5}$$

Here, $w_1(t)$ and $w_2(t)$ are the nodal displacements in time. This is a natural assumption since the finite element edge is straight in cases of simple shape functions describing linear displacement distributions in the element. In such a case, the third term of (7.1) reduces to zero. That is why we must write the Renaudot formula (7.1) at constant speed in a different form:

$$\frac{\mathrm{d}^2 w(\mathrm{v}t,t)}{\mathrm{d}t^2} = \left.\frac{\partial^2 w(x,t)}{\partial t^2}\right|_{x=\mathrm{v}t} + \mathrm{v} \left.\frac{\partial^2 w(x,t)}{\partial x \partial t}\right|_{x=\mathrm{v}t} + \mathrm{v}\frac{\mathrm{d}}{\mathrm{d}t}\left[\left.\frac{\partial w(x,t)}{\partial x}\right|_{x=\mathrm{v}t}\right]\,. \tag{8.6}$$

The third term of (8.6) is developed in its Taylor series in terms of the time increment $\Delta t = h$

$$
\begin{aligned}
\left[\left.\frac{\partial w(x,t)}{\partial x}\right|_{x=\mathrm{v}t}\right]^{t+h} &= \left[\left.\frac{\partial w(x,t)}{\partial x}\right|_{x=\mathrm{v}t}\right]^{t} + \\
&+ \left\{\frac{\mathrm{d}}{\mathrm{d}t}\left[\left.\frac{\partial w(x,t)}{\partial x}\right|_{x=\mathrm{v}t}\right]\right\}^{t}(1-\gamma)h + \left\{\frac{\mathrm{d}}{\mathrm{d}t}\left[\left.\frac{\partial w(x,t)}{\partial x}\right|_{x=\mathrm{v}t}\right]\right\}^{t+h}\gamma h.
\end{aligned}
\tag{8.7}
$$

The upper indices indicate the time at which the respective terms are defined. We assume the backward difference formula ($\gamma = 1$). In this case we have

$$
\left\{\frac{\mathrm{d}}{\mathrm{d}t}\left[\left.\frac{\partial w(x,t)}{\partial x}\right|_{x=\mathrm{v}t}\right]\right\}^{t+h} = \frac{1}{h}\left[\left.\frac{\partial w(x,t)}{\partial x}\right|_{x=\mathrm{v}t}\right]^{t+h} - \frac{1}{h}\left[\left.\frac{\partial w(x,t)}{\partial x}\right|_{x=\mathrm{v}t}\right]^{t}. \tag{8.8}
$$

Using (8.5) and (8.8), the equation (8.6) is given by the difference formula

$$
\begin{aligned}
\frac{\mathrm{d}^2 w}{\mathrm{d}t^2} &= \left(1 - \frac{x_0 + \mathrm{v}t}{b}\right)\ddot{w}_1^{i+1} + \frac{x_0 + \mathrm{v}t}{b}\ddot{w}_2^{i+1} - \frac{\mathrm{v}}{b}\dot{w}_1^{i+1} + \frac{\mathrm{v}}{b}\dot{w}_2^{i+1} - \\
&- \frac{\mathrm{v}}{bh}w_1^{i+1} + \frac{\mathrm{v}}{bh}w_2^{i+1} + \frac{\mathrm{v}}{bh}w_1^{i} - \frac{\mathrm{v}}{bh}w_2^{i}.
\end{aligned}
\tag{8.9}
$$

The upper index denotes the time layer. The energy (8.3), with respect to (8.4) and (8.9) can be written in quadratic form, which, after a classical minimization, results in the matrix equation (8.2), where

$$
\mathbf{M}_m = m\begin{bmatrix} (1-\kappa)^2 & 0 & \kappa(1-\kappa) & 0 \\ 0 & 0 & 0 & 0 \\ \kappa(1-\kappa) & 0 & \kappa^2 & 0 \\ 0 & 0 & 0 & 0 \end{bmatrix}, \tag{8.10}
$$

$$
\mathbf{C}_m = \frac{m\mathrm{v}}{b}\begin{bmatrix} -(1-\kappa) & 0 & 1-\kappa & 0 \\ 0 & 0 & 0 & 0 \\ -\kappa & 0 & \kappa & 0 \\ 0 & 0 & 0 & 0 \end{bmatrix}, \tag{8.11}
$$

$$
\mathbf{K}_m = \frac{m\mathrm{v}}{bh}\begin{bmatrix} -(1-\kappa) & 0 & 1-\kappa & 0 \\ 0 & 0 & 0 & 0 \\ -\kappa & 0 & \kappa & 0 \\ 0 & 0 & 0 & 0 \end{bmatrix}, \tag{8.12}
$$

and

$$\mathbf{e}_m = \frac{m\mathrm{v}}{bh} \begin{bmatrix} (1-\kappa)(w_2-w_1) \\ 0 \\ \kappa(w_2-w_1) \\ 0 \end{bmatrix}, \tag{8.13}$$

with coefficient $\kappa = (x_0 + \mathrm{v}h)/b\ ,\ \ 0 < \kappa \leq 1$. κ is a parameter which defines the position of the mass in the element at the beginning of the time increment.

This determines the position of the mass at time $t = h$, related to the finite element length b. The different terms describe the transverse inertia force related to the vertical acceleration, the Coriolis force, and the centrifugal force. The matrix factors $\mathbf{M}_m$, $\mathbf{C}_m$, and $\mathbf{K}_m$ can be called the mass, the damping, and the stiffness matrices. The last term $\mathbf{e}_m$ describes the nodal forces at the beginning of the time interval $[t_i, t_i + \Delta t]$. We must emphasize here that the matrices (8.10)–(8.12) and the vector (8.13) contribute only the moving inertial particle effect. The matrices of the mass influence in a finite element of a structure must be added to the global system of equations. We notice that the matrices (8.10)–(8.12) differ from the matrices that result from the solution for the case of direct differentiation of (7.1).

8.2 The Newmark Method in the Vibrations of String

The finite element of a string that carries an inertial particle was tested with the use of the Newmark method. In each time-step, the global matrices **M**, **C**, and **K** must be computed since the contributions of (8.10)–(8.12) vary. The string being tested has dimensionless length $l = 1$, tensile force $N = 1$, cross-sectional area $A = 1$, and mass density $\rho = 1$. The travelling mass $m = 1$ was accompanied by the force $P = -1$ (Figure 8.2). Figure 8.3 depicts the mass trajectory at various speeds v. This diagram can be compared with the semi-analytical results depicted in Figure 3.3. We note good coincidence for a whole range of speeds. Moreover, the discontinuity at the final support is exhibited in the numerical results. This discontinuity was reported in [48] and was also obtained by the space–time finite element method [25]. The mid-span deflections are depicted in Figure 8.4. The method can also be applied to over-critical speed (Figures 8.5 and 8.6).

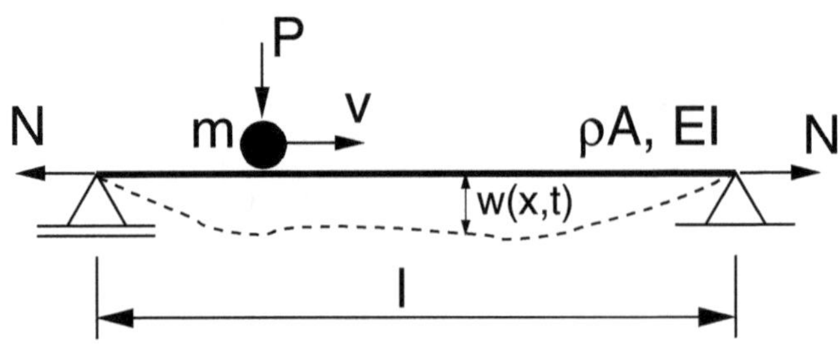

Fig. 8.2 The scheme of the tested system.

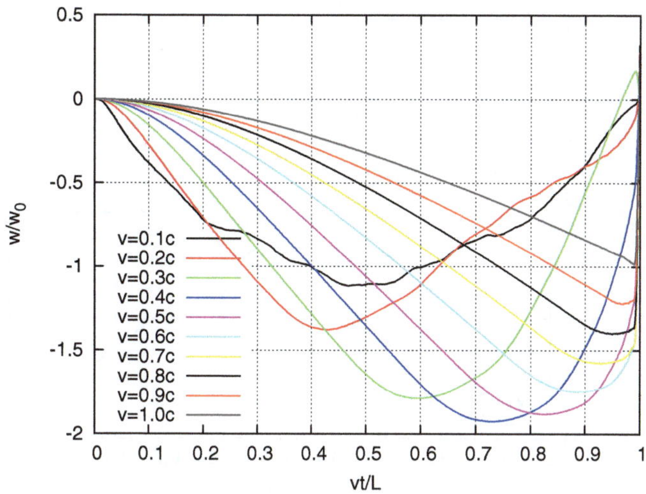

Fig. 8.3 The mass trajectory at velocities 0.1–1.0 times the wave speed c, computed numerically by the Newmark method.

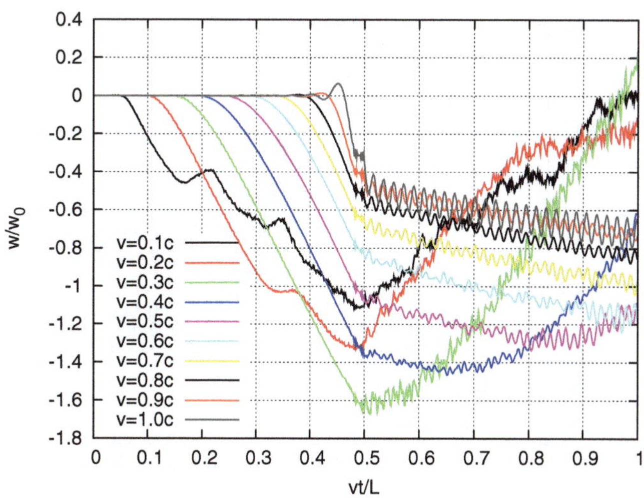

Fig. 8.4 The displacement of the middle of the span at velocities 0.1–1.0 times the wave speed c.

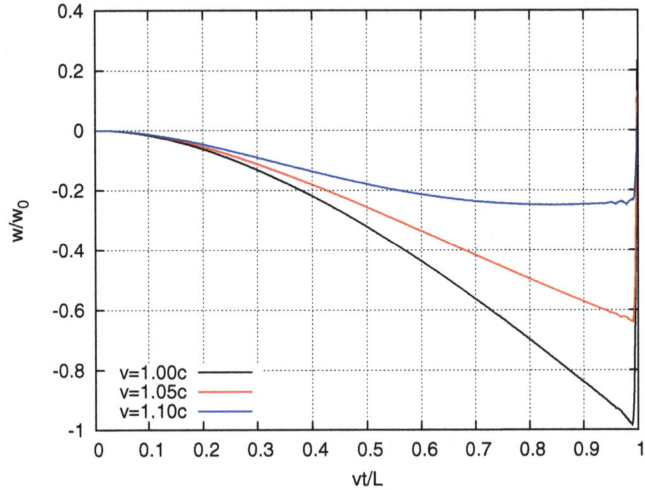

Fig. 8.5 The mass trajectory at velocities 1.0–1.1 times the wave speed c, computed numerically by the Newmark method.

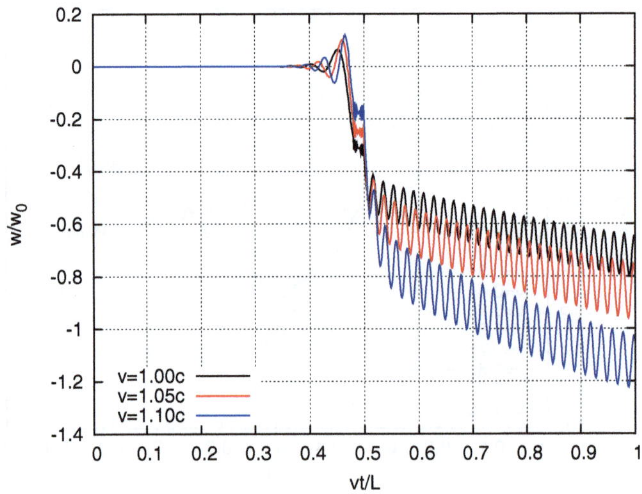

Fig. 8.6 The displacement of the middle of the span at velocities 1.0–1.1 times the wave speed c.

8.3 The Newmark Method in Vibrations of the Bernoulli–Euler Beam

The solution procedure is relatively simple in the case of the Bernoulli–Euler beam with third order shape functions. The characteristic matrices can be easily derived (8.14)–(8.16). We use them in verification of various results from the literature:

$$\mathbf{M}_m = m \begin{bmatrix} (2\kappa^3-3\kappa^2+1)^2 & b\kappa(\kappa^2-2\kappa+1)(2\kappa^3-3\kappa^2+1) & \kappa^2(3-2\kappa)(2\kappa^3-3\kappa^2+1) & b\kappa^2(\kappa-1)(2\kappa^3-3\kappa^2+1) \\ b\kappa(\kappa^2-2\kappa+1)(2\kappa^3-3\kappa^2+1) & b^2\kappa^2(\kappa^2-2\kappa+1)^2 & b\kappa^3(3-2\kappa)(\kappa^2-2\kappa+1) & b^2\kappa^3(\kappa-1)(\kappa^2-2\kappa+1) \\ \kappa^2(3-2\kappa)(2\kappa^3-3\kappa^2+1) & b\kappa^3(3-2\kappa)(\kappa^2-2\kappa+1) & \kappa^4(2\kappa-3)^2 & b\kappa^4(1-\kappa)(2\kappa-3) \\ b\kappa^2(\kappa-1)(2\kappa^3-3\kappa^2+1) & b^2\kappa^3(\kappa-1)(\kappa^2-2\kappa+1) & b\kappa^4(1-\kappa)(2\kappa-3) & b^2\kappa^4(\kappa-1)^2 \end{bmatrix} \tag{8.14}$$

$$\mathbf{C}_m = 2m\mathrm{v} \begin{bmatrix} 6\kappa(\kappa-1)(2\kappa^3-3\kappa^2+1)/b & (3\kappa^2-4\kappa+1)(2\kappa^3-3\kappa^2+1) & 6\kappa(1-\kappa)(2\kappa^3-3\kappa^2+1)/b & \kappa(3\kappa-2)(2\kappa^3-3\kappa^2+1) \\ 6\kappa^2(\kappa-1)(\kappa^2-2\kappa+1) & b\kappa(\kappa^2-2\kappa+1)(3\kappa^2-4\kappa+1) & 6\kappa^2(1-\kappa)(\kappa^2-2\kappa+1) & b\kappa^2(\kappa^2-2\kappa+1)(3\kappa-2) \\ 6\kappa^3(1-\kappa)(2\kappa-3)/b & \kappa^2(3-2\kappa)(3\kappa^2-4\kappa+1) & 6\kappa^3(\kappa-1)(2\kappa-3)/b & \kappa^3(3-2\kappa)(3\kappa-2) \\ 6\kappa^3(\kappa-1)^2 & b\kappa^2(\kappa-1)(3\kappa^2-4\kappa+1) & -6\kappa^3(\kappa-1)^2 & b\kappa^3(\kappa-1)(3\kappa-2) \end{bmatrix}, \tag{8.15}$$

$$\mathbf{K}_m = m\mathrm{v}^2 \begin{bmatrix} 6(2\kappa-1)(2\kappa^3-3\kappa^2+1)/b^2 & 2(3\kappa-2)(2\kappa^3-3\kappa^2+1)/b & 6(1-2\kappa)(2\kappa^3-3\kappa^2+1)/b^2 & 2(3\kappa-1)(2\kappa^3-3\kappa^2+1)/b \\ 6\kappa(\kappa^2-2\kappa+1)(2\kappa-1)/b & 2\kappa(\kappa^2-2\kappa+1)(3\kappa-2) & 6\kappa(1-2\kappa)(\kappa^2-2\kappa+1)/b & 2\kappa(\kappa^2-2\kappa+1)(3\kappa-1) \\ 6\kappa^2(1-2\kappa)(2\kappa-3)/b^2 & 2\kappa^2(3-2\kappa)(3\kappa-2)/b & 6\kappa^2(2\kappa-1)(2\kappa-3)/b^2 & 2\kappa^2(3-2\kappa)(3\kappa-1)/b \\ 6\kappa^2(\kappa-1)(2\kappa-1)/b & 2\kappa^2(\kappa-1)(3\kappa-2) & 6\kappa^2(1-\kappa)(2\kappa-1)/b & 2\kappa^2(\kappa-1)(3\kappa-1) \end{bmatrix}, \tag{8.16}$$

where the parameter $\kappa = (x_0 + \mathrm{v}h)/b$, x_0 is the initial position of the mass moving with speed v on the spatial element of length b in the time interval h.

8.4 The Newmark Method in Vibrations of a Timoshenko Beam

The study of wave phenomena is possible by using a more complex model of the Timoshenko beam in which the vibration equation takes into account the influence of lateral forces and rotatory inertia on the deflection line of the beam. The angle formed by the axis of the deformed beam is composed of the pure bending angle and the angle corresponding to the deformation of the pure shear. Independent interpolation of the displacements and rotation angles of the Timoshenko beam causes serious problems. Linear interpolation (8.5) of the nodal shape features renders impossible the determination of the centrifugal acceleration of the moving mass particle. Direct discretization of the terms (7.1), placed in the governing differential equation of motion, results in the following matrices:

$$\mathbf{M}_m = m \begin{bmatrix} (1-\kappa)^2 & 0 & \kappa(1-\kappa) & 0 \\ 0 & 0 & 0 & 0 \\ \kappa(1-\kappa) & 0 & \kappa^2 & 0 \\ 0 & 0 & 0 & 0 \end{bmatrix}, \tag{8.17}$$

$$\mathbf{C}_m = \frac{2mv}{b} \begin{bmatrix} -(1-\kappa) & 0 & 1-\kappa & 0 \\ 0 & 0 & 0 & 0 \\ -\kappa & 0 & \kappa & 0 \\ 0 & 0 & 0 & 0 \end{bmatrix}, \tag{8.18}$$

$$\mathbf{K}_m = \mathbf{0}\,, \tag{8.19}$$

with the coefficient

$$\kappa = \frac{x_0 + vh}{b}\,, \quad 0 < \kappa \leq 1. \tag{8.20}$$

We can apply these to a test problem and then compare the results with those obtained by semi-analytical methods. Unfortunately, the comparison is extremely unsatisfactory, especially if applied to strictly hyperbolic problems (see for example [48]). We must emphasize here that the matrices (8.10)–(8.12) and the vector (8.13) contribute only the moving inertial particle effect. The matrices of a mass influence in a finite element of the Timoshenko beam must be added to the global system of equations. Note that the matrices (8.17)–(8.19) differ from the matrices (8.10)–(8.13). The matrix **M** is the only matrix that is the same.

8.5 Numerical Results

There are few publications in which an inertial load moving on a Timoshenko beam are directly considered numerically. One can see that papers published in the literature describe methods which result in wrong responses. We can show results taken from the literature compared with semi-analytical computations. A comparison with the paper [148] is given in Figure 4.15. In the example, a very low relative velocity was assumed: $\mathrm{v}/c_1 = 0.002$ and $\mathrm{v}/c_2 = 0.001$ (shear wave and bending wave, respectively).

We will compare our diagrams with those of Lee [86]. Therefore, the data in the example is as follows: length $l = 1$ m, Young modulus $E = 207$ GPa, shear modulus $G = 77.6$ GPa, mass density $\rho = 7700$ kg/m^3. The velocity $\mathrm{v} = a\pi/l \cdot \sqrt{EI/\rho/A}$

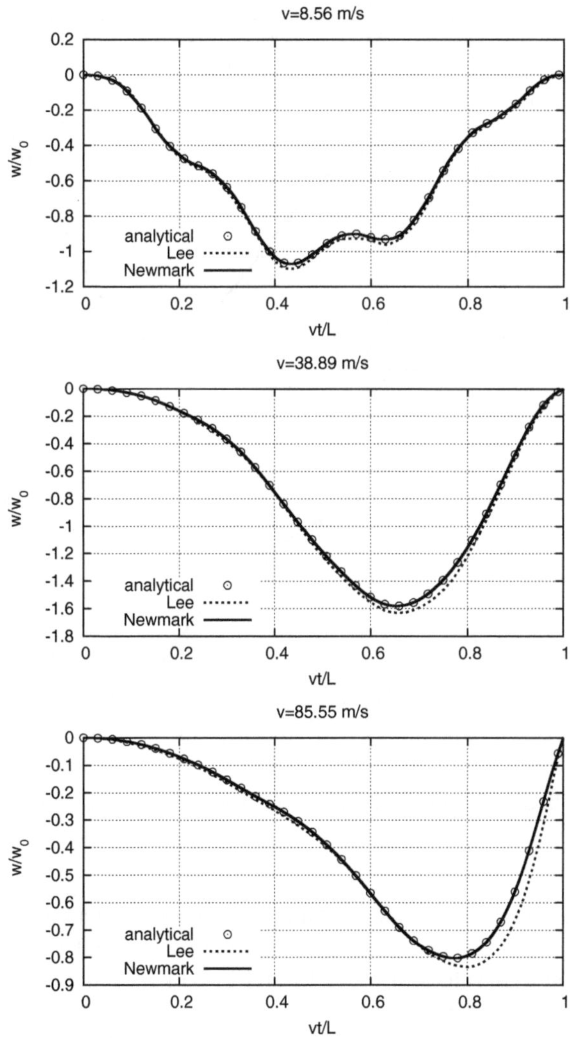

Fig. 8.7 Normalized deflections under a moving mass particle for $\beta = 0.03$: (a) $a = 0.11$, (b) $a = 0.5$ and (c) $a = 1.1$.

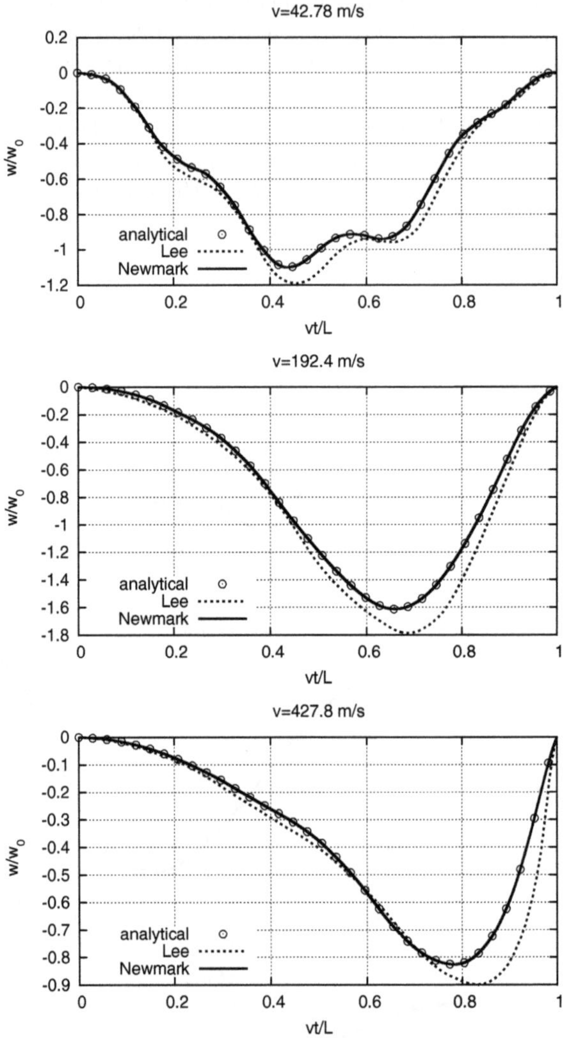

Fig. 8.8 Normalized deflections under a moving mass particle for $\beta = 0.15$: (a) $a = 0.11$, (b) $a = 0.5$ and (c) $a = 1.1$.

was determined by the parameter a. Another parameter β determines the cross sectional area $A = \beta^2 l^2/\pi$ and cross sectional inertia moment $I = \beta^4 l^4/4/\pi^3$. The moving mass m took values of 0.441 kg and 11.03 kg. Figure 8.7 exhibits the normalized deflection under the moving mass for $\beta = 0.03$ and $a = 0.11$, 0.5, and 1.1. This corresponds to a mass moving at 8.56, 38.39, 85.55 m/s on a relatively elastic beam. Figure 8.8 relates to a more rigid beam and velocities of v = 42.78, 194.4, and 427.7 m/s. Lee solved the problem semi-analytically. A fourth order differential equation was solved by the Fourier transform and finally integrated by the

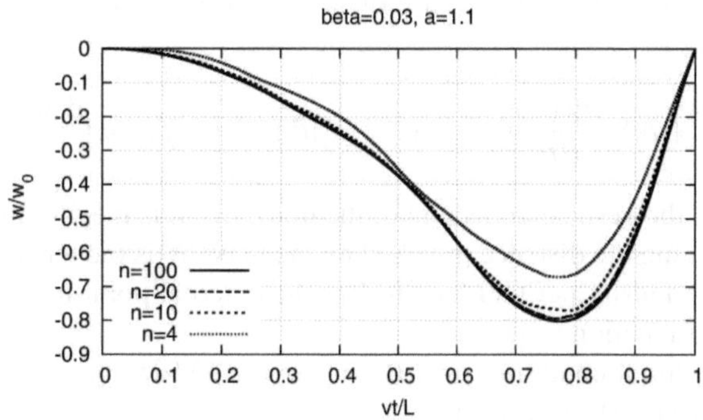

Fig. 8.9 Accuracy of the Newmark method depending on the number of finite elements.

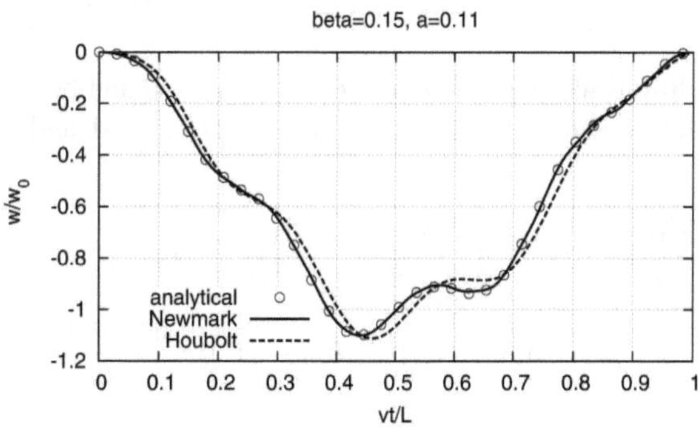

Fig. 8.10 Comparison of the Newmark and Houbolt methods in the case of a large time step.

Runge–Kutta method. In our test, we compare the results by Lee with our semi-analytically [49] obtained curves together with our Newmark time integration procedure applied to the finite element model of the Timoshenko beam. We notice a perfect coincidence of both solutions and quite good coincidence with Lee's results.

Figure 8.9 shows the accuracy, which increases with the number of elements in the structure. Ten to twenty elements is sufficient in our example.

Another comparison was carried out between the Newmark and Houbolt methods. Both methods are sufficiently accurate. However, the curve for the Newmark method perfectly coincides with our semi-analytical results (Figure 8.10).

8.6 Accelerating Mass—Numerical Approach

8.6.1 Mathematical Model

Let us consider the differential equations of structures containing a concentrated mass. We will focus our attention on the term which describes the forces induced by a moving inertial particle. In the case of a string, we can write the equation in the form

$$-N\frac{\partial^2 w(x,t)}{\partial x^2} + \rho A\frac{\partial^2 w(x,t)}{\partial t^2} = \delta(x-f(t))P - \delta(x-f(t))m\frac{\mathrm{d}^2 w(f(t),t)}{\mathrm{d}t^2} \,. \quad (8.21)$$

Here, $w(x,t)$ is the vertical deflection of the mid-line, m is the moving mass, $f(t)$ is the function giving the distance travelled by the mass, N is the tension of the string, ρA is the mass density per unit length, P is the external point force, and it usually contributes a gravitation force mg.

We impose initial conditions $w(x,0) = 0,\ \partial w(x,t)/\partial t\,|_{t=0} = 0$ and boundary conditions $w(0,t) = 0,\ w(l,t) = 0$.

The Bernoulli–Euler beam is described by the equation

$$EI\frac{\partial^4 w(x,t)}{\partial x^4} + \rho A\frac{\partial^2 w(x,t)}{\partial t^2} = \delta(x-f(t))P - \delta(x-f(t))m\frac{\mathrm{d}^2 w(f(t),t)}{\mathrm{d}t^2} \,, \quad (8.22)$$

with initial conditions $w(x,0) = 0,\ \partial w(x,t)/\partial t\,|_{t=0} = 0$ and boundary conditions $w(0,t) = 0,\ w(l,t) = 0, \partial^2 w(0,t)/\partial x^2 = 0, \partial^2 w(l,t)/\partial x^2 = 0$, and the Timoshenko beam is

$$\rho A\frac{\partial^2 w(x,t)}{\partial t^2} - \frac{GA}{k}\left(\frac{\partial^2 w(x,t)}{\partial x^2} - \frac{\partial \psi(x,t)}{\partial x}\right) =$$

$$= \delta(x-f(t))P - \delta(x-f(t))m\frac{\mathrm{d}^2 w(f(t),t)}{\mathrm{d}t^2}, \quad (8.23)$$

$$\rho I\frac{\partial^2 \psi(x,t)}{\partial t^2} - EI\frac{\partial 2\psi(x,t)}{\partial x^2} - \frac{GA}{k}\left(\frac{\partial w(x,t)}{\partial x} - \psi(x,t)\right) = 0 \,,$$

with the same boundary and initial conditions as for the Bernoulli–Euler beam. Here, EI is the bending stiffness, GA/k is the shear stiffness, ρI is the rotatory inertia of the cross section of the beam, and ψ is the angle of rotation of the cross section.

In each type of problem we have the identical inertial term $\delta(x - f(t))m\cdot \mathrm{d}^2 w(f(t),t)/\mathrm{d}t^2$. Below we will consider only this term, since the remaining parts of the equations are treated in the classical way by the finite element method.

Let us follow the direct derivation commonly carried on in the literature. The acceleration of a mass particle moving at a varying speed v in the space–time domain is described by the Renaudot formula:

$$\frac{\mathrm{d}^2 w(f(t),t)}{\mathrm{d}t^2} = \left.\frac{\partial^2 w(x,t)}{\partial t^2}\right|_{x=f(t)} + 2\mathrm{v}\left.\frac{\partial^2 w(x,t)}{\partial x\partial t}\right|_{x=f(t)} + \mathrm{v}^2\left.\frac{\partial^2 w(x,t)}{\partial x^2}\right|_{x=f(t)} +$$

$$+\dot{\mathrm{v}}\left.\frac{\partial w(x,t)}{\partial x}\right|_{x=f(t)}, \quad (8.24)$$

where $f(t)$ describes the position of the load. The above formula simply represents the chain rule of differentiation. The corresponding parts of the equation describe the lateral acceleration, the Coriolis acceleration, the centrifugal acceleration, and the acceleration associated with the change of particle velocity. These names are generally not adequate in the case of all structures. Let us compare two different problems: the vibrations of a string and the longitudinal vibrations of a bar. In both cases, we have the identical governing equation. However, in the case of longitudinal displacements we can not call the forces described by the terms of the equation either centrifugal or Coriolis.

8.6.2 The Finite Element Carrying the Moving Mass Particle

Let us consider a finite element of length b of the edge of the mass trajectory. The mass particle m passes through the finite element with a varying velocity v in the time interval h, starting at the point $x = x_0$ (Figure 8.1). The equation of virtual work which describes the motion of the inertial particle can be written in the following form

$$\Pi_m = \int_0^b w^*(x)\delta(x - f(t))\, m \frac{\mathrm{d}^2 w(f(t),t)}{\mathrm{d}t^2} \mathrm{d}x \,. \tag{8.25}$$

The virtual displacement w^* is expressed by (8.4). The position of the moving point can be described by a quadratic function in time:

$$f(t) = x_0 + \mathrm{v}t + \frac{1}{2}\dot{\mathrm{v}}t^2 \,. \tag{8.26}$$

We take first-order polynomials as the shape functions describing the interpolation of the displacements (8.5). Here, $w_1(t)$ and $w_2(t)$ are the nodal displacements in time. This is a natural assumption since the finite element edge is straight for simple shape functions describing linear displacement distributions in the element. In such cases, the third term of (8.24) reduces to zero. That is why we must write the Renaudot formula (8.24) in a different form:

$$\begin{aligned}\frac{\mathrm{d}^2 w(f(t),t)}{\mathrm{d}t^2} = {} & \left.\frac{\partial^2 w(x,t)}{\partial t^2}\right|_{x=f(t)} + \mathrm{v} \left.\frac{\partial^2 w(x,t)}{\partial x \partial t}\right|_{x=f(t)} + \dot{\mathrm{v}} \left.\frac{\partial w(x,t)}{\partial x}\right|_{x=f(t)} + \\ & + \mathrm{v}\frac{\mathrm{d}}{\mathrm{d}t}\left[\left.\frac{\partial w(x,t)}{\partial x}\right|_{x=f(t)}\right] .\end{aligned} \tag{8.27}$$

The fourth term of (8.27) is developed in a Taylor series in powers of the time increment $\Delta t = h$

$$\left[\frac{\partial w(x,t)}{\partial x} \bigg|_{x=f(t)} \right]^{t+h} = \left[\frac{\partial w(x,t)}{\partial x} \bigg|_{x=f(t)} \right]^{t} + \\ + \left\{ \frac{\mathrm{d}}{\mathrm{d}t} \left[\frac{\partial w(x,t)}{\partial x} \bigg|_{x=f(t)} \right] \right\}^{t} (1-\gamma)h + \left\{ \frac{\mathrm{d}}{\mathrm{d}t} \left[\frac{\partial w(x,t)}{\partial x} \bigg|_{x=f(t)} \right] \right\}^{t+h} \gamma h . \tag{8.28}$$

The upper indices indicate the time at which the respective terms are defined. Using (8.8), we assume the backward difference formula ($\gamma = 1$). After classical minimization of the equation (8.25) with respect to (8.27) and (8.8), we obtain

$$\mathbf{M}_m = m \begin{bmatrix} (1-\kappa)^2 & 0 & \kappa(1-\kappa) & 0 \\ 0 & 0 & 0 & 0 \\ \kappa(1-\kappa) & 0 & \kappa^2 & 0 \\ 0 & 0 & 0 & 0 \end{bmatrix} , \tag{8.29}$$

$$\mathbf{C}_m = \frac{m\mathrm{v}}{b} \begin{bmatrix} -(1-\kappa) & 0 & 1-\kappa & 0 \\ 0 & 0 & 0 & 0 \\ -\kappa & 0 & \kappa & 0 \\ 0 & 0 & 0 & 0 \end{bmatrix} , \tag{8.30}$$

$$\mathbf{K}_m = \frac{m}{b} \left(\frac{\mathrm{v}}{h} + \dot{\mathrm{v}} \right) \begin{bmatrix} -(1-\kappa) & 0 & 1-\kappa & 0 \\ 0 & 0 & 0 & 0 \\ -\kappa & 0 & \kappa & 0 \\ 0 & 0 & 0 & 0 \end{bmatrix} , \tag{8.31}$$

and

$$\mathbf{e}_m = \frac{m\mathrm{v}}{bh} \begin{bmatrix} (1-\kappa)(w_2 - w_1) \\ 0 \\ \kappa(w_2 - w_1) \\ 0 \end{bmatrix} , \tag{8.32}$$

with coefficient $\kappa = (x_0 + \mathrm{v}h + 1/2\dot{\mathrm{v}}h^2)/b$, $0 < \kappa \leq 1$. κ is a parameter which defines the position of the mass in the element at the beginning of the time increment.

This determines the position of the mass at time $t = h$, related to the finite element length b. The different terms describe the transverse inertia force related to the vertical acceleration, the Coriolis force, and the centrifugal force. The matrix factors $\mathbf{M}_m$, $\mathbf{C}_m$, and $\mathbf{K}_m$ can be called the mass, damping, and stiffness matrices. The last

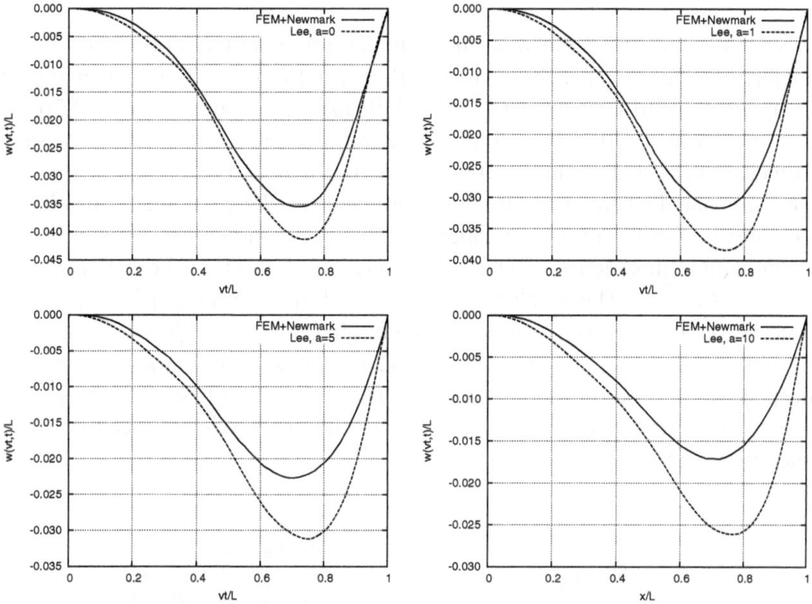

Fig. 8.11 Comparison of displacements of a Bernoulli–Euler beam under a moving contact point with those published by Lee [84]—acceleration parameter $\bar{a} = 0, 1, 5, 10$.

term $\mathbf{e}_m$ describes the nodal forces at the beginning of the time interval $[t_i, t_i + \Delta t]$. We must emphasize here that the matrices (8.29)–(8.31) and the vector (8.32) contribute only the moving inertial particle effect. The matrices of the mass influence in a finite element of a structure must be added to the global system of equations. We note that the matrices (8.29)–(8.31) differ from the matrices that result in divergence of the solution in the case of direct differentiation of (8.24)[1].

[1] Matrices that result in divergence:

$$\mathbf{M}_m = m \begin{bmatrix} (1-\kappa)^2 & 0 & \kappa(1-\kappa) & 0 \\ 0 & 0 & 0 & 0 \\ \kappa(1-\kappa) & 0 & \kappa^2 & 0 \\ 0 & 0 & 0 & 0 \end{bmatrix} \quad \mathbf{C}_m = \frac{2m\mathrm{v}}{b} \begin{bmatrix} -(1-\kappa) & 0 & 1-\kappa & 0 \\ 0 & 0 & 0 & 0 \\ -\kappa & 0 & \kappa & 0 \\ 0 & 0 & 0 & 0 \end{bmatrix},$$

$$\mathbf{K}_m = \frac{m\dot{\mathrm{v}}}{b} \begin{bmatrix} -(1-\kappa) & 0 & 1-\kappa & 0 \\ 0 & 0 & 0 & 0 \\ -\kappa & 0 & \kappa & 0 \\ 0 & 0 & 0 & 0 \end{bmatrix}, \quad \kappa = \frac{x_0 + \mathrm{v}h + \frac{1}{2}\dot{\mathrm{v}}h^2}{b}, \quad 0 < \kappa \leq 1.$$

8.6.3 Accelerating Mass—Examples

Now we will compare the displacements under a moving mass obtained from our approach with the reference results by Lee [84]. The Bernoulli–Euler beam of length $l = 6$ m, bending stiffness $EI/\rho/A = 275.4408$ m^4/s^2, moving mass $m = 0.2\rho/A/l$, initial velocity at $x = 0$ of $\mathrm{v}_0 = 6$ m/s, acceleration $a = \bar{a}EI/\rho/A/l^3$ was assumed for dimensionless coefficient $\bar{a} = 0$, 1, 5, and 10 (Figure 8.11).

The Timoshenko beam was also considered in [87]. We compare our results with those published in the reference paper [86], using the same data, already listed at the beginning of this section. The acceleration $a = \mathrm{v}/\mathrm{v}_{cr}$, where the critical velocity $\mathrm{v}_{cr} = \pi/l\sqrt{EI/\rho/A}$. The acceleration $\dot{\mathrm{v}}$ is defined by a non-dimensional parameter $\kappa = \dot{\mathrm{v}}\rho A l^3/E/I$. Two cases were considered: First the case of $\beta = 0.03$, $a = 0.11$ was computed and is depicted in Figure 8.12 for the acceleration $\kappa = 1$, for a constant speed $\kappa = 0$, and for a small retardation ($\kappa = -0.05$). Figure 8.13 presents the case for a higher initial speed $a = 0.5$ and $\beta = 0.03$, for a constant speed $\kappa = 0$, and acceleration with $\kappa = 1$.

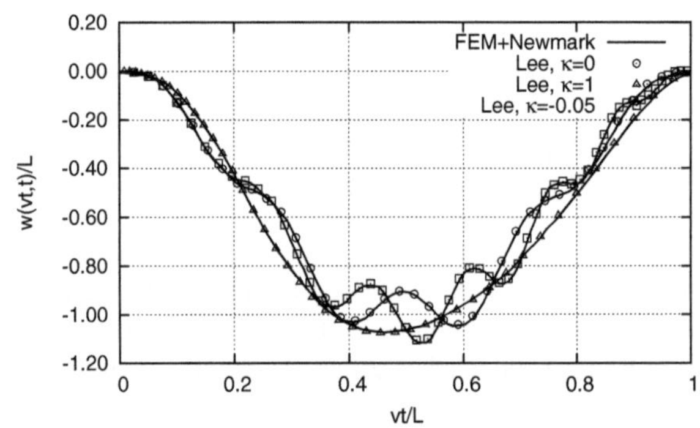

Fig. 8.12 Comparison of displacements of a Timoshenko beam under a moving contact point with those published by Lee [87]—$\beta = 0.03$, $a = 0.11$.

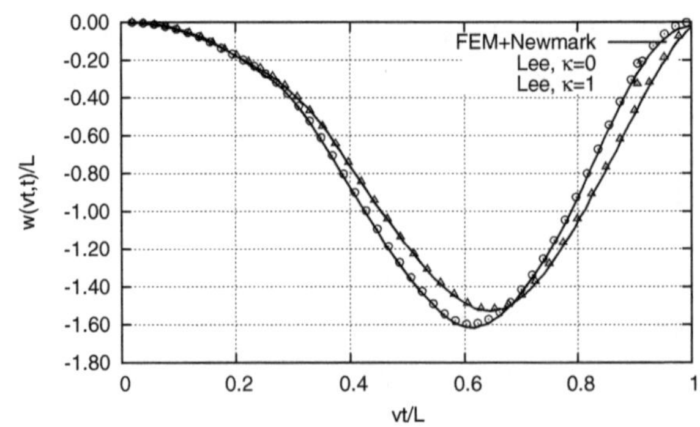

Fig. 8.13 Comparison of displacements of a Timoshenko beam under a moving contact point with those published by Lee [87]—$\beta = 0.03$, $a = 0.5$.

8.7 Conclusions

In this section we proposed a new approach to the vibration analysis of structures subjected to a moving inertial particle by use of the finite element method in space and a general time integration method, for example SSpj, in time, here represented by the Newmark and Houbolt methods. The elements describing a moving massive particle (7.29)–(7.31) can be commonly used both in the Euler beam and the Timoshenko beam. Their appearances are simpler than those of the classical matrices (8.14)–(8.16) for the Euler beam. In engineering practice, most dynamic simulations are performed by the Newmark method. An approach which extends a group of problems that can be directly solved by this commonly used method is valuable. We showed that these matrices yield accurate and stable solutions for a mass moving on a structure. Timoshenko beams or other shear resistant structures exhibit discontinuities in their solutions of the differential equations [48, 49]. Although in practice nonlinear effects smooth the trajectories, large jumps in the physical quantities are observed. The same computational result should be obtained both by semi-analytical and numerical tools. There is no reason for saying that numerical solutions converge to inaccurate results. Our finite element approach proves that even simple elemental matrices derived from a mathematically correct analysis can give perfect convergence to the analytical expressions.

There are two different ways to numerically treat differential equations in structural dynamics. The first one requires the separation of the spatial variables and time, after which two different discretizations are applied to space and to time, and finally two different solution methods are used. Commonly, the finite element method is applied to space while the central difference method or the Newmark method is applied to time. Thus the time marching procedure is established. In this case the equilibrium of forces is provided a selected time-instants, separated in time by the time step. This approach is based on a strong form of the problem. It is well elaborated for problems defined by differential equations with constant coefficients. Variable coefficients require deriving the time integration method starting from the differential equation. Classical inertia and stiffness matrices related to space can not be directly brought to play in a classical time integration method.

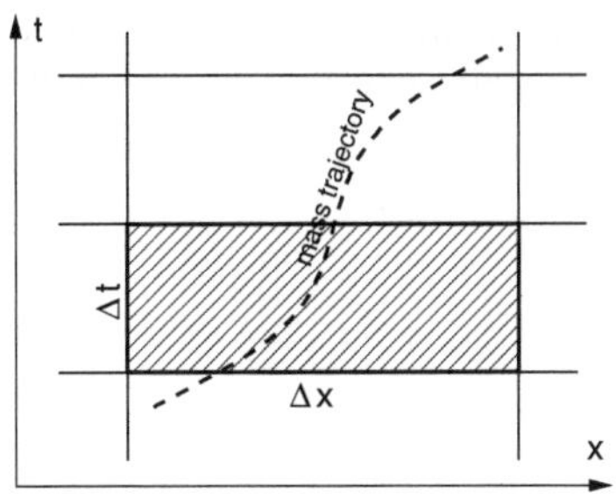

Fig. 8.14 The mass trajectory in space and time.

The second approach, called the space–time formulation [17, 9, 22], is based on the equilibrium of the energy of a structure in a time interval (Figure 8.14). It is based on the weak formulation and allows us to solve much more complicated problems, including moving concentrated physical parameters. This approach was successfully applied to the moving mass problem, solved by discrete methods [25, 24, 26].

Although the space–time approach in the case of a differential equation with constant coefficients and stationary discretization results in practically the same algorithms as the classical time integration methods, most engineers select the methods of the Newmark group for computing. A simple modification of the inertia matrix in the Newmark algorithm or direct differentiation of the acceleration of the mass particle according to the equation describing its position in time and then incorporating the resulting matrices into the solution method fails.

The practice of numerical simulations, however, requires simplicity and efficiency in the procedures. The characteristic matrices for an inertial particle should be capable of being easily incorporated into computer procedures. Thus all existing commercial codes would gain new calculating abilities. We will focus our attention on this aim.

Several classical methods for the numerical integration of the differential equations of motion can be included in one general formula, derived from the expansion of the motion function into a Taylor series. The displacements and derivatives are written in short as follows:

$$\mathbf{y}_{i+1} = \sum_{q=0}^{p-1} \frac{\Delta t^q}{q!}\mathbf{y}_i^{(q)} + \frac{\Delta t^p}{p!}\mathbf{y}_{i+\alpha}^{(p)}, \quad \dot{\mathbf{y}}_{i+1} = \sum_{q=1}^{p-1} \frac{\Delta t^{q-1}}{(q-1)!}\mathbf{y}_i^{(q)} + \frac{\Delta t^{p-1}}{(p-1)!}\mathbf{y}_{i+\alpha}^{(p)}. \quad (8.33)$$

Here, $\mathbf{y}_i$ are the known values and subsequent derivatives $\dot{\mathbf{y}}_i$, $\ddot{\mathbf{y}}_i$, etc. $\alpha_i^{(p)}$ contains unknown coefficients in terms of the remainder of the development. The above expansions allow us to write a family of methods. The general time integration method is characterized by two parameters: p, the number of terms in the Taylor series, and j, the order of the differential equation. We can construct more or less complex integration patterns, choosing the appropriate Taylor series. In the particular case $p = 2$ and $j = 2$, the method is identical with the Newmark method, and for $p = 3$ and $j = 2$, it coincides with the Houbolt method. Other well known algorithms are covered by the formula (8.33) as well. In further tests we will use the Newmark method and the Houbolt method.

Chapter 9
Meshfree Methods in Moving Load Problems

9.1 Meshless Methods (Element-Free Galerkin Method)

The idea of meshless methods is to eliminate the mesh generation stage, which is the main disadvantage of the finite element method (or other classical discrete methods). In a meshless method, the set of separated points is placed in the domain of the structure. Interpolation functions (shape functions) are then generated not in element subdomains, but in arbitrarily placed nodal points.

In the case of very large deformations it is necessary to modify the mesh of finite elements step by step. In order to minimize this significant cost of computation, researchers are trying to use meshfree methods, e.g., [27, 28, 46, 90]. These methods also seem to be appropriate for the task of a moving inertial load.

The shape functions are stretched on points which are in the neighbourhood of the given point, in the stepping subdomain Ω (Figure 9.1). We determine the shape function on the basis of the moving least squares criterion (MLS) [82]. This method consists in minimizing the differences between the exact and approximate solution. The sum of squared errors of the approximation at all the nodes ($i = 1,\ldots,m$) is a functional:

$$J = \sum_{i=1}^{m} W(x-x_i)\left[w^h(x,x_i) - w_i\right]^2 , \tag{9.1}$$

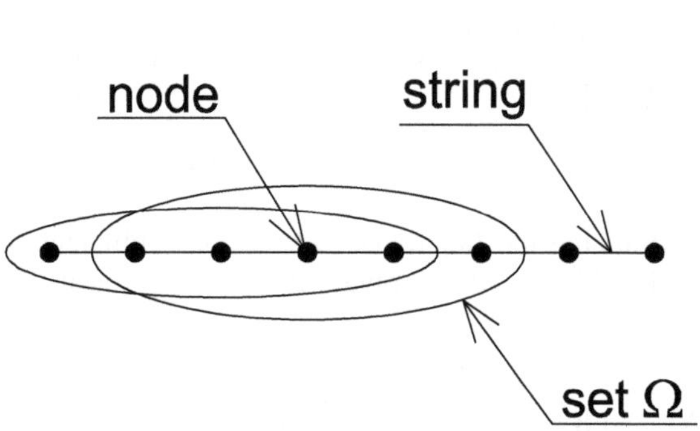

Fig. 9.1 The set Ω moving along a string.

C.I. Bajer and B. Dyniewicz: Numerical Analysis of Vibrations of Structures, LNACM 65, pp. 241–246.
springerlink.com

where $W(x-x_i)$ is the weight function, $w^h(x,x_i)$ is an approximation function, and the w_i are the nodal values. We assume exponential shape functions:

$$W(x-x_i)=\begin{cases} \mathrm{e}^{-\left(\frac{x-x_i}{\alpha}\right)^2} & \text{if } (x-x_i)\leq 1, \\ 0 & \text{if } (x-x_i)>1 \ . \end{cases} \tag{9.2}$$

The coefficient α depends on the size of the domain Ω and the number of points in the domain.

In this case, the exact solution has been approximated by an nth degree polynomial

$$w^h(x,x_i)=\sum_{j=1}^{n} p_j(x_i)a_j(x)=\mathbf{p}^{\mathsf{T}}(x_i)\mathbf{a}(x) \ , \tag{9.3}$$

where the monomials in the interpolation polynomial are

$$\mathbf{p}^{\mathsf{T}}=\left[1,x,x^2,\ldots\right] \ , \tag{9.4}$$

with approximation coefficients

$$\mathbf{a}^{\mathsf{T}}(x)=\left[a_0(x),a_1(x),a_2(x),\ldots\right] \ . \tag{9.5}$$

The functional (9.1) can be written in matrix form:

$$J=(\mathbf{Pa}-\mathbf{w})^{\mathsf{T}}\mathbf{W}(\mathbf{Pa}-\mathbf{w}) \ , \tag{9.6}$$

where $\mathbf{P}$ is a full matrix of dimension $(n\times m)$

$$\mathbf{P}=\begin{pmatrix} p_0(x_1) & p_1(x_1) & \cdots & p_m(x_1) \\ p_0(x_2) & p_1(x_2) & \cdots & p_m(x_2) \\ \vdots & \vdots & \ddots & \vdots \\ p_0(x_n) & p_1(x_n) & \cdots & p_m(x_n) \end{pmatrix} , \tag{9.7}$$

and $\mathbf{W}$ is a diagonal matrix of dimension $(n\times n)$

$$\mathbf{W}=\begin{pmatrix} W(x-x_1) & 0 & \cdots & 0 \\ 0 & W_2(x-x_2) & \cdots & 0 \\ \vdots & \vdots & \ddots & \vdots \\ 0 & 0 & \cdots & W(x-x_n) \end{pmatrix} . \tag{9.8}$$

As a result of minimizing the functional (9.6) at the coefficients of the approximating polynomial ($\partial J/\partial a_i=0$) we obtain the sought form of the vector $\mathbf{a}$ from equation (9.3)

$$\mathbf{a}(x)=\mathbf{A}^{-1}\mathbf{B}\mathbf{w} \ , \tag{9.9}$$

where $\mathbf{A}=\mathbf{P}^{\mathsf{T}}\mathbf{W}\mathbf{P}$ and $\mathbf{B}=\mathbf{P}^{\mathsf{T}}\mathbf{W}$. Eqs (9.3) and (9.9) lead to:

$$w^h(x,x_i) = \mathbf{p}^\mathsf{T}(x_i)\,\mathbf{A}^{-1}\,\mathbf{B}\,\mathbf{w} = \sum_{j=1}^{n} \phi_j^m(x)\, w_j\,, \tag{9.10}$$

where $\phi_j^m(x)$ is the shape function and m is the degree of the approximation polynomial

$$\phi_j^m(x) = \mathbf{p}^\mathsf{T}(x_i)\,\mathbf{A}^{-1}\,\mathbf{B} = [\phi_1^m(x), \phi_2^m(x), \ldots \phi_n^m(x)]\ . \tag{9.11}$$

The general criterion of MLS approximation requires differentiation of the equation $\mathbf{p}^\mathsf{T}(x_i)\mathbf{A}^{-1}\mathbf{B}$. In order to simplify this task, we apply a zero degree polynomial approximation ($m=0$). This leads to the shape functions called the Shepard functions, dependent only on a weight function of the nodes

$$\phi_i^0 = \frac{W(x-x_i)}{\sum_{j=1}^{n} W(x-x_j)}\ . \tag{9.12}$$

Selecting these functions is a significant problem when the nodes are irregularly distributed. In the general case, the designation of the matrix describing the test structure requires numerical integration, resulting in additional computational costs.

The resulting shape function (9.12) was used for discretization of a string. Using virtual work, we can obtain the stiffness and inertia matrices:

$$\begin{aligned} k_{ij} &= \int_\Omega B_i^T\, N\, B_j\, d\Omega\,, \\ m_{ij} &= \int_\Omega \phi_i^T\, \rho A\, \phi_j\, d\Omega\,, \end{aligned} \tag{9.13}$$

where $B = \partial\phi/\partial x$, N is the tensile force, and ρA is the mass of the string. The nodal matrices $\mathbf{k}$ and $\mathbf{m}$ computed for domains Ω are assembled into global matrices $\mathbf{K}$ and $\mathbf{M}$. The proper choice of the parameter α is still the fundamental problem of the weight function (9.2). In the case of a regular distribution of points in the set Ω, where the distance between the nodes is denoted by b, the coefficient α can described with high accuracy by $\alpha = b/\sqrt{\pi}$.

9.2 Results

The numerical results will be compared with the analytical results. We used the following data: length of the string $l=1$, tensile force $N=1$, mass of the a string $\rho A = 1$, external force $P=-1$ travelling with constant speed $\mathrm{v} = 0.1c$ (where c, as usual, is the wave speed in the string).

In Figure 9.2 we plot the displacements under a moving force and at the middle of the string length. The analytical solution coincides well with the one obtained by the finite element method or the central difference method. The meshless method results in a good approximation to the accurate solution. In this method we can notice the phase error, visible especially in the second half of the observation period. This means that the rigidity is higher than required. However, the smoothing effect is the main reason for the differences in the results. A significantly lower number of

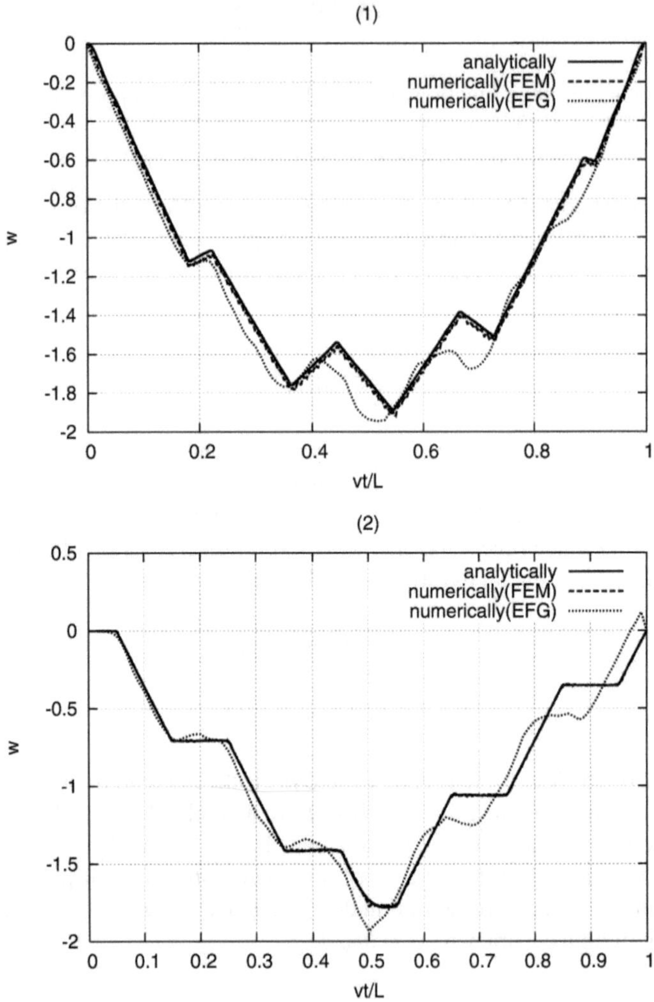

Fig. 9.2 Displacements under constant moving force: (1) under moving force, (2) at the middle of the string (number of points in Ω is 10).

points in the moving domain (2 instead of 10 as in the previous case) (Figure 9.3) results in quite similar plots. Both the amplitude and the period of the vibrations are acceptable.

The numerical approach in the case of a Winkler foundation does not differ from the analytical solution. We must emphasize here that a continuously moving force in the numerical application is replaced by a sequence of marching pairs of forces, applied to nodal points. In such a case we neglect the mixed derivatives of the formulation. We can notice that such an analysis does not introduce a significant error (Figure 9.4).

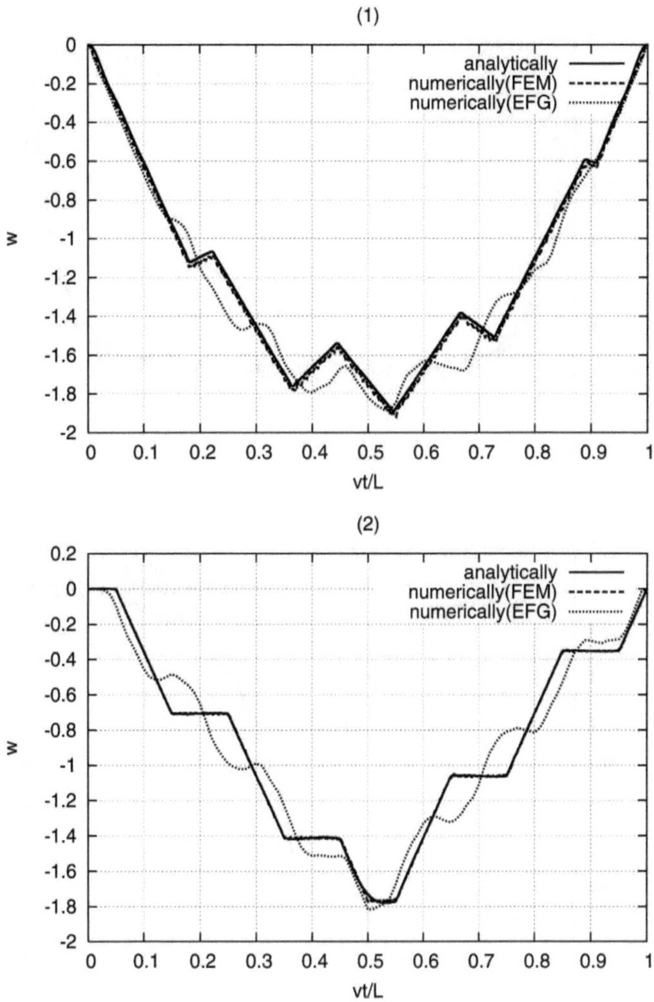

Fig. 9.3 Displacements under constant moving force: (1) under moving force, (2) at the middle of the string (number of points in Ω is 2).

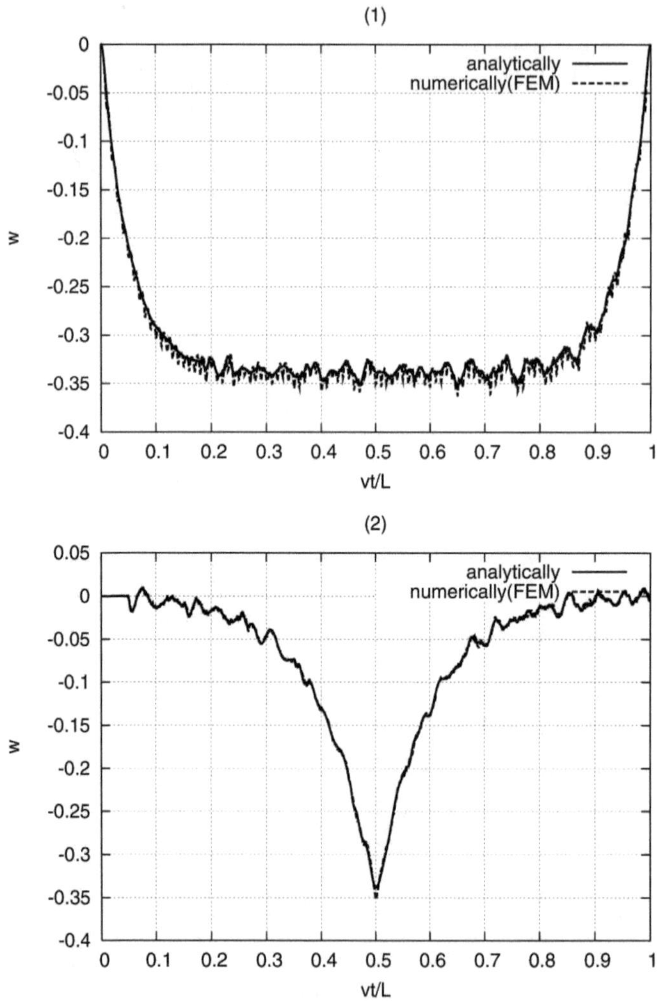

Fig. 9.4 Displacements under constant moving force—string put on a Winkler foundation: (1) under moving force, (2) at the middle of the string.

Chapter 10
Examples of Applications

The examples of the calculations of the selected engineering problems given in this Chapter demonstrate the practice of numerical solutions. In real structures we always ask questions as to what geometry and what values of the material data are appropriate to pass from the physical model of the structure to the numerical one. Real shapes are usually complex and we try to simplify them, replacing curves with straight lines, non-uniformly distributed material parameters with homogeneous material, material damping with a numerical decay of the amplitude. Let us consider, as a first example, a track subjected to a moving vehicle. We can build a detailed three-dimensional model using cubes or tetrahedra with many degrees of freedom describing the foundation, ballast, track elements, rails, wheels, and the remaining part of the vehicle. We can include contact phenomena, friction, material nonlinearities, thermo-mechanical coupling, etc. However, such a model nowadays would be a challenge even for a static problem. Calculating the solution can last even a quarter of an hour. That is relatively long considering the computational power of multi-core processors. In a dynamic analysis, such a computation must be repeated thousands of times. The duration of the task exceeds any reasonable length of time. That is why we must still simplify our numerical models and improve the computational tools. Fortunately, a coarse discretization and a simplified mesh does not influence the frequencies significantly. The amplitudes are worse.

In vibration analysis of structures under moving loads we often search for the interaction of the moving vehicle and the load-carrying structure. In the case of multi-point contact of a base with a multi degree-of-freedom vehicle, such interactions are essential.

The problem of ill-conditioned systems of algebraic equations is well known in computer analysis. Especially, the assumption of constants whose ranges vary several times may cause problems. If the commercial software recalculates the given data to scaled values, we can partly trust the solutions. This issue is essential since in dynamic simulations the computational error can occur at least once during each step of the computations and, in an initially invisible way, affect severly the final results. In complex problems, we will not be able to detect such incorrect solutions. In the case of custom-written computer programs we ourselves should ensure that

C.I. Bajer and B. Dyniewicz: Numerical Analysis of Vibrations of Structures, LNACM 65, pp. 247–270.
springerlink.com

Table 10.1 Material data of structural elements.

parameter	value [SI units]	value [cm, g, μs]
rail		
E	210 GPa	2.1
A	$7.7 \cdot 10^{-}3$ m^2	77.
ν	0.2	0.2
ρ	7860 kg/m^3	7.86
G	77 GPa	0.77
I	$30.55 \cdot 10^{-6}$ m^4	3055.
elastic pad		
E	0.25–0.50 GPa	$2.5 \cdot 10^{-3}$–$5.0 \cdot 10^{-3}$
A	$22.5 \cdot 10^{-3}$ m^2	225.
ρ	1800 kg/m^3	1.8
sleeper (concrete)		
E	30 GPa	0.3
G	7.7 GPa	0.077
A	$15 \cdot 10^{-3}$ m^2	150.
ρ	2400 kg/m^3	2.4
sleeper (wooden)		
E	11 GPa	0.11
A	$35 \cdot 10^{-3}$ m^2	350.
ρ	0.65 g/cm^3	0.65
Y-type sleeper (steel)		
E	210 GPa	2.1
A	$21.6 \cdot 10^{-3}$ m^2	216.
ρ	7860 kg/m^3	7.86
ground		
k_z (soft)	10^5 Pa	10^{-6}
k_z (sand compacted)	10^8 Pa	10^{-3}
k_z (rigid)	10^{10} Pa	10^{-1}
ν	0.3	0.3
ballast		
E	500 MPa	$5 \cdot 10^{-3}$
ρ	>1200 kg/m^3	>1.2
ν	0.3	0.3

the matrices in our systems of equations are well-conditioned. The simplest way is to deliver the data, both geometric and material, within a similar range. For this purpose, it is better to replace the SI system of units by another one, more suitable. We chose the following system of units: cm, g, μs. This results in most of data's having almost the same range, especially the Young modulus and the time step.

Table 10.2 Data assumed in simulation.

rail UIC-60 bending stiffness EI	6.42 MN/m^2
mass density per unit length ρA	60 kg/m
damping loss factor of rail η_r	0.02
stiffness of elastic pad per unit length	300 MN/m^2
damping loss factor of a pad η_p	0.2
sleeper mass per unit length of a track (per one rail)	250 kg/m
ballast stiffness per unit length of a track (one rail)	100 MN/m^2
ballast damping loss factor η_b	1.0
foundation damping loss factor η_f	0.1

Below we will give the material data useful for simulations of railway problems. All the values are given both in the SI system and in our system of units. In Table 10.1 one can find the data for the fundamental types of elements of the track. The material and structural data assumed in the further simulation is collected in Table 10.2.

10.1 Dynamics of the Classical Vehicle–Track System

Now let us have a look at the real example of vibrations of a carriage moving on a classical track. The custom computer software used here was written by the authors, implementing the numerical approaches given in this book. We use the geometric and material data from [42].

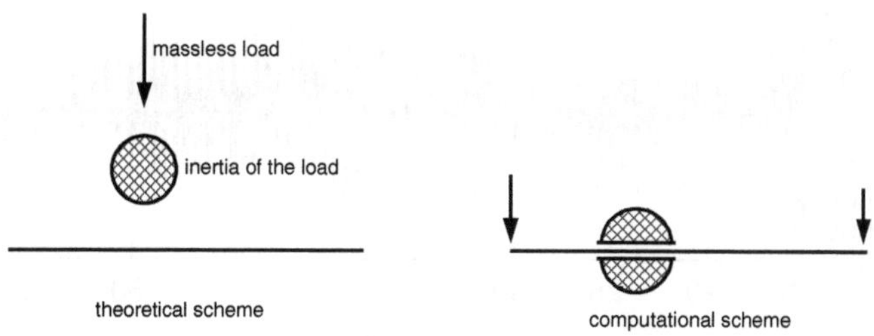

Fig. 10.1 Theoretical scheme of the problem and the scheme assumed for computations.

The replacement of the real physical load with the computational load is explained in Figure 10.1. The finite element is subjected at the intermediate point to the force with the inertia parameter, i.e., the concentrated mass. This force, usually placed in a numerical model at the right-hand side of the resulting system of algebraic equations, can be simply distributed over the neighbouring nodes. The bending moments in the case of a beam must appear at the finite element joints as well. The concentrated mass is incorporated directly into the left-hand side matrices. Their coefficients vary in each time step and this requires the solution of a

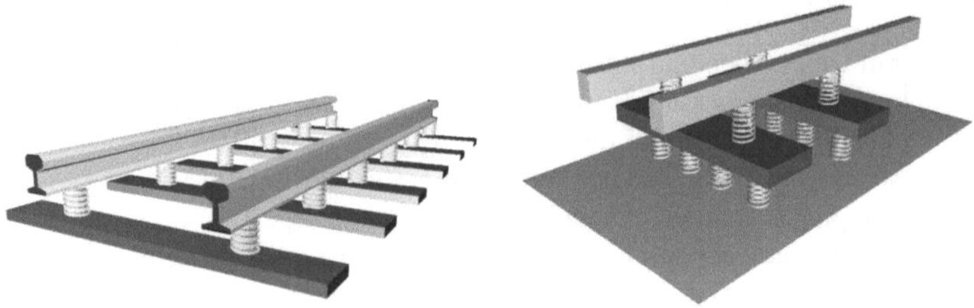

Fig. 10.2 Substructures assumed in analysis.

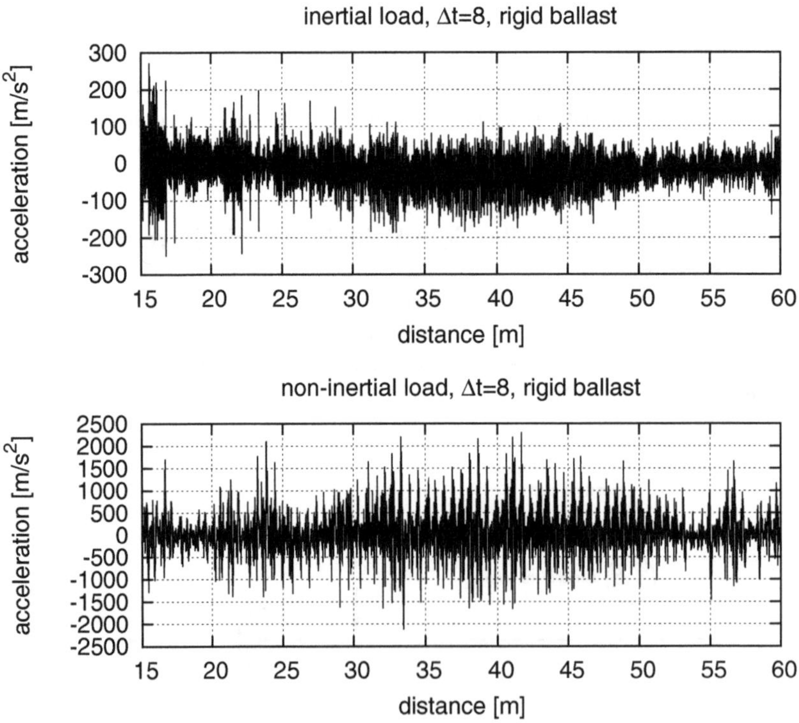

Fig. 10.3 Vertical accelerations of the axle box at a speed of 290 km/h with the inertial and non-inertial loads assumed in the model.

system of equations at every time step. No iterations are required, unless unilateral contact is assumed. There are two advantages of such a solution: accurate and faster computations.

The track model is composed of plates, beams, grid or frame elements, and springs. A simple track structure can be considered in the same manner as a

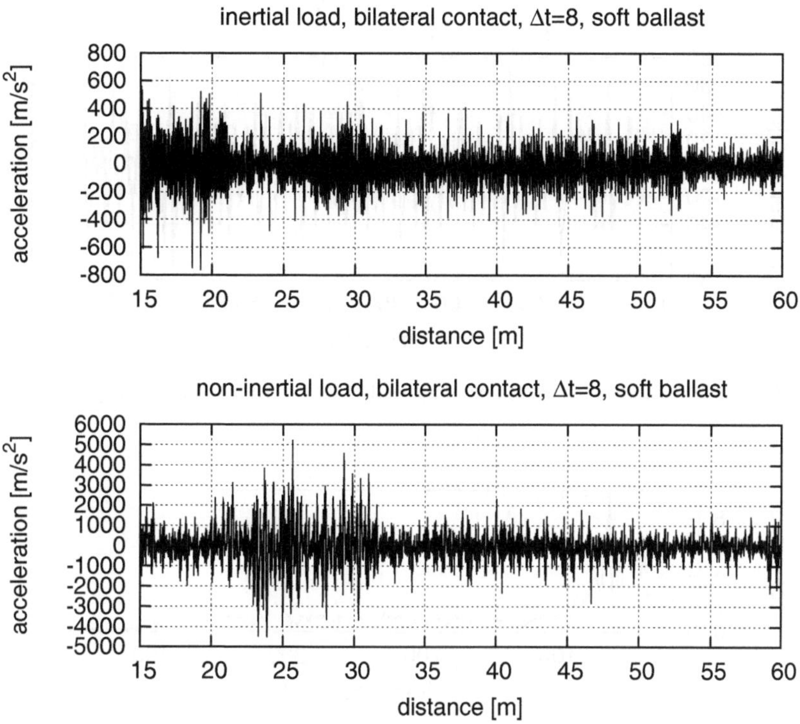

Fig. 10.4 Vertical accelerations of the axle box at a speed of 290 km/h with the inertial and non-inertial loads assumed in the model with soft ballast.

complex one. Let us look at the simplest classical track (Figure 10.2), built of sleepers as grid elements placed on an elastic Winkler foundation, springs which model elastic pads, and grid elements which describe rails, both straight and curved. In both Figures 10.3 and 10.4 in the case of a non-inertial load (lower diagrams) we can notice the strong influence of the sleepers. With an inertial load (upper plots), this influence is moderate and the dynamic response is more realistic.

We can compare our results with the reference paper [42] (Figure 10.5). Both Figures 10.3 and 10.5, obtained for an inertial load, exhibit a similar range of accelerations of the axle box. The signal in Figure 10.5 shows a low frequency mode which is difficult to explain. The response of our numerical simulation has the same magnitude of accelerations and has more realistic higher frequency oscillations. The model analysis of the plotted signal is depicted in Figure 10.6. Figure 10.7 shows the accelerations of the car's body at a speed of 290 km/h.

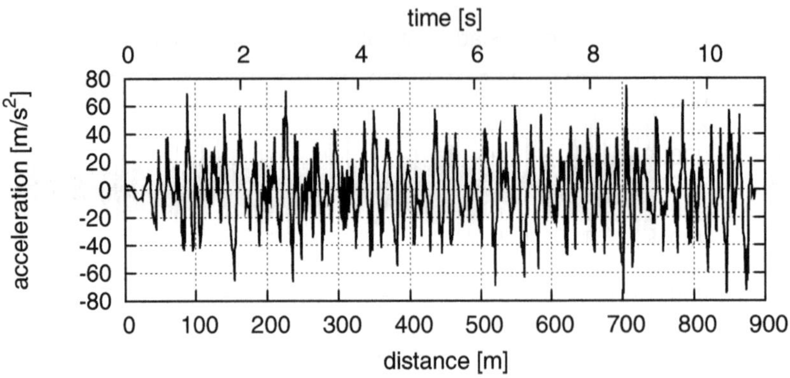

Fig. 10.5 Accelerations of the axle box at a speed of 290 km/h taken from [42].

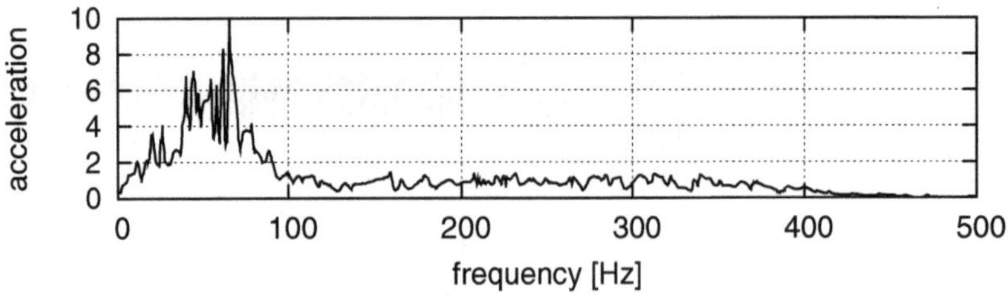

Fig. 10.6 Modal analysis of the acceleration signal from Figure 10.5 [42].

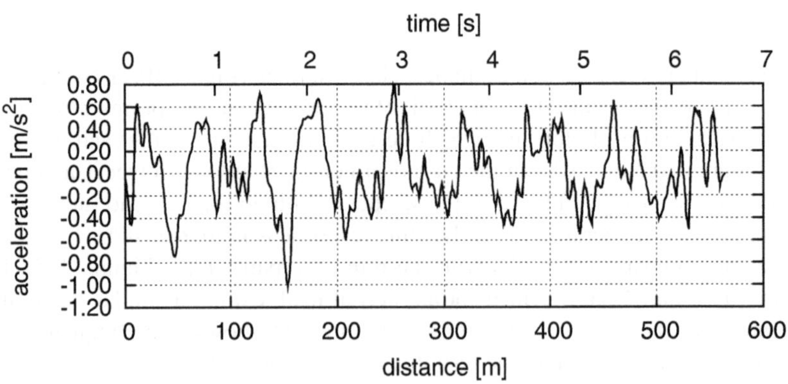

Fig. 10.7 Accelerations of the body of an ETR500 car at a speed of 290 km/h taken from [42]—deformable bodies model.

10.2 Dynamics of the System Vehicle—Y-Type Track

Ballastless tracks and steel sleeper wedge tracks (also called Y-type sleepers) are an example of a dynamic track–vehicle system. Modernization, for reasons of the security of the suspension system and due to the high demands of railway rules, is rarely applied and requires a long and costly research of its prototypes. The modernization of old railway tracks using new technologies gives a noticeable improvement in the dynamic collaboration of the wheel–rail system. The line Plaszow Cracov–Auschwitz is an example where part of the track was made with steel Y-type sleepers. This type of sleeper is also used by the funicular railway to Gubałówka in Zakopane (Figure 10.8) and in Krynica Górska. This scheme compared with a classical track is depicted in Figure 10.9.

Fig. 10.8 Y-type track.

The advantage of Y-type sleepers is an increase in the lateral stiffness and inertia of the track by including ballast which collaborates with the steel sections (Figure 10.10). This is especially important on curves and in mountain areas. The disadvantage is its higher implementation cost. Measurements on experimental sections show a significant reduction in noise at train crossings.

While the construction of trial road maintenance sections is time-consuming, computer simulations can be done quickly. They are valuable when the analysed model corresponds to the physical model as closely as possible [132, 51]. In its development, the issue is to determine what portion of the wheel weight in contact

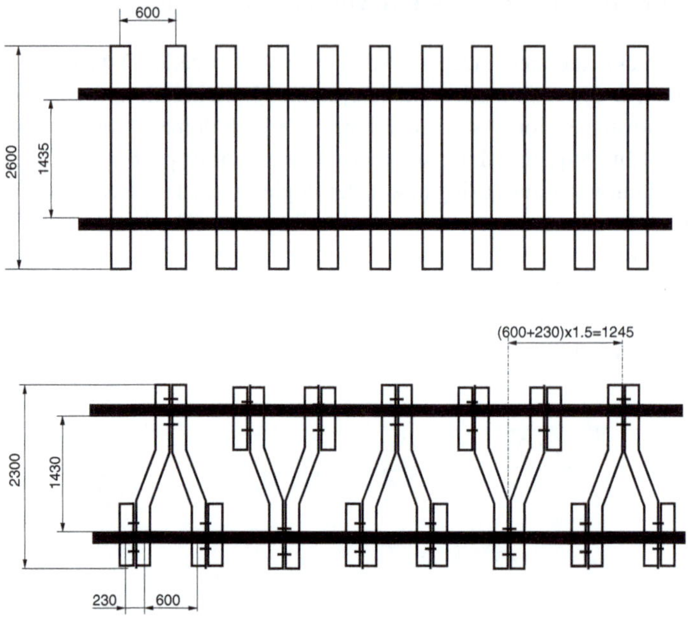

Fig. 10.9 The track with classical and Y-type sleepers.

Fig. 10.10 Wedge steel sleeper.

with the rail dynamically interacts with it. Other parameters of the model can be relatively well-chosen on the basis of the technical documentation. It is certain that at high speeds, the track should not be loaded by only nonlinertial spring elements, and the mass should be taken into account, on which we can then place even a complex vehicle model.

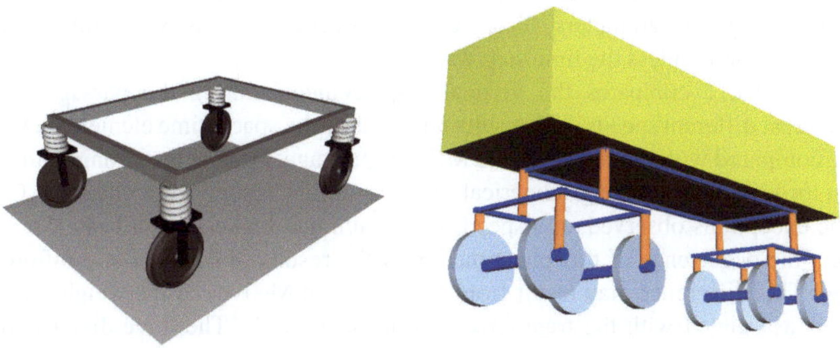

Fig. 10.11 Model of the track and rail vehicle scheme adopted in numerical computations.

The task of passing a rail vehicle over the track of a classic or Y-type is considered as a problem of two subsystems. The first one is a track consisting of rails, sleepers, viscoelastic pads, and viscoelastic soil. The second subsystem is a rail vehicle, built in a simple way of oscillators, connected by a deformable frame (Figure 10.11). We assume that vibrations of both rails are coupled by the wheelsets, and that the element which couples the vibrations that propagate along the rail is the rail vehicle boogie frame. Rails, sleepers, and the boogie frame, are taken as the grid elements, with three degrees of freedom at a node. Elastic spacers were adopted for the finite elements of a rod. The soil was taken as an inertial Winkler foundation. The frame rigidity and inertia of the individual elements were suitably chosen. In the case of a simplified model of the vehicle in which the wheels are granulated masses, the mass of the wheel, which is accompanied by transverse displacements of the rails, were chosen within the range of 15%–50% by weight of the wheel. Both dynamical systems were solved independently, by building and solving the corresponding systems of algebraic equations. The grid nodes of both discrete systems moved relative to each other and therefore a simple iterative procedure of balancing the forces in both systems was natural to begin with. In the first stage, the track was subjected at the contact points of the wheels with the rails, to forces corresponding to the pressure of the dynamic forces of the boogie wheels. The result was a vertical displacement of a discretized rail element. This allowed of determining the vertical displacements of the rail at the points of contact with the wheels. These displacements were taken into account as the boundary conditions in the solution of the boogie subjected to external forces, such as its own weight and the weight of the wagon body. The reactions at the points of contact with the rails were the result of solving this stage. These reactions, with opposite signs, were used again to load the track. Iteratively repeating the procedure conducted us in several steps to balance the static system and then let go to the next while in a dynamic process. The procedure worked correctly within a certain range of parameters. In our case, in practice, no loss of stability of the solution occured unless the time step was too large.

Figure 10.12 compares the vertical displacements during the passage of a wheelset at different speeds. The results obtained by the space–time element method were compared with the results of the Medyna package. Despite the strongly differing approaches for creating numerical models, similar results were obtained. One of the exceptions observed, at a speed of 72 km/h and higher, was a beat. We can expect the occurrence of this phenomenon in the results of Medyna at a different speed. This difference may result from a rejection in Medyna of the weight of the wheels associated with the transverse motion of the rails. The three-dimensional image of distorted rail axes at successive moments are shown in Figure 10.13. We see higher amplitudes of displacements of the classical track, especially visible at a distance from the rail vehicle wheels. In order to make a comparison, the results of the measurements recorded in Germany are depicted in Figure 10.14. In addition to the displacements in time showing the average values, similar to the quasistatic case, the values diminished by static deflection are also presented.

Two characteristic cases of the analysis of Y-type sleepers were selected. The rolling of a vehicle with regular, perfect wheels was the first. A vehicle with polygonised, corrugated wheels, which subjected the track to oscillating forces, was the second. The initial stage of rolling on the rails was an excitation of the system. The response of the wheelset/track system depends on the velocity. In the higher velocity range, in the case of perfect wheels, the sleeper type has a significant influence (Figure 10.15).

The second case of the problem is depicted in Figure 10.16. We notice a higher level of oscillations for the Y-type sleepers. However, the average level of displacements is considerably lower. Moreover, with increasing train velocity the amplitudes in both cases start to become similar.

In addition to the improved dynamic properties registered at the rail level, one can see that there are smaller amplitudes of the selected point on the frame of a rail vehicle. The differences between passage on the classical track and the Y-type track is significant and amounts to tens of percent (Figure 10.17).

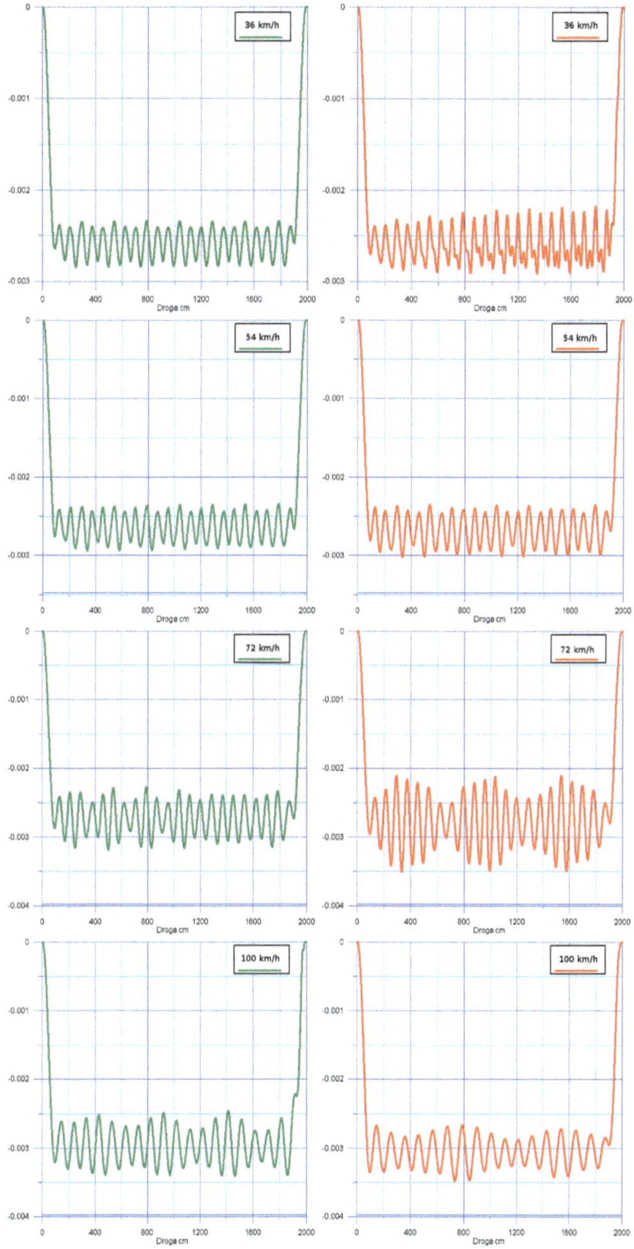

Fig. 10.12 Vertical displacements of the contact point of the wheel and the rail computed with the package Medyna (left column) and by the space–time element method (right column) at 36, 54, 72 and 100 km/h.

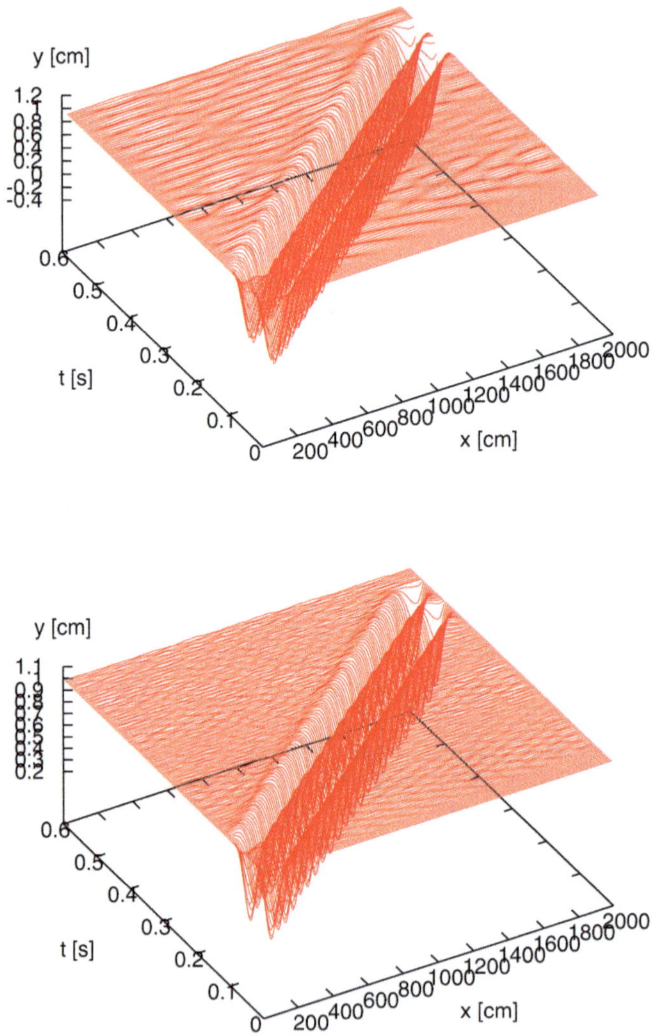

Fig. 10.13 Vertical displacements of a classical (left) and Y-type track (right) in time-space, loaded by a boogie moving at a speed of 40 m/s.

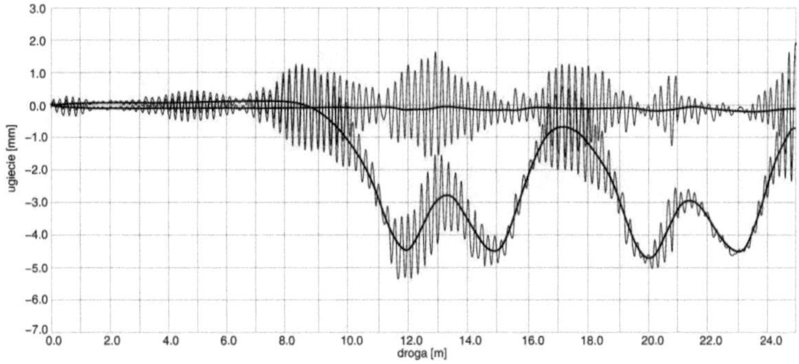

Fig. 10.14 Measurements of vertical displacements of rails in passage at a speed of 60 km/h and diminished by static deflection.

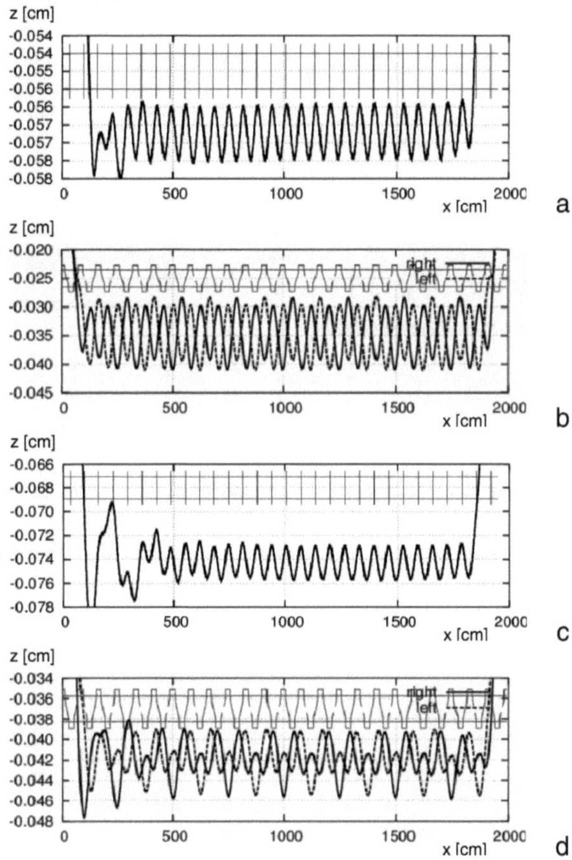

Fig. 10.15 The vertical displacement of contact points while driving at a speed of 30 m/s (a and b) and 50 m/s (c and d).

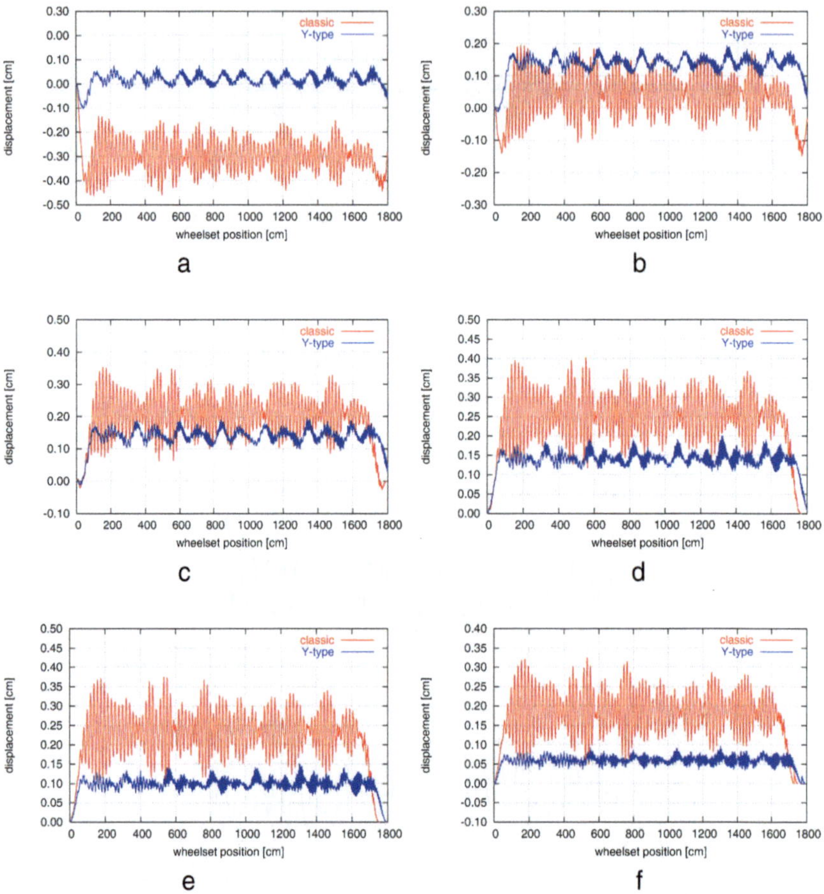

Fig. 10.16 Vertical displacements recorded at distance of (a) 120 cm, (b) 140 cm, (c) 160 cm, (d) 180 cm, (e) 200 cm, and (f) 220 cm from the contact point of the first wheelset of the boogie with a classical track and with Y-type sleepers at a speed of 40 m/s.

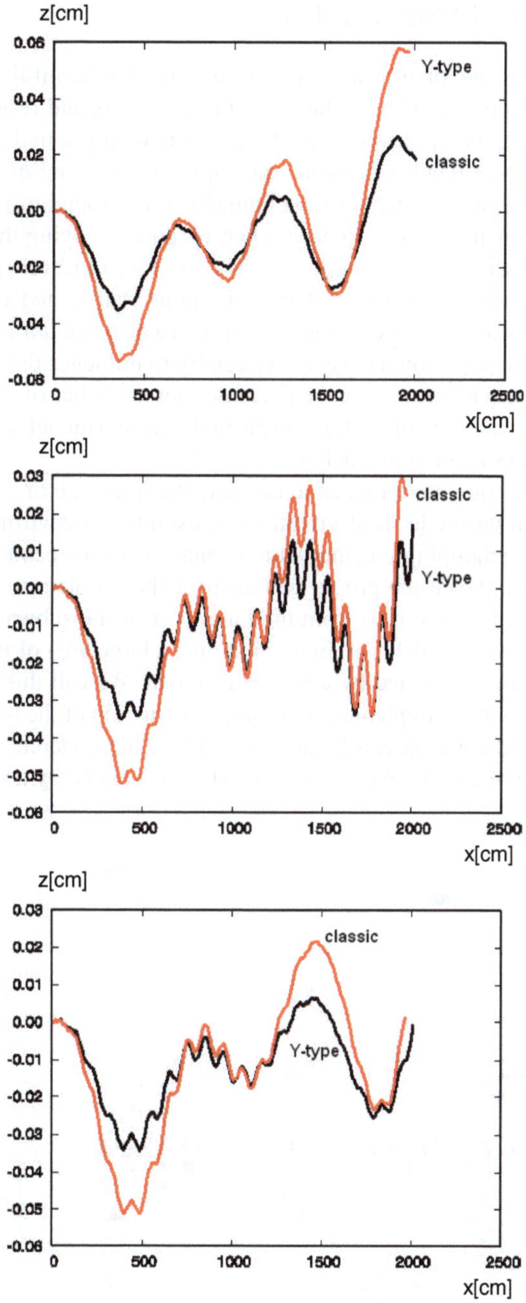

Fig. 10.17 The vertical displacement of the vehicle frame at a speed of 30, 34, and 36 m/s.

10.3 Dynamics of Subway Track

Analysis of the passage of subways aims at assessing the harmful effects of vibrations on the environment. This applies to both passengers and residents of nearby buildings. Harm can be on the one hand loud and annoying sounds at low frequencies, and on the other hand, vibrations causing uneven settling of the foundations of buildings and plaster scratch. For the human ear, the sound of two modal components, with frequencies close to each other, is more annoying than a single frequency, even one with a higher amplitude. The increase in travel speed of subway carriages increases the amplitudes of the vibrations transmitted to the surroundings. Existing subway tunnels can not be rebuilt. We also can not institute any profound vibroacoustic isolation. However, we can try to influence the dynamics of the vehicle–track system by modifying the dynamic characteristics of the rail vehicle or by changing the foundation of the track in the underground tunnel. One such attempt to modify the tracks is presented below.

The idea of the computations was to estimate the influence of different types of vibration isolation on the level of vibrations transmitted to environment. The efficiency of vibration damping together with an analysis of the economic cost of the foundation could indicate the proper treatment of the problem to engineers. Two methods of isolation of the rails from the outer part of the subway tunnel are depicted in Figures 10.18 and 10.19. In the first one, a large area of the cross section of the concrete base is isolated by a viscoelastic mat. We call this, deep isolation. This solution can only be implemented during construction of the tunnel. Later than that, only a second solution can be applied: part of the concrete base can be replaced with a longitudinally placed reinforced beam, isolated by a viscoelastic mat.

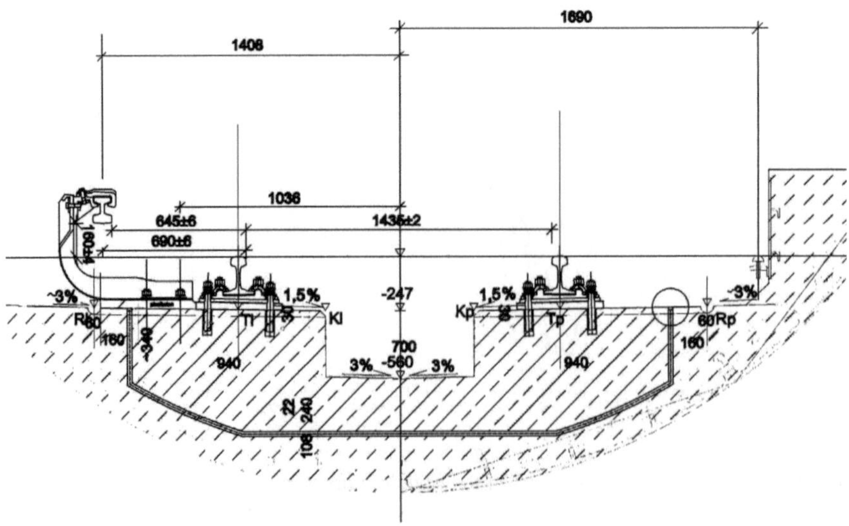

Fig. 10.18 Subway track with deep isolation.

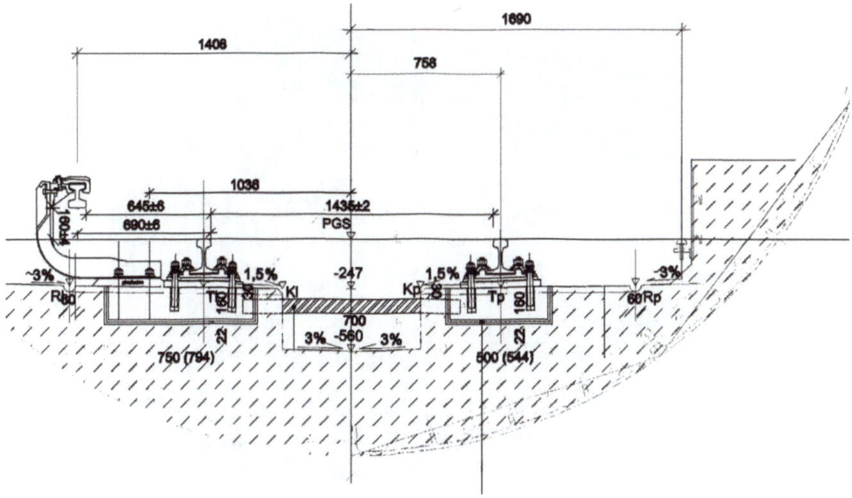

Fig. 10.19 Subway track with shallow isolation.

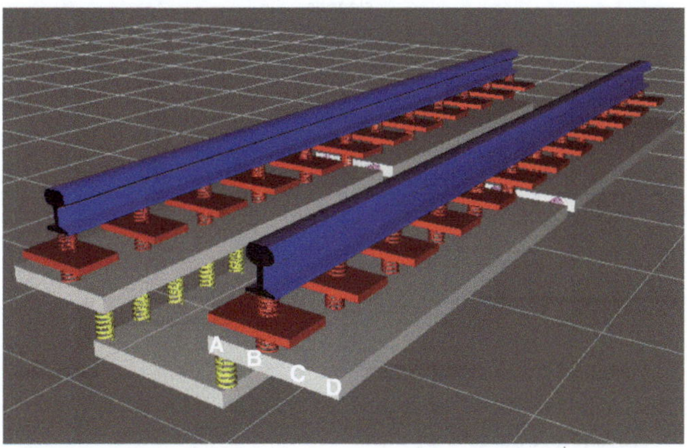

Fig. 10.20 Model of subway track.

This method is called shallow isolation. The idea is to transfer the dynamic effects from the points of local contact of the wheels with the rails to a larger area of the track. The arrangement of the longitudinal beams under the sleepers so as to be isolated from the foundation further increases the inertia of the vehicle–rail–sleepers interfaces and is a kind of dynamic absorber. The selection of the cross-section of the longitudinal beam and the shallow insulation should take into account, in addition to the dynamic properties of the system components (especially the suspension, mass distribution, stiffness of elastic pads, or rail fastening), also the speed of travel on particular intervals of track.

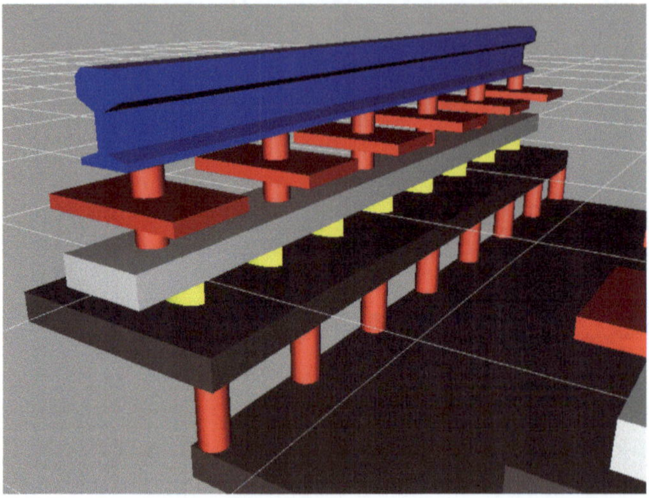

Fig. 10.21 Shallow isolation of the subway track.

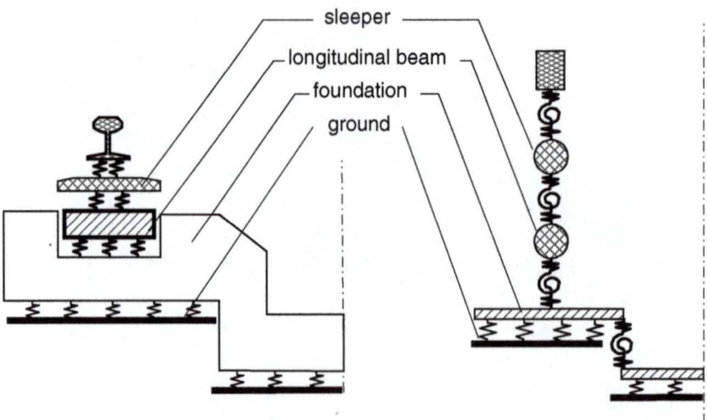

Fig. 10.22 Modification of the subway track.

The basic simplified numerical model is built as shown in Figure 10.20. Then the model is modified according to the type of isolation. The scheme has been extended to a bar situated under the sleepers, insulated from the substrate with an elastic layer. Such a scheme has allowed of analysing the possibility of isolating the basic layout of the track from the environment. The vibrations caused by passing vehicles, and especially the coupling of the vibrations caused by successive axles, can thus be significantly reduced. In the particular case of adopting a little stiffness of the longitudinal beam, one can reduce the extended problem to the basic scheme. The numerical model was described by the following types of elements: the grid—applied to the rails and sleepers, beams—as applied to the longitudinal beams, plates—describing the foundation and the elastic-viscous Winkler substrate

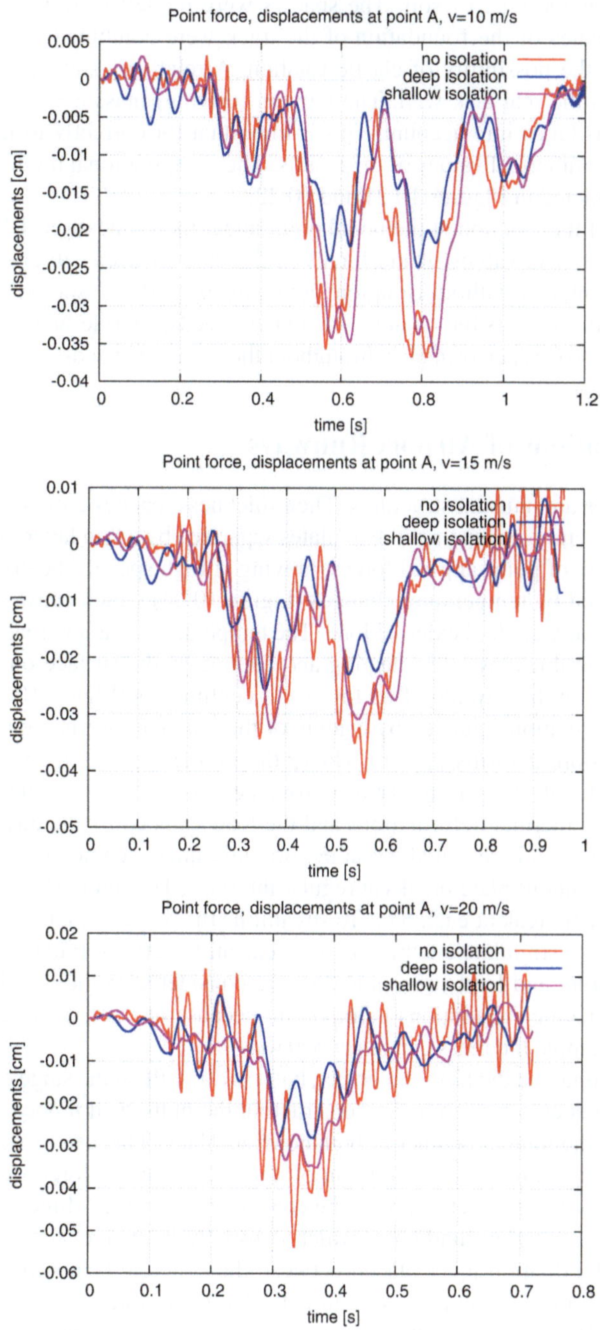

Fig. 10.23 The reaction under the foundation slab with various types of vibration isolation, at speeds of 10, 15, and 20 m/s.

which describes the type of soil. The spacers were modelled by visco-elastic elements. The plates of the foundation of the track were combined with the spring elements with the possibility of elastic rotation. A relatively simple model of the track describes the real task well. The moving (driving) mass associated with the beam (rail) was taken into account. This is important for correctly formulating the model and the numerical values of the results. The computational model of shallow isolation is depicted in Figures 10.21 and 10.22.

Examples of the reaction at the nodal point under the foundation slab are shown in Figure 10.23. The values of the forces in the case of deep vibration isolation, shallow isolation, and without isolation, are compared. At low speeds, the shallow mat does not make a visible change. At higher speeds, its role begins to emerge. A deep mat improves performance throughout the range of speeds.

10.4 Vibrations of Airport Runways

Runways plates are unusual structures. Their thickness reaches 2 m. Moreover, this is a layered structure, consisting of two plates separated by a thin layer of insulation. The load is a set of point-applied forces, moving at high speed. The problem is the distortion caused by temperature changes (Figure 10.24) which cause stresses and by the dynamically applied external load. The temperature gradient reaches 0.08 °C per mm of plate thickness. The heating and cooling of the surface on sunny days causes changes in the altitude of up to several centimeters (Figure 10.25). A plate loaded by deformations induced by heating of the top layer, located on a unilateral elastic foundation, deforms upward. Under the influence of its own weight, long plates fall and break at a certain distance from their ends because of the stress accumulation due to thermal deformations and the load induced by the aircraft (Figure 10.26). This process is repeated, breaking off subsequent segments at the ends. In this way, a continuous plate breaks at regular intervals. The figure shows that a continuous plate will break at each 6 m. To streamline the process, plates are incised at fixed distances to stimulate cracking in the weakened sections. Finally, short plates resist temperature and load stresses and so are considered in the simulations. The problem of dynamic calculations is important because of the large aircraft takeoff weight (for example the mass can reach 600 tons).

It's easy to imagine the physical object, but it is difficult, at the stage of numerical description, to decide how to begin constructing the numerical model. One should consider, e.g., the mutual contact of both concrete slabs. This can be dealt with locally, in the neighbourhood of the point of application of the force. A thick layer of the surface of the concrete slab can be considered as a three-dimensional block. Then, too, there is the ground base, which is usually treated as a set of unilateral constraints. The dilation layer placed between the concrete layers may be grafted adhesively with plates. The plate system is loaded by a system of moving concentrated forces. We should also take into account the air pressure, which is especially important when trying to bond off sealed plate edges from each other and from the ground. We will not go into details of the study of the problem, but will only include

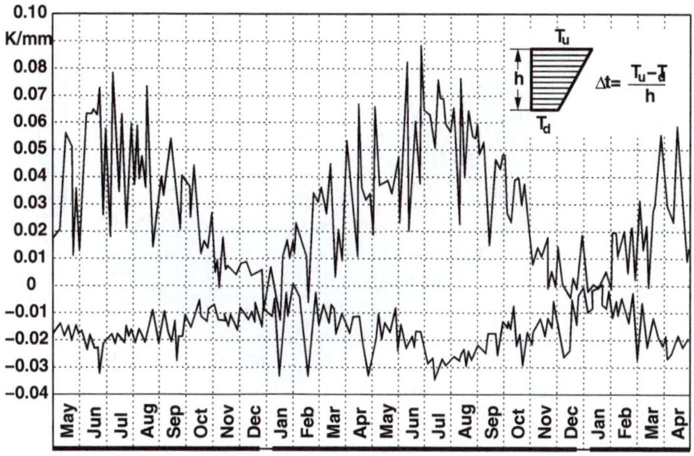

Fig. 10.24 Annual changes in temperature gradients in the airfield plate during the day (upper graph) and during the night (lower graph).

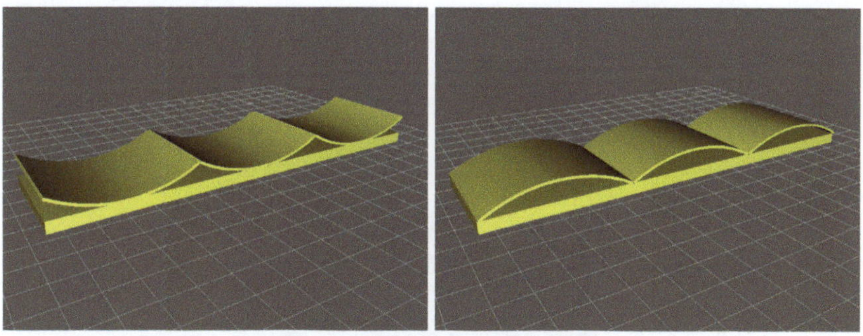

Fig. 10.25 Deformation of plate segments due to diurnal changes in temperature.

the range of accepted data. The task was described as a pair of plates of medium thickness, bonded with viscoelastic dampers, resting on a Winkler viscoelastic foundation. Vibrations of plates of dimensions 8×8 m were studied. A symmetric half of the task was considered. The loading was performed by an oscillator subjected to a force of 1 MN, moving at speeds of 180–360 km/h (Figure 10.27). The top plate had a thickness of 40 cm, and the lower, 130 cm. Figure 10.28 presents the casting process of the upper plate. The viscoelastic foundation has a stiffness of $k_z = 4 \cdot 10^7$ N/m^2, and the dilation mat E =(0.2–2.0)$\cdot 10^{11}$ N/m^2. The replacement thickness of the mat equal to 8 cm was assumed in order to take into account the deformation of the concrete block panels in the vertical direction. Figure 10.29 shows the initial state of the simulation. The wave propagation from the source is visible. Let us recall that the symmetric half being considered has dimensions 4×8 m and is relatively narrow. Our figures have disproportionate dimensions. Figure 10.30 shows

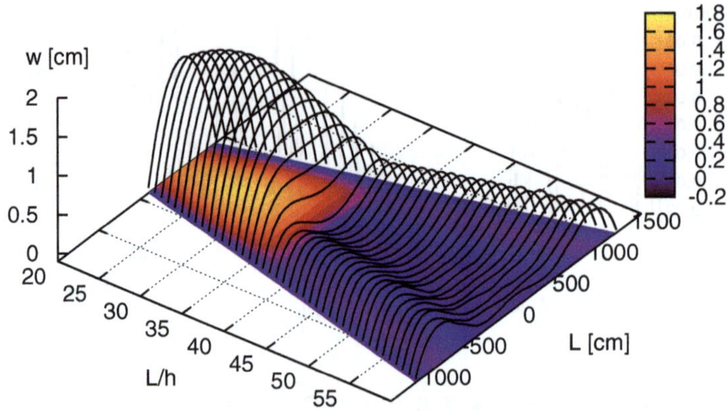

Fig. 10.26 The deflection line w of the plate deformed by a temperature gradient depending on the length l and ratio of length to thickness l/h.

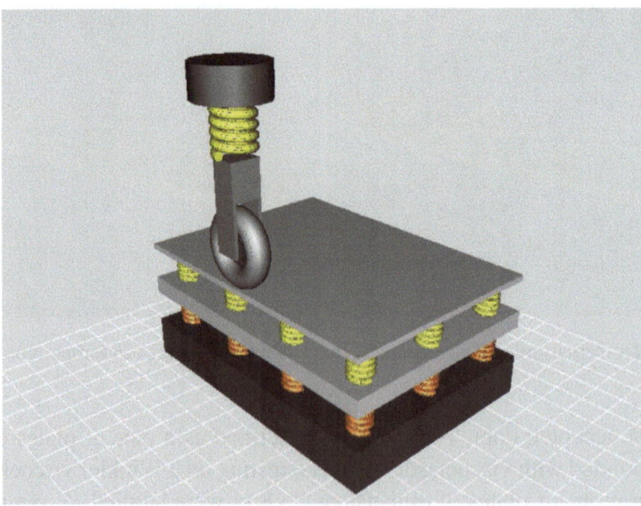

Fig. 10.27 Diagram of the airfield plate.

Fig. 10.28 Upper plate construction.

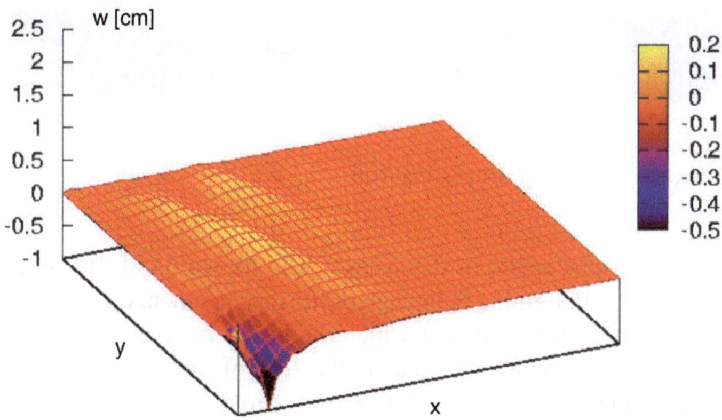

Fig. 10.29 Wave in the plate in the initial state of the motion.

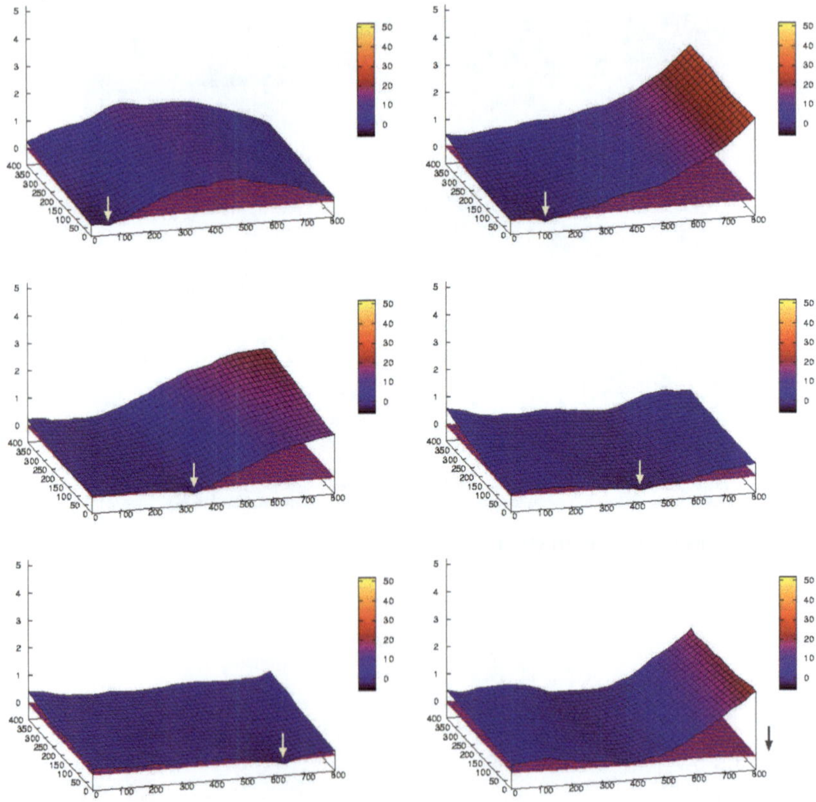

Fig. 10.30 Successive stages of the passing load.

six phases of the displacement of the system. The loosening of the upper plate is apparent. Two factors determine the image of the displacements: the strength of the dilation layer, i.e., the stress value at which the bonds between the two plates are broken, and the way in which the contact in the zone of force application and stress concentration is modelled. The factors that play a role in a zone limited to a very small loaded area, virtually decide the outcome. That is why the task can not be analysed by limiting attention to a pair of plates. Proper selection of the parameters describing the two decisive factors must be preceded by a three-dimensional static analysis of the strain state.

Appendix A
Computer Programs

The computer programs for our test simulations and for real track and airfield plate simulations were written in Fortran, C, and the C# programming language. In further sections, listings of the programs will be given in Fortran 77, since this seems to be the most efficient tool for numerical analyses. The important feature in favour of this language is a free Lapack library with highly efficient procedures for linear algebra. For our purpose, the solution of the system of equations must be efficient and should allow operating with a minimum number of coefficients. A band matrix enables solving with a highly discretized spatial domain in a reasonable time. We must apply procedures for non-symmetric matrices since the contribution of the non-symmetric matrix **C** of the moving point mass makes the symmetric matrices of the structure non-symmetric.

These computer programs are written in a relatively clear form. The global matrices are collected as band matrices. The code can be shortened and the execution made more efficient if we rearrange some procedures. However, we intend here to keep a clear form of the code.

A.1 String—Space–Time Element Method

The program for string analysis uses the space–time finite element method. The parameters alpha and gamma control the space–time element method with a Dirac virtual function. The next data describe the length of the string (xl), tensile force (xf), mass density (ro), cross sectional area (a), external force (q0), and the moving mass (xm0). The number of points ne is declared as a parameter, the speed of the mass vm and time step h are declared in the text. The procedure for solving the system of equations DGESV is taken from the Lapack library. The subroutine maska computes the moving mass matrices.

```
      program string
      implicit real*8(a-h,o-z)
      parameter(np=161)
      dimension sklok(2,4),
     x     ska(np,np),ska0(np,np),skb(np,np),skb0(np,np),
     x     f(np),u(2,np),v(2,np),e(np),ipom(np)
      data alpha,gamma,  xl,  xf,  ro,  a,    q0,  xm0
     x     /1.d0, .01d0, 1d0, 1d0, 1d0, 1d0, -1d0, 1d0/


c------------------------------------------------------------
c     string with the moving mass - space-time element method
c------------------------------------------------------------
c       np - number of points
c       ne - number of elements (ne=np-1)
c       alpha, gamma - parameters of the space-time element method
c                     0.707 < alpha <= 1.0
c                        gamma=0 --- no damping
c       vm - velocity of the mass
c       h -time step
c       xl - length of the string
c       xf - tensile force
c       ro - mass density
c       a - cross section area
c       q0 - massless fpoint force
c       xm0 - moving mass
c       ia, ip - indices: ia=1 and ip=2 or ia=2 and ip=1
c

      vm=0.5
      h=xl/vm/(np-1)/40

      ip=1
      ne=np-1
      dx=xl/ne

      do 701 i3=1,np
         e(i3)=0d0
         f(i3)=0d0
         u(1,i3)=0d0
         v(1,i3)=0d0
         u(2,i3)=0d0
         v(2,i3)=0d0
         do 701 j3=1,np
            ska0(i3,j3)=0d0
            skb0(i3,j3)=0d0
 701  continue

c     .................................... element matrix
      call elem(sklok,xf,ro,a,dx,h,alpha,1)

c     .................................... global matrix assembly
      do 10 i=1,ne
         call dod(i,ska0,skb0,sklok,np)
 10   continue

c       tmax - final time of the experiment
c       nkrok - number of time steps

      tmax=xl/vm*1.0
      nkrok=int(tmax/h)

c.................................................... time step loop
      do 100 it=1,nkrok
```

```
      ip=3-ip
      ia=3-ip
      czas=it*h

c................................ nodal forces from the previous step
      do 11 i=1,np
 11      f(i)=-e(i)

c................................. the load located in element 'iel'
      iel=int(czas*vm/dx-1d-10)+1
      if(iel.gt.ne)iel=0
      x0=(czas-h)*vm-iel*dx+dx
      xkappa=(x0+vm*alpha*h)/dx

      if(iel.gt.0)then
c......................................... moving load is added
         sk2=(czas*vm-iel*dx+dx)/dx
         sk1=1d0-sk2
         xm0s1=sk1*xm0
         xm0s2=sk2*xm0
         if(czas*vm.gt.0.)then
            f(iel  )=f(iel  )+sk1*q0
            f(iel+1)=f(iel+1)+sk2*q0
         endif
      endif

c................................ recuperation of matrix ska, skb
      do 72 ii=1,np
         do 72 jj=1,np
            ska(ii,jj)=ska0(ii,jj)
 72         skb(ii,jj)=skb0(ii,jj)

c......................... mass matrix
      call maska(sklok,xm0,alpha,vm,h,x0,xf,a,dx,xkappa)
      if(it.gt.0)call dod(iel,ska,skb,sklok,np)

c...................................... multiplication: ska*v1
c                           f:=F-A*v1  (f - right-hand side vector)
      do 31 i=1,np
        do 31 j=1,np
 31        f(i)=f(i)-ska(i,j)*v(ia,j)

      call warbrz(skb,f,np)

      call DGESV(np,1,skb,np,ipom,f,np,info)
      if(info.ne.0)stop

      do 66 ii=1,np
 66      v(ip,ii)=f(ii)

      beta=1.-alpha/(1.+gamma)

      do 51 ii=1,np
 51      u(ip,ii)=u(ia,ii)+h*((1-beta)*v(ia,ii)+beta*v(ip,ii))

      call elem(sklok,xf,ro,a,dx,h,alpha,2)
      do 83 ii=1,np
 83      e(ii)=0d0

      do 84 n=1,ne
         do 84 ii=1,2
            do 84 jj=1,2
 84            e(n-1+ii)=e(n-1+ii)+sklok(ii,jj)*u(ip,n-1+jj)
```

```
      xmnoz=xm0*vm**2/dx

      ul=u(ip,iel)
      up=u(ip,iel+1)
      vl=v(ip,iel)
      vp=v(ip,iel+1)

c...................... nodal forces
      e(iel  )=e(iel  )+xmnoz*( ul-up)/dx
      e(iel+1)=e(iel+1)+xmnoz*(-ul+up)/dx

      if(iel.eq.0)then
          ylok=0d0
      else
          ylok=sk1*u(ip,iel)+sk2*u(ip,iel+1)
      endif

      write(*,1001)czas*vm/xl,ylok,u(ip,(np-1)/2)
1001  format(3(1pe15.4))
100   continue
      end


c.................................................... mass marix
      subroutine maska(ste,xm0,alpha,v,h,x0,xf,a,b,xkappa)
      implicit real*8(a-h,o-z)
      dimension ste(2,4)

      do 9 i=1,2
         do 9 j=1,4
 9          ste(i,j)=0d0

c----------------stiffness
      xmnoz=xm0*v*v*h/b/b
      al=1-alpha
      ste(1,1)=ste(1,1)+xmnoz*al
      ste(1,2)=ste(1,2)-xmnoz*al
      ste(2,1)=ste(2,1)-xmnoz*al
      ste(2,2)=ste(2,2)+xmnoz*al

      ste(1,3)=ste(1,3)+xmnoz*(1.-al)
      ste(1,4)=ste(1,4)-xmnoz*(1.-al)
      ste(2,3)=ste(2,3)-xmnoz*(1.-al)
      ste(2,4)=ste(2,4)+xmnoz*(1.-al)

c-----------------damping
      xmnoz=xm0*v/b
      xkappa2=xkappa
      ste(1,1)=ste(1,1)-xmnoz*(1.-xkappa2)*(1-al)
      ste(1,2)=ste(1,2)+xmnoz*(1.-xkappa2)*(1-al)
      ste(2,1)=ste(2,1)-xmnoz*(xkappa2)*(1-al)
      ste(2,2)=ste(2,2)+xmnoz*(xkappa2)*(1-al)

      ste(1,3)=ste(1,3)-xmnoz*(1.-xkappa2)*al
      ste(1,4)=ste(1,4)+xmnoz*(1.-xkappa2)*al
      ste(2,3)=ste(2,3)-xmnoz*(xkappa2)*al
      ste(2,4)=ste(2,4)+xmnoz*(xkappa2)*al

c----------------inertia
      xmnoz=xm0/h
      ste(1,1)=ste(1,1)-xmnoz*(1-xkappa)**2
      ste(1,2)=ste(1,2)-xmnoz*xkappa*(1-xkappa)
```

```
      ste(2,1)=ste(2,1)-xmnoz*xkappa*(1-xkappa)
      ste(2,2)=ste(2,2)-xmnoz*xkappa**2

      ste(1,3)=ste(1,3)+xmnoz*(1-xkappa)**2
      ste(1,4)=ste(1,4)+xmnoz*xkappa*(1-xkappa)
      ste(2,3)=ste(2,3)+xmnoz*xkappa*(1-xkappa)
      ste(2,4)=ste(2,4)+xmnoz*xkappa**2
      end


c..................................... string element
      subroutine elem(ste,f,ro,a,dx,h,alpha,itryb)
      implicit real*8(a-h,o-z)
      dimension ste(2,4)

      xr=f*a/dx

      ste(1,1)= xr
      ste(1,2)=-xr
      ste(2,1)=-xr
      ste(2,2)= xr

      if(itryb.eq.2)return

      xmnoz=h*alpha**2/2.d0

      do 16 i=1,2
         do 16 j=3,4
 16         ste(i,j)=ste(i,j-2)*xmnoz

      xmnoz=(1.d0-alpha/2.d0)*alpha*h

      do 17 i=1,2
         do 17 j=1,2
 17         ste(i,j)=ste(i,j)*xmnoz

c     inertia matrix.............................................
      xmas=ro*a*dx
      ste(1,1)=ste(1,1)-xmas/3d0/h
      ste(1,2)=ste(1,2)-xmas/6d0/h
      ste(2,1)=ste(2,1)-xmas/6d0/h
      ste(2,2)=ste(2,2)-xmas/3d0/h

      ste(1,3)=ste(1,3)+xmas/3d0/h
      ste(1,4)=ste(1,4)+xmas/6d0/h
      ste(2,3)=ste(2,3)+xmas/6d0/h
      ste(2,4)=ste(2,4)+xmas/3d0/h

      end


c........................................ boundary conditions
      subroutine warbrz(sk,r,np)
      implicit real*8(a-h,o-z)
      dimension sk(np,1),r(1),izam(2)
      data izam/1,0/
      izam(2)=np

      do 111 iii=1,2
      nrowb=izam(iii)

      do 430 j=1,np
```

```
      sk(nrowb,j)=0.d0
430   sk(j,nrowb)=0.d0
      r(nrowb)=0.d0
111   sk(nrowb,nrowb)=1.d0
      end

      subroutine dod(n,ska,skb,sklok,lukl)
      implicit real*8(a-h,o-z)
      dimension ska(lukl,1),skb(lukl,1),sklok(2,4)

      do 1 i=1,2
       do 1 j=1,2
         ska(n-1+i,n-1+j)=ska(n-1+i,n-1+j)+sklok(i,j)
1        skb(n-1+i,n-1+j)=skb(n-1+i,n-1+j)+sklok(i,j+2)
      end
```

A.2 Timoshenko Beam—Newmark Method

The Timoshenko beam is computed with the Newmark method. Subintegrated finite element matrices were assumed. The program beam is built up in the following way. First, the elemental beam stiffness and inertia matrices are computed (elemk and elemm). Then the global matrices are assembled (dodkm). The time stepping scheme is started and the element carrying the mass is determined (iel). The portion of the mass assigned to both nodes of the element iel is computed. The procedure matk computes the matrix of coefficients for the Newmark method. In the same part, the moving mass is added to the global matrix and to the right-hand side vector. The boundary conditions are taken into account in warbrz and the system of equations is solved by solve. Finally, the velocity and acceleration vector is computed and the execution returns to the beginning of the time loop.

The program is autonomous, i.e., it does not require external library procedures. The Gauss procedure is included.

```
      program beam
      implicit real*8(a-h,o-z)
      parameter(nbd=4,nbd2p1=2*nbd+1,ne=100,nszf=(ne+1)*2)
      dimension sklok(4,4),smlok(4,4),sk(nszf,nbd2p1),
     x     sm(nszf,nbd2p1),skkon(nszf,nbd2p1),t(nszf),f(nszf),
     x     u1(nszf),v1(nszf),p1(nszf)
c  input data [centymeter, gram microsekond]
c  Young modulus, cross sec. moment, Poissone coeff., cross sec. ar
ea
c  height of the cross section, mass density, length, travelling ma
ss,
c  gravity acceleration, shape coefficient
      data xxe, xxi, xni,  xxa,  xxro,  xxl0, xxm0,  graw,  xk
     x    /1d0,.2d0,.3d0, 2.5d0, 7.8d0, 200d0, 5d2, 10d-10, 1.2d0/

c   time step, velocity, Newmark beta, Newmark gamma
      data h, v, beta, gamma /50.d0, 1d-4, 0.25d0, 0.5d0/
c-------------------------------------------------------
c     Timoshenko with the moving mass - Newmark method
c-------------------------------------------------------
        data iel,ielmem/1,1/
      open(2,file='results.dat')

c     sklok - stiffness element matrix
c     smlok - mass element matrix
c     sk    - global stiffness matrix
c     sm    - global mass matrix

      dx=xxl0/ne

      do 701 i3=1,nszf
         f(i3)=0d0
         u1(i3)=0d0
         v1(i3)=0d0
         p1(i3)=0d0
         do 701 j3=1,2*nbd+1
            sk(i3,j3)=0d0
            sm(i3,j3)=0d0
 701     continue

c     elemental matrices: stiffness and inertia
      call elemk(sklok,xxe,xxa,xxi,dx,xni,xk)
      call elemm(smlok,xxro,xxa,xxi,dx)

c     global matrices assembly - loop over elements
      do 10 i=1,ne
         call dodkm(i,sk,sklok,sm,smlok,nszf,nbd)
 10   continue

      tmax=xxl0/v*1.2d0
c...................................... starting time loop
      do 100 it=1,int(tmax/h)
      time=it*h

      do 7 i=1,nszf
 7       f(i)=0.d0
c     'iel' -element that carry the mass
      iel=int(time*v/dx)+1
      if(iel.gt.ne)iel=0
        xkappa=time*v/dx-iel+1d0

      do 11 i=1,nszf
 11      f(i)=0.d0
      if(iel.gt.0)then
         sk2=(time*v-iel*dx+dx)/dx
         sk1=1d0-sk2
         xxm0s1=sk1*xxm0
         xxm0s2=sk2*xxm0
c...................................... external moving load
         f(2*iel-1)=f(2*iel-1)-xxm0s1*graw
         f(2*iel+1)=f(2*iel+1)-xxm0s2*graw
      endif

      call matk(beta,gamma,h,sk,sm,skkon,f,t,nbd,nszf,u1,v1,p1,
     x   xkappa,xxm0,dx,iel,ielmem,v)
        call warbrz(skkon,f,nszf,nbd)
      call solve(skkon,f,nszf,nbd)
      call weknas(beta,gamma,u1,v1,p1,h,f,nszf)
      if(iel.eq.0)then
           ylok=0d0
      else
           ylok=sk1*u1(2*iel-1)+sk2*u1(2*iel+1)
      endif

      write(*,7001)time*v/xxl0,ylok,u1(ne+1)
      write(2,7001)time*v/xxl0,ylok,u1(ne+1)
 7001 format(f10.7,1pe15.4,1pe15.4)
        ielmem=iel
 100  continue
      end

c................................................................
      subroutine elemk(sklok,xe,a,xi,xl,xni,xk)
      implicit real*8(a-h,o-z)
      dimension sklok(4,4)

      xg=xe/2.d0/(1d0+xni)
      s1=xe*xi/xl
      s2=xk*xg*a/xl
      sklok(1,1)=s2
      sklok(1,2)=s2*xl/2.d0
      sklok(1,3)=-s2
      sklok(1,4)=s2*xl/2.d0
      sklok(2,1)=sklok(1,2)
      sklok(2,2)=s2*xl**2/4.d0+s1
      sklok(2,3)=-s2*xl/2.d0
      sklok(2,4)=s2*xl**2/4.-s1
      sklok(3,1)=sklok(1,3)
      sklok(3,2)=sklok(2,3)
      sklok(3,3)=s2
      sklok(3,4)=-s2*xl/2.d0
      sklok(4,1)=sklok(1,4)
      sklok(4,2)=sklok(2,4)
      sklok(4,3)=sklok(3,4)
      sklok(4,4)=s2*xl**2/4.d0+s1
      end

c................................................................
      subroutine elemm(smlok,ro,a,xi,xl)
      implicit real*8(a-h,o-z)
      dimension smlok(4,4)

      s=ro*a*xl
      ss=ro*xi*xl
      smlok(1,1)=s/3.d0
      smlok(1,2)=0.d0
      smlok(1,3)=s/6.d0
      smlok(1,4)=0.d0
```

```
      smlok(2,1)=0.d0
      smlok(2,2)=ss/3.d0
      smlok(2,3)=0.d0
      smlok(2,4)=ss/6.d0
      smlok(3,1)=s/6.d0
      smlok(3,2)=0.d0
      smlok(3,3)=s/3.d0
      smlok(3,4)=0.d0
      smlok(4,1)=0.d0
      smlok(4,2)=ss/6.d0
      smlok(4,3)=0.d0
      smlok(4,4)=ss/3.d0

      end

c...................................... global matrix assembly
      subroutine dodkm(n,sk,sklok,sm,smlok,nszf,nbd)
      implicit real*8(a-h,o-z)
      dimension sk(nszf,-nbd:nbd),sm(nszf,-nbd:nbd),
     x  sklok(4,4),smlok(4,4)

      do 1 i=1,4
        do 1 j=1,4
          sk(2*n-2+i,j-i)=sk(2*n-2+i,j-i)+sklok(i,j)
1         sm(2*n-2+i,j-i)=sm(2*n-2+i,j-i)+smlok(i,j)
      end

c........... solution of the system of equations (Gauss procedure)
      subroutine solve(s,f,n,nbd)
      implicit real*8(a-h,o-z)
      dimension s(n,-nbd:nbd),f(1),x(10,10)

        do 91 i=1,10
        do 91 j=1,10
        x(i,j)=0d0
91      if(j-i.ge.-nbd.and.j-i.le.nbd)x(i,j)=s(i,-i+j)

      do 3 i=1,n-1
        do 3 j=i+1,min0(i+nbd,n)
         do 4 k=1,nbd
          if(dabs(s(i,0)).lt.1e-20)then
            write(*,*)'solve - singular matrix=',i
            stop
          endif
4         s(j,k+i-j)=s(j,k+i-j)-s(j,i-j)*s(i,k)/s(i,0)
3       f(j)=f(j)-s(j,i-j)/s(i,0)*f(i)

      do 6 i=n,1,-1
        do 5 k=1,min0(nbd,n-i)
5        f(i)=f(i)-s(i,k)*f(i+k)
6       f(i)=f(i)/s(i,0)
      end

c........................................ boundary conditions
      subroutine warbrz(sk,r,nszf,nbd)
      implicit real*8(a-h,o-z)
      dimension sk(nszf,-nbd:nbd),r(1),izam(2)
      data izam/1,0/

      izam(2)=nszf-1

      do 111 iii=1,2
        nrowb=izam(iii)
        r(nrowb)=0.d0
        do 430 j=-nbd,nbd
          sk(nrowb,j)=0.d0
          if(nrowb-j.ge.1.and.nrowb-j.le.nszf)sk(nrowb-j,j)=0.d0
430     continue
        sk(nrowb,0)=1.d0
111   continue
      end

c................................ Newmark procedure - first stage
      subroutine matk(beta,gamma,h,sk,sm,skkon,f,t,nbd,nszf,
     x    u1,v1,p1,xkappa,xm0,b,iel,ielmem,v)
      implicit real*8(a-h,o-z)
      dimension sk(nszf,-nbd:nbd),sm(nszf,-nbd:nbd),
     x     skkon(nszf,-nbd:nbd),t(1),f(1),u1(1),v1(1),p1(1)

      a0=1.d0/beta/h/h
      a1=gamma/beta/h
      a2=1.d0/beta/h
      a3=1.d0/2.d0/beta-1.d0
      a4=gamma/beta-1.d0
      a5=h/2.d0*(gamma/beta-2.d0)
      a6=h*(1.d0-gamma)
      a7=gamma*h
      iel1=iel*2-1
      iel2=iel*2+1

      do 2 i=1,nszf
        do 2 j=-nbd,nbd
 2        skkon(i,j)=sk(i,j)+a0*sm(i,j)
c................................. point mass addition
      wsp11m= xm0*(1.0d0-xkappa)*(1.0d0-xkappa)
      wsp13m= xm0*xkappa*(1.0d0-xkappa)
      wsp31m= xm0*xkappa*(1.0d0-xkappa)
      wsp33m= xm0*xkappa*xkappa

      wsp11c=-xm0*v/b*(1.0d0-xkappa)
      wsp13c= xm0*v/b*(1.0d0-xkappa)
      wsp31c=-xm0*v/b*xkappa
      wsp33c= xm0*v/b*xkappa

      wsp11k=-xm0*v/b/h*(1.0d0-xkappa)
      wsp13k= xm0*v/b/h*(1.0d0-xkappa)
      wsp31k=-xm0*v/b/h*xkappa
      wsp33k= xm0*v/b/h*xkappa

      wsp11 = wsp11m*a0 + wsp11c*a1 + wsp11k
      wsp13 = wsp13m*a0 + wsp13c*a1 + wsp13k
      wsp31 = wsp31m*a0 + wsp31c*a1 + wsp31k
      wsp33 = wsp33m*a0 + wsp33c*a1 + wsp33k

      skkon(iel1, 0)=skkon(iel1, 0) + wsp11
      skkon(iel1, 2)=skkon(iel1, 2) + wsp13
      skkon(iel2,-2)=skkon(iel2,-2) + wsp31
      skkon(iel2, 0)=skkon(iel2, 0) + wsp33

      do 3 i=1,nszf
 3      t(i)=a0*u1(i) + a2*v1(i) + a3*p1(i)

      do 31 i=1,nszf
        do 31 j=-nbd,nbd
        if(i+j.ge.1.and.i+j.le.nszf) f(i)=f(i) + sm(i,j)*t(i+j)
31      continue

      f(iel1)=f(iel1) + wsp11m*t(iel1) + wsp13m*t(iel2)
```

```
      f(iel2)=f(iel2) + wsp31m*t(iel1) + wsp33m*t(iel2)

      do 4 i=1,nszf
4        t(i)=a1*u1(i) + a4*v1(i) + a5*p1(i)

      t_iel1 = a1*u1(iel1) + a4*v1(iel1) + a5*p1(iel1)
      t_iel2 = a1*u1(iel2) + a4*v1(iel2) + a5*p1(iel2)
c............................................ f:=f+C*t
      f(iel1)=f(iel1) + wsp11c*t_iel1 + wsp13c*t_iel2
      f(iel2)=f(iel2) + wsp31c*t_iel1 + wsp33c*t_iel2
c............................................ nodal forces of the mass
      eps=u1(2*ielmem+1)-u1(2*ielmem-1)
      f(iel1)=f(iel1)+(xm0*v/b/h)*(1.0d0-xkappa)*eps
      f(iel2)=f(iel2)+(xm0*v/b/h)*         xkappa *eps
      end

c.................................. Newmark procedure - second stag
e
      subroutine weknas(beta,gamma,u1,v1,p1,h,r,nszf)
      implicit real*8(a-h,o-z)
      dimension u1(1),v1(1),p1(1),r(1)

      a0=1.d0/beta/h/h
      a1=gamma/beta/h
      a2=1.d0/beta/h
      a3=1.d0/2.d0/beta-1.d0
      a4=gamma/beta-1.
      a5=h/2.d0*(gamma/beta-2.d0)
      a6=h*(1.d0-gamma)
      a7=gamma*h

      do 1 i=1,nszf
         u2=r(i)
         p2=a0*(u2-u1(i))-a2*v1(i)-a3*p1(i)
         v2=v1(i)+a6*p1(i)+a7*p2
         u1(i)=u2
         v1(i)=v2
         p1(i)=p2
1     continue
      end
```

A.3 Mindlin Plate—Space–Time Element Method

The theory of this model of the plate is given in Section 7.4.2, page 213. The program is written in Fortran 77. It uses the Lapack library procedure dpbsv. It solves the system of algebraic equations. The coefficient matrix is formed as a band non-symmetric matrix.

The input data set is listed below. Two first lines with values are read by the program as a file plate.cmm (lines 23–26). The remaining lines are comments and are omitted.

```
20    10   1200.   1200.   10.   .5d0   .01    9.81d-3   0.08   10d6
0.30d0   0.2d0   2.4d0   40.d0   1d-6

nx    ny   xlx    yly     dt    alfa vjazdy  qtotal   gamma xmasa
 e      xni       ro       t      xkz
```

The data is given in the [cm, g, μs] system.

```
      program plate
      implicit real*8(a-h,o-z)
      parameter(ist=4500000)
c***************************************************************
c       Mindlin plate analysis under a movin mass;
c***************************************************************
      dimension ma(ist)
      data nbd,ir,lwe,lss,lsswsp/190,2,4,3,2/

      nbd1=nbd+1
      open(8,file='results.dat',status='unknown')
c
c     nbd   - bandwith of the stiffness matrix
c     ir=2  - length of Real/length of integer:  R*8/I*4
c
      do 1 i=1,ist
1       ma(i)=0

c     reading initial data ...................................
      open(1,file='plate.cmm',status='old')
      read(1,*)nx,ny,xlx,yly,dt,alfa,vjazdy,qtotal,gamma,xmasa
      read(1,*)e,xni,ro,t,xkz
      close(1)

      np=(nx+1)*(ny+1)
      ne=nx*ny
      nwb=(nx+ny)*2
      nszf=np*lss

c     nx, ny     - No. of spatial elements in x and y direction
c     nwb        - No. of fixed nodes
c     dt,h       - time step
c     e          - Young modulus
c     xni        - Poissone coefficient
c     ro         - mass density
c     alfa       - parametr [0,1] procedury calkowania rown. ruchu
c     vjazdy     - velocity of the moving mass
c     gamma      - coefficient of numerical damping
c     kz         - stiffness of foundation

c     memory partition..........................................
c
c     1     - wpw
c     n1    - npe
c     n2    - ityp
c     n3    - iwb
c     n4    - r                   (right-had side vector)
c     n5    -
c     n6    - ef
c     n7    - q                   (velocity vector)
c     n8    - u                   (displacement vector)
c     n13   - skd | ind           (stiffness matrix) | temp. matrix
c     n14   - xmater
c     n15   - ind

      n1= 1+np*lsswsp*ir
      n2=n1+lwe*ne
      n3=n2+ne

      call dane(ma(1),ma(n1),ma(n3),ma(n2),np,ne,nx,ny,
     x     xlx,yly,dx,dy,nwbakt,lwe,lss,nbd00)

      nwb=nwbakt
```

```
      if(nbd00.gt.nbd)then
         write(*,*)'increase nbd ',nbd,' -->',nbd00
         stop
      else
         write(*,*)'decrease nbd ',nbd,' -->',nbd00
      endif

      n4=n3+2*nwb
      n6=n4+nszf*ir
      n7=n6+nszf*ir
      n8=n7+2*nszf*ir
      n13=n8+nszf*ir
      n14=n13+nszf*nbd1*ir
      n15=n14+20*ir
      n16=n15+ne
      n20=n16

      write(*,710)n16,ist
 710  format(' mem/ist=',i10,'/',i10)
      if(n16.gt.ist)stop 'lack of memory'
c-------------------------------------------------------------------
      iq1=1
      iq2=2
      it=0

100   it=it+1
      czas=it*dt
      write(*,*)'step, time=',it,czas
      call load(ma(1),ma(n4),np,lss,
     x     czas,vjazdy,qtotal,nx,ny,xlx,yly,dt,alfa,xmasa)

      call loadnodal(ma(n6),ma(n4),np,lss)

      call formk(ma(n13),ma(1),ma(n1),np,ne,dx,dy,ma(n4),lwe,lss,
     1     nszf,nbd,nbd1,dt,alfa,ma(n6),ma(n7),iq1,iq2,ma(n2),
     2     xmasa,ximasa,vjazdy,ma(n8),xlx,yly,
     3      e,xni,ro,t,xkz)

      call modyfwb(ma(n13),nbd,nbd1,np,lss,ma(n4),nwb,ma(n3))

      call dpbsv('U',nszf,nbd,1,ma(n13),nbd1,ma(n4),nszf,info)
      if(info.ne.0)stop 'singular matrix'

      call przemieszczenia(ma(n4),ma(n7),ma(n8),iq1,iq2,alfa,dt,
     x     gamma,np,lss)

      call silywezlowe(ma(1),ma(n1),np,ne,dx,dy,ma(n4),lwe,lss,
     1     nszf,nbd,nbd1,dt,alfa,ma(n6),ma(n2),ma(n8),ma(n14),
     2     xlx,yly,xmasa,vjazdy,e,xni,ro,t,xkz)

      call wydruki(ma(1),ma(n1),ma(n8),ma(n7),iq1,iq2,np,ne,lss,
     x     nx,ny,czas,vjazdy,xlx,wspx,e,xi)

      iq1=3-iq1
      iq2=3-iq2
      if(wspx.lt.1.2)go to 100
      end


      subroutine dane(wpw,npe,iwb,ityp,np,ne,nx,ny,xlx,yly,
     x     dx,dy,nwb,lwe,lss,nbd00)
      implicit real*8(a-h,o-z)
      dimension wpw(2,1),npe(lwe,1),iwb(2,1),ityp(1)
```

```
      dx=xlx/nx
      dy=yly/ny

      l=0
      do 9 i=0,nx-1
         do 9 j=0,ny-1
            l=l+1
            npe(1,l)=i*(ny+1)+j+1
            npe(2,l)=i*(ny+1)+j+2
            npe(3,l)=(i+1)*(ny+1)+j+1
            npe(4,l)=(i+1)*(ny+1)+j+2
            ityp(l)=1
    9 continue

      nwb=0
      do 10 i=1,ny+1
          nwb=nwb+1
          iwb(1,nwb)=i
          iwb(2,nwb)=100
          nwb=nwb+1
          iwb(1,nwb)=(nx)*(ny+1)+i
   10     iwb(2,nwb)=100

      do 11 i=2,nx
          nwb=nwb+1
          iwb(1,nwb)=(ny+1)*(i-1)+1
          iwb(2,nwb)=100
          nwb=nwb+1
          iwb(1,nwb)=(ny+1)*(i)
   11     iwb(2,nwb)=100

      ll=0
      do 12 i=0,nx
         do 12 j=0,ny
            ll=ll+1
            wpw(1,ll)=i*dx
            wpw(2,ll)=j*dy
   12 continue

      nbd00=0
      do 91 i=1,ne
         n1=iabs(npe(1,i)-npe(2,i))
         nbd00=max0(nbd00,n1)
         if(npe(3,i).gt.0)then
            n1=iabs(npe(2,i)-npe(3,i))
            nbd00=max0(nbd00,n1)
            n1=iabs(npe(3,i)-npe(4,i))
            nbd00=max0(nbd00,n1)
            n1=iabs(npe(4,i)-npe(1,i))
            nbd00=max0(nbd00,n1)
         endif
   91 continue

      nbd00=(nbd00+1)*lss
      write(*,*)'bandwidth =',nbd00

      do 71 i=1,nwb
         if(iwb(1,i).gt.np)stop 7
   71 continue

      end
```

```
      subroutine FORMK(skd,WPW,NPE,np,ne,dx,dy,r1,lwe,lss,
     1     nszf,nb,nbd1,h,alfa,ef,q,iq1,iq2,ityp,
     2     xmasa,ximasa,vm,u,xlx,yly,e,xni,ro,t,xkz)
      implicit real*8(a-h,o-z)

      dimension skd(nbd1,1),wpw(2,1),r1(6),dd(12,24),npe(lwe,1),
     1     q(2,1),wsp(2,4),nel(4),ef(1),ityp(1),u(1)
      common /wezly/x0,ktory_wez1,ktory_wez2,xkappa,sk1,sk2,ine

      lsh=lwe*lss

      do 300 j=1,nszf
         do 300 i=1,nbd1
  300       skd(i,j)=0d0

      DO 400 N=1,NE
         do 81 ii=1,12
            do 81 jj=1,24
   81          dd(ii,jj)=0d0

      nel(1)=npe(1,n)
      nel(2)=npe(2,n)
      nel(3)=npe(3,n)
      nel(4)=npe(4,n)
c-----------------------------------------------
      call plate_elem(dd,dx,dy,h,alfa,1,e,xni,ro,t,xkz)
      lwelok=4

      do 7 i=1,lwelok
         nw1=nel(i)
         do 7 ii=1,lss
            il=nw1*lss-lss+ii
            ix=i*lss-lss+ii

            do 7 j=1,lwelok
               nw2=nel(j)
               do 7 jj=1,lss
                  ik=nw2*lss-lss+jj
                  iy=j*lss-lss+jj
                  r1(il)=r1(il)-dd(ix,iy)*q(iq1,nw2*lss-lss+jj)
                  if(il.le.ik)skd(nbd1+il-ik,ik)=
     x                   skd(nbd1+il-ik,ik)+dd(ix,iy+lsh)
    7          continue
  400  CONTINUE

      nel(1)=ktory_wez1
      nel(2)=ktory_wez2
      lsh=6

      call mass(dd,dx,dy,h,alfa,xmasa,vm)

      lwelok=2
      do 71 i=1,lwelok
         nw1=nel(i)
         do 71 ii=1,lss
            il=nw1*lss-lss+ii
            ix=i*lss-lss+ii

            do 71 j=1,lwelok
               nw2=nel(j)
               do 71 jj=1,lss
```

```
                  ik=nw2*lss-lss+jj
                  iy=j*lss-lss+jj
                  r1(il)=r1(il)-dd(ix,iy)*q(iq1,nw2*lss-lss+jj)
                  if(il.le.ik)skd(nbd1+il-ik,ik)=
     x                skd(nbd1+il-ik,ik)+dd(ix,iy+lsh)
 71             continue
        end

      subroutine modyfwb(b,nbd,nbd1,np,lss,r,nwb,iwb)

      implicit real*8(a-h,o-z)
      dimension b(nbd1,1),r(6),iwb(2,1)

      do 1 iii=1,nwb
         iwezel=iwb(1,iii)
         kod=iwb(2,iii)
         do 2 jj=1,lss
            icyfra=int(kod/10**(lss-jj))
            kod=kod-icyfra*10**(lss-jj)

            if(icyfra.eq.1)then
               nrl=iwezel*lss-lss+jj
               al=0d0

               do 400 ii=1,nbd1
                  lin=nrl-nbd1+ii
                  if(lin.gt.np*lss)stop 2
                  if(ii.gt.np*lss)stop 3
                  r(lin)=r(lin)-al*b(ii,nrl)
                  b(ii,nrl)=0d0
 400           continue

               do 401 ii=1,min0(nbd,np*lss-nrl)
                  r(nrl+ii)=r(nrl+ii)-al*b(nbd1-ii,nrl+ii)
                  b(nbd1-ii,nrl+ii)=0d0
 401           continue

               b(nbd1,nrl)=1.d0
               r(nrl)=al
            endif

 2       continue
 1    continue

      end


      subroutine load(wpw,fzewn,np,lss,
     x      czas,vjazdy,qtotal,nx,ny,xlx,yly,h,alfa,xm0)
      implicit real*8(a-h,o-z)

      dimension wpw(2,1),fzewn(1)
      common /wezly/x0,ktory_wez1,ktory_wez2,xkappa,sk1,sk2,ine
      common sw,ktory_wez1m,ktory_wez2m

c     zeroing of external load vector

      do 2 i=1,np*lss
 2       fzewn(i)=0d0

      dx=xlx/nx
      dy=yly/ny
```

```
c     nodes which are subjected by a load: ktory_wez1, ktory_wez2
      x=vjazdy*czas
      if(x.le.xlx)then
         ine=int(czas*vjazdy/dx-1d-10)+1
         x0=(czas-h)*vjazdy-ine*dx+dx
         xkappa=(x0+vjazdy*alfa*h)/dx

         ktory_wez1=(ny+1)*(ine-1)+ny/2+1
         ktory_wez2=(ny+1)*(ine  )+ny/2+1

         sk1=1d0-xkappa
         sk2=xkappa

         fzewn(ktory_wez1*3-2)=-qtotal*sk1
         fzewn(ktory_wez2*3-2)=-qtotal*sk2
      nnn=(ny+1)*(nx/2+1)-ny/2

      else
         w1=0.
         w2=0.
         ktory_wez1=1
         ktory_wez2=1
      endif

      fzewn(3*ktory_wez1-2)=fzewn(3*ktory_wez1-2)-sw*(1-xkappa)
      fzewn(3*ktory_wez2-2)=fzewn(3*ktory_wez2-2)-sw*xkappa
      end



      subroutine loadnodal(ef,r,np,lss)
      implicit real*8(a-h,o-z)
      dimension ef(1),r(1)

c     computation of right-hand side vector

      do 1 i=1,np*lss
 1       r(i)=r(i)-ef(i)
      end



      subroutine przemieszczenia(r,q,u,iq1,iq2,alfa,dt,gamma,np,lss)
      implicit real*8(a-h,o-z)
      dimension r(1),q(2,1),u(1)

      beta=1d0-alfa/(1d0+gamma)

      do 9 i=1,np*lss
         q(iq2,i)=r(i)
         u(i)=u(i)+(q(iq1,i)*(1d0-beta)+q(iq2,i)*beta)*dt
 9    continue
      end


      subroutine silywezlowe(WPW,NPE,np,ne,dx,dy,r1,lwe,lss,
     1     nszf,nbd,nbd1,h,alfa,ef,ityp,u,xmater,xlx,yly,
     2     xxm0,vm,e,xni,ro,t,xkz)
      implicit real*8(a-h,o-z)

      dimension wpw(2,1),r1(6),dd(12,24),npe(lwe,1),
     2 wsp(2,4),nel(4),ef(1),ityp(1),u(1),xmater(20)
      common /wezly/x0,ktory_wez1,ktory_wez2,xkappa,sk1,sk2,ine
```

```
      common sw,ktory_wez1m,ktory_wez2m

      lsh=lwe*lss
      itryb=2

      do 17 i=1,nszf
 17      ef(i)=0d0

      DO 400 N=1,NE
         nel(1)=npe(1,n)
         nel(2)=npe(2,n)
         nel(3)=npe(3,n)
         nel(4)=npe(4,n)

c     element matrix.........................................
      call plate_elem(dd,dx,dy,h,alfa,itryb,e,xni,ro,t,xkz)
         lwelok=4
         do 7 i=1,lwelok
            nw1=nel(i)
            do 7 ii=1,lss
               il=nw1*lss-lss+ii
               ix=i*lss-lss+ii

               do 7 j=1,lwelok
                  nw2=nel(j)
                  do 7 jj=1,lss

                     ik=nw2*lss-lss+jj
                     iy=j*lss-lss+jj
                     ef(il)=ef(il)+dd(ix,iy)*u(nw2*lss-lss+jj)
 7       continue
 400  CONTINUE

c............. nodal forces of the mass .................
      u1=u(3*ktory_wez1m-2)
      u2=u(3*ktory_wez2m-2)

      u3=u(3*ktory_wez1-2)
      u4=u(3*ktory_wez2-2)
      ktory_wez1m=ktory_wez1
      ktory_wez2m=ktory_wez2

      sw=xxm0*vm/dx/h*(u4-u3-u2+u1)

      end



      subroutine wydruki(wpw,npe,u,q,iq1,iq2,np,ne,lss,nx,ny,czas,
     x     vjazdy,xlx,wspx,e,xi)
      implicit real*8(a-h,o-z)
      character*24 dysk
      dimension wpw(2,1),npe(4,1),u(1),q(2,1)
      common /wezly/x0,ktory_wez1,ktory_wez2,xkappa,sk1,sk2,ine

      wspx=czas*vjazdy/xlx
      dx=xlx/nx

      nnn=(ny+1)*(nx/2+1)-ny/2
      w_sle=u(ktory_wez1*3-2)*sk1+u(ktory_wez2*3-2)*sk2
      write(8,991)czas*vjazdy/xlx,w_sle,u(nnn*3-2)
 991     format(1pe15.6,1pe15.6,1pe15.6)

      end
```

```
      subroutine plate_elem(dd,dxbok,dybok,h,alfa,itryb,e,xni,ro,t,x
kz)
      implicit real*8(a-h,o-z)
      dimension iksi(4),ieta(4)
      dimension dd(12,24)
      data iksi,ieta/1,-1,1,-1,-1,-1,1,1/
c.........................................
      xkappa=5./6.
      g=e/2./(1+xni)
      a=dxbok/2.
      b=dybok/2.
c.........................................
      wsp1=e*t**3/48./a/b/(1.-xni**2)
      wsp2=g*xkappa*t

      do 711 i=1,4
         do 711 j=1,4
            dd(3*i-2,3*j-2)=0d0
            dd(3*i-2,3*j-1)=0d0
            dd(3*i-2,3*j)=0d0
            dd(3*i-1,3*j-2)=0d0
            dd(3*i,3*j-2)=0d0

            dd(3*i-1,3*j-1)=((2*a**2*ieta(i)*ieta(j)*
     x      (iksi(i)*iksi(j)+3)+b**2*iksi(i)*iksi(j)*
     x      (1-xni)*(ieta(i)*ieta(j)+3))/6)*wsp1
            dd(3*i-1,3*j)=(-a*b*(2*ieta(i)*xni*iksi(j)+ieta(j)*
     x      iksi(i)*(1-xni))/2)*wsp1
            dd(3*i,3*j-1)=(a*b*(ieta(i)*iksi(j)*(xni-1)-2*
     x      ieta(j)*xni*iksi(i))/2)*wsp1
            dd(3*i,3*j)=((2*b**2*iksi(i)*iksi(j)*
     x      (ieta(i)*ieta(j)+3)-a**2*ieta(i)*ieta(j)*(xni-1)*
     x      (iksi(i)*iksi(j)+3))/6)*wsp1
c----------------------
            dd(3*i-2,3*j-2)=dd(3*i-2,3*j-2)+((a**2*ieta(i)*ieta(j)*
     x      (iksi(i)*iksi(j)+3)+b**2*iksi(i)*iksi(j)*
     x      (ieta(i)*ieta(j)+3))/(12*a*b))*wsp2
            dd(3*i-2,3*j-1)=dd(3*i-2,3*j-1)+(-a*ieta(i)*
     x      (iksi(i)*iksi(j)+3)/12)*wsp2
            dd(3*i-2,3*j)=dd(3*i-2,3*j)+(b*iksi(i)*
     x      (ieta(i)*ieta(j)+3)/12)*wsp2
            dd(3*i-1,3*j-2)=dd(3*i-1,3*j-2)+(-a*ieta(j)*(iksi(i)*
     x      iksi(j)+3)/12)*wsp2
            dd(3*i-1,3*j-1)=dd(3*i-1,3*j-1)+
     x      (a*b*(ieta(i)*ieta(j)+3)*(iksi(i)*iksi(j)+3)/36)*wsp2
            dd(3*i-1,3*j)=dd(3*i-1,3*j)+0d0
            dd(3*i,3*j-2)=dd(3*i,3*j-2)+(b*iksi(j)*
     x      (ieta(i)*ieta(j)+3)/12)*wsp2
            dd(3*i,3*j-1)=dd(3*i,3*j-1)+0d0
            dd(3*i,3*j)=dd(3*i,3*j)+(a*b*(ieta(i)*ieta(j)+3)*
     x      (iksi(i)*iksi(j)+3)/36)*wsp2
711      continue

c........................... elastic foundation
      do 73 i=1,12,3
73      dd(i,i)=dd(i,i)+xkz*dxbok*dybok/4.

c........................... further dynamic stiffness matrix
      if(itryb.eq.2)return

c     formulation of entire matrix [A|B] ..................
```

```
      xmnoz=h*alfa**2/2.d0
      do 16 i=1,12
         do 16 j=13,24
 16         dd(i,j)=dd(i,j-12)*xmnoz

      xmnoz=(1.d0-alfa/2.d0)*alfa*h
      do 17 i=1,12
         do 17 j=1,12
 17         dd(i,j)=dd(i,j)*xmnoz

c     addition of the mass matrix ........... [-1/h*...|+1/h*...]

      xmasa=dxbok*dybok*t*ro/4./h/108.

      do 71 ii=1,4
         do 71 jj=1,4
            if(ii.eq.jj)mnoz=48
            if(iabs(ii-jj).eq.1)mnoz=24
            if(iabs(ii-jj).eq.2)mnoz=12
            if(iabs(ii-jj).eq.3)mnoz=24
            dd(3*ii-2,   3*jj-2)=dd(3*ii-2,   3*jj-2)-xmasa*mnoz
            dd(3*ii-2,12+3*jj-2)=dd(3*ii-2,12+3*jj-2)+xmasa*mnoz
 71   continue

      xmasa=dxbok*dybok*t**3*ro/4./h/108.

      do 72 ii=1,4
         do 72 jj=1,4
            if(ii.eq.jj)mnoz=4
            if(iabs(ii-jj).eq.1)mnoz=2
            if(iabs(ii-jj).eq.2)mnoz=1
            if(iabs(ii-jj).eq.3)mnoz=2

            dd(3*ii-1,   3*jj-1)=dd(3*ii-1,   3*jj-1)-xmasa*mnoz
            dd(3*ii  ,   3*jj  )=dd(3*ii  ,   3*jj  )-xmasa*mnoz
            dd(3*ii-1,12+3*jj-1)=dd(3*ii-1,12+3*jj-1)+xmasa*mnoz
            dd(3*ii  ,12+3*jj  )=dd(3*ii  ,12+3*jj  )+xmasa*mnoz
 72   continue
      end


      subroutine mass(dd,dxbok,dybok,h,alfa,xxm0,vm)
      implicit real*8(a-h,o-z)
      dimension dd(12,24),xm(6,12),xc(6,12),xk(6,12)
      common /wezly/x0,ktory_wez1,ktory_wez2,xkappa,sk1,sk2,ine

c     formulation of matrices of mass element

      data xm,xc,xk/72*0d0,72*0d0,72*0d0/

      do 11 i=1,12
         do 11 j=1,24
 11         dd(i,j)=0d0

      alfa2=0.
c...........................formulation of matrix M
      xm(1,1)=-xxm0/h*(1.-xkappa)**2
      xm(1,4)=-xxm0/h*xkappa*(1.-xkappa)
      xm(4,1)=-xxm0/h*xkappa*(1.-xkappa)
      xm(4,4)=-xxm0/h*xkappa**2

      xm(1,7)= xxm0/h*(1.-xkappa)**2
      xm(1,10)=xxm0/h*xkappa*(1.-xkappa)
      xm(4,7)= xxm0/h*xkappa*(1.-xkappa)
      xm(4,10)=xxm0/h*xkappa**2

c...........................formulation of matrix C
      w=2d0*xxm0*vm/dxbok

      xc(1,1)=-w*(1-xkappa)*(1-alfa2)
      xc(1,4)= w*(1-xkappa)*(1-alfa2)
      xc(4,1)=-w*xkappa*(1-alfa2)
      xc(4,4)= w*xkappa*(1-alfa2)

      xc(1,7)= -w*(1-xkappa)*alfa2
      xc(1,10)= w*(1-xkappa)*alfa2
      xc(4,7)= -w*xkappa*alfa2
      xc(4,10)= w*xkappa*alfa2

      do 1 i=1,6
      do 1 j=1,12
1        dd(i,j)=xm(i,j)+xc(i,j)

      end
```

A.4 Kirchhoff Plate — Space-Time Element Method

Below we show the code for thin plate element, siutable for the program from Appendix A.3. The matrices for thin plate element are given in the section 7.4.1, page 204.

```
      subroutine plate(dd,dxbok,dybok,h,alfa,itryb,e,xni,ro,t,xkz)

      implicit real*8(a-h,o-z)
      dimension k1(12,12),k2(12,12),k3(12,12),k4(12,12)
      dimension dd(12,24)
c
c      rectangular element of the thin plate
c
c      numbering of nodes
c
c      ^ y
c      |
c      2-------4
c      |       |
c      1-------3--->x
c
      data k1/
     x    60, 30, 0,30, 15, 0,-60,30, 0,-30,15, 0,
     x      30, 20, 0,15, 10, 0,-30,10, 0,-15, 5, 0,
     x      0, 0, 0, 0, 0, 0, 0, 0, 0, 0, 0, 0,
     x      30, 15, 0,60, 30, 0,-30,15, 0,-60,30, 0,
     x      15, 10, 0,30, 20, 0,-15, 5, 0,-30,10, 0,
     x      0, 0, 0, 0, 0, 0, 0, 0, 0, 0, 0, 0,
     x      -60,-30,0,-30,-15,0,60, -30,0,30, -15,0,
     x      30, 10, 0,15, 5, 0,-30,20, 0,-15,10, 0,
     x      0, 0, 0, 0, 0, 0, 0, 0, 0, 0, 0, 0,
     x      -30,-15,0,-60,-30,0,30, -15,0,60, -30,0,
     x      15, 5, 0,30, 10, 0,-15,10, 0,-30,20, 0,
     x      0, 0, 0, 0, 0, 0, 0, 0, 0, 0, 0, 0/


      data k2/
     x  60 ,0, -30, -60, 0, -30 ,30  ,0, -15, -30, 0, -15 ,
     x  0  ,0,  0,    0 , 0,  0  , 0  ,0,  0,    0 , 0,  0  ,
     x -30 ,0, 20,   30 , 0, 10  ,-15 ,0, 10,   15 , 0,  5  ,
     x -60 ,0, 30,   60 , 0, 30  ,-30 ,0, 15,   30 , 0, 15  ,
     x  0  ,0,  0,    0 , 0,  0  , 0  ,0,  0,    0 , 0,  0  ,
     x -30 ,0, 10,   30 , 0, 20  ,-15 ,0,  5,   15 , 0, 10  ,
     x 30  ,0, -15, -30, 0, -15 ,60  ,0, -30, -60, 0, -30 ,
     x  0  ,0,  0,    0 , 0,  0  , 0  ,0,  0,    0 , 0,  0  ,
     x -15 ,0, 10,   15 , 0,  5  ,-30 ,0, 20,   30 , 0, 10  ,
     x -30 ,0, 15,   30 , 0, 15  ,-60 ,0, 30,   60 , 0, 30  ,
     x  0  ,0,  0,    0 , 0,  0  , 0  ,0,  0,    0 , 0,  0  ,
     x -15 ,0,  5,   15 , 0, 10  ,-30 ,0, 10,   30 , 0, 20  /

      data k3/
     x      30, 15, -15,-30,-15, 0, -30, 0, 15, 30, 0, 0,
     x      15, 0, -15,-15, 0, 0, 0, 0, 0, 0, 0, 0,
     x      -15,-15, 0, 0, 0, 0, 15, 0, 0, 0, 0, 0,
     x      -30,-15, 0, 30, 15, 15, 30, 0, 0, -30, 0, -15,
     x      -15, 0, 0, 15, 0, 15, 0, 0, 0, 0, 0, 0,
     x      0, 0, 0, 15, 15, 0, 0, 0, 0, -15, 0, 0,
     x      -30, 0, 15, 30, 0, 0, 30, -15,-15,-30,15, 0,
     x      0, 0, 0, 0, 0, 0, -15, 0, 15, 15, 0, 0,
     x      15, 0, 0, 0, 0, 0, -15,15, 0, 0, 0, 0,
     x      30, 0, 0, -30, 0, -15,-30,15, 0, 30, -15,15,
     x      0, 0, 0, 0, 0, 0, 15, 0, 0, -15, 0, -15,
     x      0, 0, 0, -15, 0, 0, 0, 0, 0, 15, -15, 0/


      data k4/
     x      84, 6,-6,-84,-6,-6,-84, 6, 6,84, -6, 6,
     x      6, 8, 0,-6, -8, 0,-6, -2, 0, 6, 2, 0,
     x      -6, 0, 8, 6, 0,-2, 6, 0,-8,-6, 0, 2,
```

```
     x      -84,-6, 6,84, 6, 6,84, -6,-6,-84, 6,-6,
     x      -6, -8, 0, 6, 8, 0, 6, 2, 0,-6, -2, 0,
     x      -6, 0,-2, 6, 0, 8, 6, 0, 2,-6, 0,-8,
     x      -84,-6, 6,84, 6, 6,84, -6,-6,-84, 6,-6,
     x      6, -2, 0,-6, 2, 0,-6, 8, 0, 6, -8, 0,
     x      6, 0,-8,-6, 0, 2,-6, 0, 8, 6, 0,-2,
     x      84, 6,-6,-84,-6,-6,-84, 6, 6,84, -6, 6,
     x      -6, 2, 0, 6, -2, 0, 6, -8, 0,-6, 8, 0,
     x      6, 0, 2,-6, 0,-8,-6, 0,-2, 6, 0, 8/

c.........................................

      ddx =e*t**3/12./(1.-xni**2)
      ddy =ddx
      dd1 =ddx*xni
      ddxy=ddx*(1.-xni)/2.

      p2=(dxbok/dybok)**2

c       write(*,*)'dxdy=',dxbok,dybok

      do 10 i=1,12
         do 10 j=1,12
 10         dd(i,j)=(ddx*k1(i,j)/p2+ddy*k2(i,j)*p2+dd1*k3(i,j)+
     x              ddxy*k4(i,j))/15./dxbok/dybok


      do 11 i=1,12
         do 13 j=2,12,3
            dd(i,j)=dd(i,j)*dxbok
 13         dd(j,i)=dd(j,i)*dxbok
         do 14 j=3,12,3
            dd(i,j)=dd(i,j)*dybok
 14         dd(j,i)=dd(j,i)*dybok
 11   continue

c      elastic foundation
      do 73 i=1,12,3
 73     dd(i,i)=dd(i,i)+xkz*dxbok*dybok/4.

c......................

      if(itryb.eq.2)return

c     computation of entire matrix [A|B] ..................

      xmnoz=h*alfa**2/2.d0
      do 16 i=1,12
         do 16 j=13,24
 16         dd(i,j)=dd(i,j-12)*xmnoz

      xmnoz=(1.d0-alfa/2.d0)*alfa*h
      do 17 i=1,12
         do 17 j=1,12
 17         dd(i,j)=dd(i,j)*xmnoz



c     addition of mass matrix ...........................
c     [-1/h*...|+1/h*...]

      xmasa=dxbok*dybok*t*ro/4./h/108.

      do 71 ii=1,4
         do 71 jj=1,4
            if(ii.eq.jj)mnoz=48
            if(iabs(ii-jj).eq.1)mnoz=24
            if(iabs(ii-jj).eq.2)mnoz=12
            if(iabs(ii-jj).eq.3)mnoz=24
            dd(3*ii-2,    3*jj-2)=dd(3*ii-2,    3*jj-2)-xmasa*mnoz
            dd(3*ii-2,12+3*jj-2)=dd(3*ii-2,12+3*jj-2)+xmasa*mnoz
 71   continue

      xmasa=dxbok*dybok*t**3*ro/4./h/108.

      do 72 ii=1,4
         do 72 jj=1,4
            if(ii.eq.jj)mnoz=4
            if(iabs(ii-jj).eq.1)mnoz=2
            if(iabs(ii-jj).eq.2)mnoz=1
            if(iabs(ii-jj).eq.3)mnoz=2

            dd(3*ii-1,    3*jj-1)=dd(3*ii-1,    3*jj-1)-xmasa*mnoz
            dd(3*ii  ,    3*jj  )=dd(3*ii  ,    3*jj  )-xmasa*mnoz

            dd(3*ii-1,12+3*jj-1)=dd(3*ii-1,12+3*jj-1)+xmasa*mnoz
            dd(3*ii  ,12+3*jj  )=dd(3*ii  ,12+3*jj  )+xmasa*mnoz
 72   continue

      end
```

References

1. Achenbach, I.D., Sun, C.T.: Moving load on flexibly supported Timoshenko beam. Int. J. Solid Struct. 1, 353–370 (1965)
2. Akin, J.E., Mofid, M.: Numerical solution for response of beams with moving mass. J. of Struc. Engng. 115(1), 120–131 (1989)
3. Andrianov, I.V., Awrejcewicz, J.: Dynamics of a string moving with time-varying speed. J. Sound and Vibr. 292, 935–940 (2006)
4. Antosik, P., Mikusiński, J., Sikorski, R.: Theory of distributions. The sequential approach. Elsevier-PWN, Amsterdam-Warszawa (1973)
5. Argyris, J.H., Chan, A.S.L.: Application of the finite elements in space and time. Ing. Archiv. 41, 235–257 (1972)
6. Argyris, J.H., Scharpf, D.W.: Finite elements in space and time. Nucl. Engng. Design 10, 456–469 (1969)
7. Argyris, J.H., Scharpf, D.W.: Finite elements in time and space. Aeron. J. Roy. Aeron. Soc. 73, 1041–1044 (1969)
8. Ayre, R.S., Jacobsen, L.S., Hsu, C.S.: Transverse vibration of one- and two-span beams under the action of a moving mass load. In: ASME (ed.) Proc. of the First US Nat. Congr. of Appl. Mech., pp. 81–90 (1951)
9. Bajer, C.: Space–time finite element formulation for the dynamical evolutionary process. Appl. Math. and Comp. Sci. 3(2), 251–268 (1993)
10. Bajer, C., Bogacz, R.: Dynamic contact problem by means of the space-time element method. In: Gruber, R., Periaux, J., Shaw, R. (eds.) Proc. 5th Int. Symp. Numer. Meth. in Engng., vol. 1, pp. 313–318. Springer, Lausanne (1989)
11. Bajer, C., Bogacz, R.: On the space–time element method applied to the dynamics of vehicle–road interaction. ZAMM 71, T213–T216 (1991)
12. Bajer, C., Bohatier, C.: Solution of thermomechanical problems by the space–time finite elements. In: Chenot, J.L., Wood, R., Zienkiewicz, O. (eds.) Numerical Methods in Industrial Forming Processes, pp. 215–219. A.A. Balkema, Valbonne (1992)
13. Bajer, C., Bohatier, C.: Modélisation des systèmes mécaniques sous sollicitations dynamiques et résolution numérique des problèmes non linéaires associés. In: Méth. Inform. de la Conception Industr., Méc. des Struc., Marseille, pp. 79–90 (1993)
14. Bajer, C., Bohatier, C.: Solution of thermomechanical problems by the space–time finite elements. In: Numerical Methods in Industrial Forming Processes, Balkema, pp. 215–219 (1993)
15. Bajer, C., Kosiński, W.: Numerical modelling of thermal waves via internal state variable approach. Comp. Ass. Mech. Eng. Sci. 2, 307–319 (1995)

16. Bajer, C., Podhorecki, A.: Space–time element method in structural dynamics. Arch. of Mech. 41, 863–889 (1989)
17. Bajer, C.I.: Triangular and tetrahedral space–time finite elements in vibration analysis. Int. J. Numer. Meth. Engng. 23, 2031–2048 (1986)
18. Bajer, C.I.: Notes on the stability of non–rectangular space–time finite elements. Int. J. Numer. Meth. Engng. 24, 1721–1739 (1987)
19. Bajer, C.I.: Adaptive mesh in dynamic problem by the space–time approach. Comput. and Struct. 33(2), 319–325 (1989)
20. Bajer, C.I.: The space-time approach to rail/wheel contact and corrugations problem. Comp. Ass. Mech. Eng. Sci. 5(2), 267–283 (1998)
21. Bajer, C.I., Bogacz, R., Bonthoux, C.G.: Adaptive space–time elements in the dynamic elastic–viscoplastic problem. Comput. and Struct. 39, 415–423 (1991)
22. Bajer, C.I., Bohatier, C.: The soft way method and the velocity formulation. Comput. and Struct. 55(6), 1015–1025 (1995)
23. Bajer, C.I., Bonthoux, C.G.: State–of–the–art in true space–time finite element method. Shock Vibr. Dig. 20, 3–11 (1988)
24. Bajer, C.I., Dyniewicz, B.: Space-time approach to numerical analysis of a string with a moving mass. Int. J. Numer. Meth. Engng. 76(10), 1528–1543 (2008)
25. Bajer, C.I., Dyniewicz, B.: Numerical modelling of structure vibrations under inertial moving load. Arch. Appl. Mech. 79(6-7), 499–508 (2009)
26. Bajer, C.I., Dyniewicz, B.: Virtual functions of the space-time finite element method in moving mass problems. Comput. and Struct. 87, 444–455 (2009)
27. Belytschko, T., Krongauz, Y., Organ, D., Fleming, M., Krysl, P.: Meshless methods: an overview and recent developments. Comput. Meth. Appl. Mech. Engng. 139(1-4), 3–47 (1996)
28. Belytschko, T., Lu, Y.Y., Gu, L.: Element free Galerkin methods. Int. J. Numer. Meth. Engng. 37, 229–256 (1994)
29. Biondi, B., Muscolino, G.: New improved series expansion for solving the moving oscillator problem. J. Sound and Vibr. 281, 99–117 (2005)
30. Bogacz, R., Krzyzynski, T., Popp, K.: On the generalization of mathew's problem of the vibration of a beam on elastic foundation. Z. Angew. Math. Mech. 69, 243–252 (1989)
31. Bogacz, R., Szolc, T.: On methods of solution for the discrete–continuous system under moving load. Z. Angew. Math. Mech. 72(4), T16–T19 (1992)
32. Bogacz, R., Szolc, T.: Analysis of dynamic interaction between the continuous string and moving oscillator. Eng. Trans. 41(3-4), 361–380 (1993)
33. Bohatier, C.: A large deformation formulation and solution with space–time finite elements. Arch. Mech. 44, 31–41 (1992)
34. Bohatier, C., Bajer, C.: Kinematic approach for dynamic contact problems — the geometrical soft way method. Engng. Trans. 43(1-2), 101–111 (1995)
35. Bolotin, W.: Problems of bridge vibration under the action of the moving load (in Russian). Izvestiya AN SSSR, Mekhanika i Mashinostroenie 4, 109–115 (1961)
36. Bolotin, W.W.: On the influence of moving load on bridges (in Russian). Reports of Moscow University of Railway Transport MIIT 74, 269–296 (1950)
37. Cheng, Y.S., Au, F.T.K., Cheung, Y.K.: Vibration of railway bridges under a moving train by using bridge-track-vehicle element. Engng. Struct. 23(12), 1597–1606 (2001)
38. Cifuentes, A.O.: Dynamic response of a beam excited by a moving mass. Finite Elem. Anal. and Des. 5(3), 237–246 (1989)
39. Cook, R., Malkus, D.S., Plesha, M.E.: Concepts and applications of finite element analysis, 3rd edn. John Wiley & Sons (1989)

40. Courant, R.: Variational methods for the solution of problems of equilibrium and vibrations. Bull. Am. Math. Soc. 49, 1–23 (1943)
41. Delgado, R.M., dos Santos, S.M.: Modelling of railway bridge-vehicle interaction on high speed tracks. Comput. and Struct. 63(3) (1997)
42. Diana, G., Cheli, F., Bruni, S., Collina, A.: Dynamic interaction between rail vehicles and track for high speed train. Vehicle System Dynamics 24, 15–30 (1995)
43. Dmitrijev, A.S.: The analysis of solutions of problems with lateral oscillatory vibrations of various beam structures under the motion of non spring point load (in Russian). Machine Dynamics Problems 24, 18–28 (1985)
44. Dokainish, M.A., Subbaraj, K.: A survey of direct time–integration methods in computational structural dynamics – I. Explicit methods. Comput. and Struct. 32, 1371–1386 (1989)
45. Dokainish, M.A., Subbaraj, K.: A survey of direct time–integration methods in computational structural dynamics – II. Implicit methods. Comput. and Struct. 32, 1387–1401 (1989)
46. Dolbow, J., Belytschko, T.: An introduction to programming the meshless element free Galerkin method. Arch. Comput. Meth. Engng. 5(3), 207–241 (1998)
47. Dyniewicz, B., Bajer, C.I.: Discontinuous trajectory of the mass particle moving on a string or a beam. Machine Dyn. Probl. 31(2), 66–79 (2007)
48. Dyniewicz, B., Bajer, C.I.: Paradox of the particle's trajectory moving on a string. Arch. Appl. Mech. 79(3), 213–223 (2009)
49. Dyniewicz, B., Bajer, C.I.: New feature of the solution of a Timoshenko beam carrying the moving mass particle. Arch. Mech. 62(5), 327–341 (2010)
50. Esmailzadeh, E., Gorashi, M.: Vibration analysis of beams traversed by uniform partially distributed moving masses. J. Sound Vibr. 184, 9–17 (1995)
51. Esveld, C.: Modern railway track. MRT-Productions (2001)
52. Filho, F.V.: Finite element analysis of structures under moving loads. The Shock and Vibration Digest 10(8), 27–35 (1978)
53. Florence, A.L.: Traveling force on a Timoshenko beam. J. Appl. Mech. 32, 351–358 (1965)
54. Flügge, W., Zajac, E.: Bending impact waves in beams. Ingenieur-Archiv. 28(2), 59–70 (1959)
55. Fried, I.: Finite element analysis of time–dependent phenomena. AIAA J. 7, 1170–1173 (1989)
56. Frýba, L.: Vibrations of solids and structures under moving loads. Thomas Telford House (1999)
57. Gavrilov, S.N.: The effective mass of a point mass moving along a string on a Winkler foundation. J. Appl. Math. and Mech. 70(4), 641–649 (2006)
58. Gavrilov, S.N., Indeitsev, D.A.: The evolution of a trapped mode of oscillations in a string on an elastic foundation – moving inertial inclusion system. J. Appl. Math. and Mech. 66(5), 833–852 (2002)
59. Gear, C.W.: The numerical integration of ordinary differential equations. Math. Comp. 21, 146–156 (1967)
60. Gurtin, M.E.: Variational principles for linear elastodynamics. Arch. Rat. Mech. Anal. 16, 34–50 (1964)
61. Gurtin, M.E.: Variational principles for linear initial–value problems. Quart. Appl. Math. 22, 252–256 (1964)
62. Han, R.P.S., Lu, J., Chen, H., Houlston, R.: SUPRA — a novel space–time finite element. In: Robinson, J. (ed.) FEM Today and the Future (1993)

63. Herrera, I., Bielak, J.: A simplified version of Gurtin's variational principles. Arch. Rat. Mech. Anal. 53, 131–149 (1974)
64. Hilber, H.M., Hughes, T.J.R., Taylor, R.L.: Improved numerical dissipation for time integration algorithms in structural dynamics. Earthquake Engng. and Struct. Dyn. 5, 283–292 (1977)
65. Hino, J., Yoshimura, T., Konishi, K., Ananthanarayana, N.: A finite element prediction of the vibration of a bridge subjected to a moving vehicle load. J. Sound Vibr. 96(1), 45–53 (1984)
66. Hinton, E., Rock, T., Zienkiewicz, O.C.: A note on mass lumping and related processes in the finite element method. Earthquake Eng. Struct. Dyn. 4(3), 245–249 (1976)
67. Hoff, C., Hughes, T.J.R., Hulbert, G., Pahl, P.J.: Comparison of the Hilbert–Hughes–Taylor α-method and the θ_1-method. Comput. Meth. Appl. Mech. Engng. 76, 87–93 (1989)
68. Hughes, T.J.R., Levit, I., Winget, J.: An element by element solution algorithm for problems of structural and solid mechanics. Comput. Meth. Appl. Mech. Engng. 36, 241–254 (1983)
69. Hutton, D.V., Counts, J.: Deflection of a beam carrying a moving mass. J. Appl. Mech. 41(3), 803–804 (1984)
70. Inglis, C.E.: A mathematical treatise on vibrations in railway bridges. Cambridge University Press (1934)
71. Jeffcott, H.H.: On the vibration of beams under the action of moving loads. Philosoph. Mag. Series 7 8(48), 66–97 (1929)
72. Kacprzyk, Z.: Space–time superelement (in Polish). Archives of Civil Engineering 28(1-2), 47–55 (1982)
73. Kacprzyk, Z., Lewiński, T.: Comparison of some numerical integration methods for the equations of motion of systems with a finite number of degrees of freedom. Eng. Trans. 31(2), 213–240 (1983)
74. Kaplunov, Y.D.: The torsional oscillations of a rod on a deformable foundation under the action of a moving inertial load (in Russian). Izv. Akad. Nauk SSSR, MTT 6, 174–177 (1986)
75. Kriloff, A.N.: Über die erzwungenen Schwingungen von gleichförmigen elastischen Stäben. Mathematische Annalen 61, 211–234 (1905)
76. Kuang, Z.B., Atluri, S.N.: Temperature field due to a moving heat source. J. Appl. Mech. Trans. ASME 52, 274–280 (1985)
77. Kączkowski, Z.: The method of finite space–time elements in dynamics of structures. J. Tech. Phys. 16(1), 69–84 (1975)
78. Kączkowski, Z.: The method of time dependent finite elements (in Polish). Archives of Civil Engineering 22(3), 365–378 (1976)
79. Kączkowski, Z.: General formulation of the stiffness matrix for the space–time finite elements. Archives of Civil Engineering 25(3), 351–357 (1979)
80. Kączkowski, Z.: On application of non-rectangular space-time elements (in Polish). J. Theoret. and Appl. Mech. 21(4), 531–542 (1983)
81. Kączkowski, Z., Langer, J.: Synthesis of the space-time finite element method. Archives of Civil Engineering 26(1), 11–17 (1980)
82. Lancaster, P., Salkauskas, K.: Surfaces generated by moving least squares methods. Math. Comput. 37(155), 141–158 (1981)
83. Langer, J., Klasztorny, M.: Interpretation of Jakushev's description of concentrated mass moving along Euler-Bernoulli beam. Archives of Civil Engineering 41(1), 5–11 (1995)
84. Lee, H.P.: On the dynamic behaviour of a beam with an accelerating mass. Archive of Applied Mechanics 65(8), 564–571 (1995)

85. Lee, H.P.: Dynamic response of a beam with a moving mass. J. Sound Vibr. 191(2), 289–294 (1996)
86. Lee, H.P.: The dynamic response of a Timoshenko beam subjected to a moving mass. J. Sound Vibr. 198(2), 249–256 (1996)
87. Lee, H.P.: Transverse vibration of a Timoshenko beam acted on by an accelerating mass. Applied Acoustics 47(4), 319–330 (1996)
88. Lewiński, T.: Stability analysis of a difference scheme for the vibration equation with a finite number of degrees of freedom. Zastosowania Matematyki 18(3), 473–486 (1984)
89. Lin, Y.H., Tretheway, M.W.: Finite element analysis of elastic beams subjected to moving dynamic loads. J. Sound Vibr. 136(2), 323–342 (1990)
90. Liu, G.R.: Mesh Free Methods: Moving Beyond the Finite Element Method. CRC Press (2002)
91. Lou, P., Dai, G., Zeng, Q.: Dynamic analysis of a Timoshenko beam subjected to moving concentrated forces using the finite element method. Shock and Vibr. 14, 459–468 (2007)
92. Lou, P., Zeng, Q.Y.: Formulation of equations of motion of finite element form for vehicle-track-bridge interaction system with two types of vehicle model. Int. J. Numer. Meth. Engng. 62(3), 435–474 (2005)
93. Lowan, A.N.: On transverse oscillations of beams under the action of moving variable loads. Philosoph. Mag. Series 7 19(127), 708–715 (1935)
94. Ludwig, K.: Deformation of rail elastically supported of infinite length by loads moving at a constant horizontal velocity. In: Proc. of the 5th Int. Congr. on Appl. Mech., pp. 650–655 (1938)
95. Mackertich, S.: Response of a beam to a moving mass. J. Acoust. Soc. Am. 92(3), 1766–1769 (1992)
96. Mathews, P.M.: Vibrations of a beam on elastic foundation. ZAMM - Journal of Applied Mathematics and Mechanics 38(3-4), 105–115 (1958)
97. Metrikine, A.V., Verichev, S.N.: Instability of vibration of a moving oscillator on a flexibly supported Timoshenko beam. Archive of Applied Mechanics 71(9), 613–624 (2001)
98. Michaltsos, G., Sophianopoulos, D., Kounadis, A.N.: The effect of a moving mass and other parameters on the dynamic response of a simply supported beam. J. Sound Vibr. 191, 357–362 (1996)
99. Michaltsos, G.T.: Dynamic behaviour of a single-span beam subjected to loads moving with variable speeds. J. Sound and Vibr. 258(2), 359–372 (2002)
100. Mofid, M., Akin, J.E.: Discrete element response of beams with traveling mass. Advances in Engng. Software 25, 321–331 (1996)
101. Morgaevskii, A.B.: Critical velocities calculation in the case of a beam under moving load (in Russian). Mekhanika i mashinostroenie, Izvestiya AN SSSR, OTN 3, 176–178 (1959)
102. Mullen, R., Belytschko, T.: An analysis of an unconditionally stable explicit method. Comput. and Struct. 16, 691–696 (1983)
103. Newmark, N.M.: A method of computation for structural dynamics. ASCE J. of Engng. Mech. Div. 85, 67–94 (1959)
104. Nowacki, W.: Dynamics of structures (in Polish). Arkady, Warszawa (1972)
105. Oden, J.T.: A generalized theory of finite elements, II. Applications. Int. J. Numer. Meth. Engng. 1, 247–259 (1969)
106. Panovko, J.: Historical outline of the theory of dynamic influence of moving load (in Russian). Engineering Academy of Air Forces 17, 8–38 (1948)
107. Park, K.C.: An improved stiffly stable method for direct integration of nonlinear structural dynamic equations. Journal of Applied Mechanics 42, 464–470 (1975)

108. Park, K.C.: Practical aspects of numerical time integration. Comput. and Struct. 7, 343–353 (1977)
109. Park, K.C., Felippa, C.A.: Adaptive finite element method and its applications. In: Partitioned Analysis of Coupled Systems. North-Holland, Amsterdam (1983)
110. Park, K.C., Housner, J.M.: Semi-implicit transient analysis procedures for structural dynamics analysis. Int. J. Numer. Meth. Engng. 18, 609–622 (1982)
111. Pestel, E.: Tragwerksauslenkung unter bewegter last. Ingenieur-Archiv. 19, 378–383 (1951)
112. Pesterev, A.V., Bergman, L.A., Tan, C.A., Tsao, T.C., Yang, B.: On asymptotics of the solution of the moving oscillator problem. J. Sound and Vibr. 260, 519–536 (2003)
113. Petyt, M.: Introduction to Finite Element Vibration Analysis. Cambridge University Press (1990)
114. Pietrzakowski, M.: Simulation investigation of damping in nonlinear torsional discrete–continuous system. Machine Dyn. Probl. 3, 65–77 (1992)
115. Podhorecka, A.: The space–time finite element method in the geometrically nonlinear problems (in Polish). J. Theoret. and Appl Mech. 26(4), 683–699 (1988)
116. Podhorecki, A.: The viscoelastic space–time element. Comput. and Struct. 23, 535–544 (1986)
117. Podhorecki, A.: The space-time element method in geommetrically nonlinear viscoelasticity (in Polish). Tech. Rep. 45, Zeszyty Nauk. ATR, Bydgoszcz, Rozprawy (1991)
118. Renaudot, A.: Etude de l'influence des charges en mouvement sur la resistance, des ponts metallique a poutres droites. Annales des Ponts et Chausses 1, 145–204 (1861)
119. Richtmyer, R.D., Morton, K.W.: Difference methods for initial–value problems. John Wiley & Sons (1967)
120. Rieker, J.R., Lin, Y.H., Trethewey, M.W.: Discretization considerations in moving load finite element beam models. Finite Elem. Anal. and Des. 21, 129–144 (1996)
121. Rodeman, R., Longcope, D.B., Shampine, L.F.: Response of a string to an accelerating mass. J. Appl. Mech. 98(4), 675–680 (1976)
122. Sadiku, S., Leipholz, H.H.E.: On the dynamics of elastic systems with moving concentrated masses. Ingenieur-Archiv. 57, 223–242 (1987)
123. Saller, H.: Influence of moving load on railway track and bridges (in German, orig.: Einfluss bewegter Last auf Eisenbahnoberbau und Brücken). Kreidels Verlag, Berlin und Wiesbaden (1921)
124. Schallenkamp, A.: Schwingungen von Trägern bei bewegten Lasten. Ingenieur-Archiv. 8, 182–198 (1937)
125. Schwartz, L.: Théorie des distributions I, Paris (1950)
126. Seaborn, J.: Hypergeometric functions and their applications. Springer, New York (1991)
127. Smith, C.E.: Motion of a stretched string carrying a moving mass particle. J. Appl. Mech. 31(1), 29–37 (1964)
128. Stanisic, M.M., Euler, J.A., Montgomery, S.T.: On a theory concerning the dynamical behavior of structures carrying moving masses. Ingenieur-Archiv. 43, 295–305 (1974)
129. Steele, C.R.: The Timoshenko beam with a moving load. J. Appl. Mech. 35, 481–488 (1968)
130. Steuding, H.: Die Schwingung von Trägern bei bewegten Lasten. Ingenieur-Archiv. 6(4), 265–270 (1935)
131. Stokes, G.G.: Discussion of a differential equation relating to the breaking of railway bridges. Trans. Cambridge Philosoph. 8, 707–735 (1849)
132. Strzyżakowski, Z.: Modelling of dynamic phenomena in tranportation systems (in Polish). Politechnika Radomska, Radom (2006)

133. Szcześniak, W.: Inertial moving loads on beams (in Polish). Scientific Reports, Warsaw University of Technology, Civil Engineering 112 (1990)
134. Szcześniak, W.: Comparison of the mathematical models of the moving inertial load on Euler and Timoshenki beams (in Polish). In: VIII Symp. Dyn., Konstr. Zeszyty Naukowe Politechniki Rzeszowskiej, Mechanika, vol. 117(38), pp. 363–368 (1993)
135. Taheri, M.R., Ting, E.C.: Dynamic response of plates to moving loads: finite element method. Comput. and Struct. 34(3), 509–521 (1990)
136. Taltello, F., Burkhardt, G.: Zur Anwendung der Methode der Raum–Zeit–Elemente auf die dynamische Untersuchung von Stahlbetonbauteilen. Tech. Rep. 68, Hochschule fűr Architektur und Bauwesen, Weimar (1988)
137. Timoshenko, S.P., Young, D.H., Weaver, W.: Vibration Problems in Engineering. Wiley, New York (1974)
138. Ting, E.C., Genin, J., Ginsberg, J.H.: A general algorithm for moving mass problems. J. Sound Vib. 33(1), 49–58 (1974)
139. Trujillo, D.M.: An unconditionally stable explicit algorithm for structural dynamics. Int. J. Numer. Meth. Engng. 11, 1579–1592 (1977)
140. Willis, R.: Preliminary Essay to the Appendix B. Report of the Commissioners Appointed to Inquire into the Application of Iron to Railway Structures. W. Clowes and Sons (1849)
141. Witkowski, M.: On the time space in structural dynamics (in Polish). Prace Naukowe, Budownictwo 80. Warsaw University of Technology (1983)
142. Wood, W.L.: A unified set of single step algorithms. Part II: Theory. Int. J. Numer. Meth. Engng. 20, 2303–2309 (1984)
143. Wood, W.L., Bossak, M., Zienkiewicz, O.C.: An alpha modification of Newmark's method. Int. J. Numer. Meth. Engng. 15, 1562–1566 (1981)
144. Wu, J.J.: Dynamic analysis of an inclined beam due to moving loads. J. Sound and Vibr. 288, 107–131 (2005)
145. Wu, J.J., Whittaker, A.R., Cartmell, M.P.: The use of finite element techniques for calculating the dynamic response of structures to moving loads. Comput. and Struct. 78, 789–799 (2000)
146. Yakushev, N.Z.: Certain problems of dynamics of the beam under moving load (in Russian). Studies in Plates and Shells Theory 12, 199–220 (1974)
147. Yang, Y.B., Yau, J.D., Wu, Y.S.: Vehicla-bridge interaction dynamics. World Scientific (2004)
148. Yavari, A., Nouri, M., Mofid, M.: Discrete element analysis of dynamic response of Timoshenko beams under moving mass. Advances in Engineering Software 33(3), 143–153 (2002)
149. Yavari, A., Nouri, M., Mofid, M.: Discrete element analysis of dynamic response of Timoshenko beams under moving mass. Advances in Engng. Software 33(3), 143–153 (2002)
150. Yoshida, D.M., Weaver, W.: Finite-element analysis of beams and plates with moving loads. Intl. Assoc. Bridge Struc. Engr. 31(1), 179–195 (1971)
151. Zagustin, E.A., Young, D.H.: Dynamic response of beams to a traveling mass. Intl. Assoc. Bridge Struc. Engr. 32(1), 219–235 (1972)
152. Zemanian, A.H.: Distribution Theory and Transform Analysis: An Introduction to Generalized Functions, with Applications. Dover Publications (1987)
153. Zienkiewicz, O.C.: The finite element method in engineering science. McGraw-Hill, London (1971)
154. Zienkiewicz, O.C., Wood, W.L., Hine, N.W., Taylor, R.L.: A unified set of single step algorithms. Part I: General formulation and applications. Int. J. Numer. Meth. Engng. 20, 1529–1552 (1984)

Index

accuracy, 140
  central difference method, 108
accuracy of the method, 153
Adams
  closed method, 112
  explicit method, 110
  implicit method, 112
  open method, 110
Adams method, 109
approximate methods, 9

bar element, 169
beam, 170
  Bernoulli–Euler, 46, 85, 188
  Timoshenko, 198, 230
Bernoulli–Euler beam, 46, 85, 188
Bossak method, 117

central difference method, 105

discontinuity, 26
displacement formulation of STEM, 129

efficiency, 167
elastic foundation, 215
element
  bar, 169
  beam, 170
  simplex, 161, 169, 176
  string, 144, 226
  Timoshenko beam, 230
  triangular, 176
examples, 18, 247
  classical track, 249
  dynamics of track, 249, 253
  plate, 266
  Timoshenko beam, 59, 89
  track dynamics, 262

Fourier solution
  Bernoulli–Euler beam, 47
  string, 32

inertial load, 84
information flow, 138

Lagrange approach, 37
Lagrange equation, 50
literature, 5, 31, 46, 78

mass
  zero, 218
meshfree method, 241
method
  Adams, 109
  Bossak, 117
  central difference, 105
  Cranck–Nicolson, 101
  Houbolt, 102
  mean acceleration, 101
  Newmark, 114, 160, 223, 226, 230
  Park, 118
  Park–Housner, 118
  predictor-corrector, 99
  Runge–Kutta, 99
  space-time element, 123
  SSpj, 102
  trapezoidal, 101

Trujillo, 121
Wilson, 102
methods
approximate, 9
Mindlin plate, 213, 277

Newmark method, 114, 160, 223, 226, 230

one-degree-of-freedom, 140
oscillator, 79

Park method, 118
plate, 67, 204
on elastic foundation, 215
Mindlin, 213, 277
thick, 213
thin, 204, 283
plate element, 172
program
Mindlin plate, 277
string, 271
thin plate, 283
Timoshenko beam, 274

Renaudot approach, 71
Runge–Kutta method, 99

semi-analytical methods, 31
simplex element, 161, 169
space–time element
Bernoulli–Euler beam, 188
plate, 204
string, 182
Timoshenko beam, 198
SSpj method, 102
stability
central difference method, 107
Park–Housner method, 119
string, 32, 182
under oscillator, 79
string element, 144, 226

The Park–Housner method, 118
thick plate, 213, 277
thin plate, 204, 283
Timoshenko beam, 55, 198, 230
track
classical, 249
subway, 262
Y-type, 253
triangular elements, 176
Trujillo method, 121

velocity formulation, 140
virtual function, 151
Dirac, 153
hat function, 151
roof, 152
triangular, 152

Yakushev approach, 72

MIX
Papier aus verantwortungsvollen Quellen
Paper from responsible sources
FSC® C105338

If you have any concerns about our products,
you can contact us on
ProductSafety@springernature.com

In case Publisher is established outside the EU,
the EU authorized representative is:
Springer Nature Customer Service Center GmbH
Europaplatz 3, 69115 Heidelberg, Germany

Printed by Libri Plureos GmbH
in Hamburg, Germany